Urban Tree Management

Urban Tree Management

Editor:

Dr. Hazem Fouda, Ph.D.

Delve Publishing
www.delvepublishing.com

Urban Tree Management

Editor: Dr. Hazem Fouda, Ph.D.

708 3rd Avenue, 6th Floor
New York, NY 10017
United States of America
www.delvepublishing.com

ISBN: 978-1-68095-775-4

Library of Congress Control Number: 2016951531

About the Editor

Dr. Hazem Fouda, Ph.D.

Hazem Shawky Fouda has a PhD. In Agriculture Sciences, obtained his PhD. From the Faculty of Agriculture, Alexandria University in 2008, He is working in Cotton Arbitration & Testing General Organization (CATGO).

8. In-House Urban Forestry Research ... 43
9. Conclusions and Future Research ... 43
References ... 44

Chapter 5 Urban Forest Inventory ... 47
1. Introduction ... 47
2. Study Area and Data ... 49
3. Methodology ... 50
4. Results and Discussion ... 52
Conclusions ... 58
Acknowledgments ... 58
Author Contributions ... 58
Conflicts of Interest ... 59
References ... 59

Chapter 6 Monitoring of Recently-Planted Urban Trees ... 61
1. Introduction ... 61
2. The Tree-Planting Organization Context ... 62
3. Citizen Science ... 62
4. Measuring Urban Tree Outcomes ... 63
5. Factors That Influence Urban Tree Outcomes ... 64
6. The Planted Tree Re-Inventory Protocol ... 67
Conclusion ... 71
Appendix ... 71
Acknowledgements ... 71
References ... 71

Chapter 7 A Technical Guide to Urban and Community Forestry ... 75
1. Introduction ... 75
2. Values of Urban Trees ... 76
3. Basic Horticultural Principles ... 80
4. Urban Forestry Planning ... 83
5. Tree Selection and Site Design ... 85
6. Planting and Early Care ... 91
7. Maintenance of Established Trees ... 95
8. Preserving Trees Affected by Development ... 100
9. Utility Arboriculture ... 107
10. Coordination ... 109
11. The Politics of Trees ... 109

Chapter 8 Evaluating Urban Forestry Performance ... 113
1. Introduction and Literature Review ... 113
2. Literature Review ... 114
3. Benefits of Urban Trees ... 115
4. Urban Forest Performance Measures and Community Demographics ... 116
5. Tree Inventories and Management Plans ... 116
6. Professional Staff ... 116
7. Tree Ordinances ... 117
8. Advocacy and Advisory Organizations ... 117
References ... 117

Chapter 9 **Sustainability for Urban Forests** ... **121**
1. Introduction ... 121
2. Three Conceptual Domains ... 121
3. Interpreting Sustainability for Urban Forests ... 124
Conclusions ... 126
References ... 126

Chapter 10 **Urban Tree Canopy Cover and Environmental Justice** ... **129**
1. Introduction ... 129
2. Materials and Methods ... 130
3. Results ... 133
4. Discussion ... 135
5. Supporting Information ... 137
Conclusion ... 142
Author Contributions ... 142
References ... 142

Chapter 11 **Enhancement of Tree Productivity from Forest to Urban Land Covers** ... **145**
1. Introduction ... 145
2. Materials and Methods ... 146
3. Results ... 149
4. Discussion ... 152
5. Supporting Information ... 154
Acknowledgments ... 155
Author Contributions ... 155
References ... 155

Chapter 12 **Quantifying and Mapping the Supply of and Demand for Carbon Storage and Sequestration Service from Urban Trees** ... **159**
1. Introduction ... 159
2. Overview of Carbon Storage and Sequestration Modeling ... 161
3. Methods ... 162
4. Results ... 168
5. Discussion ... 171
Conclusions ... 174
Supporting Information ... 174
Author Contributions ... 175
References ... 175

Chapter 13 **Urban Tree Effects on Soil Organic Carbon** ... **179**
1. Introduction ... 179
2. Methods ... 180
3. Results ... 181
4. Discussion ... 182
5. Supporting Information ... 184
Acknowledgments ... 186
Author Contributions ... 186
References ... 186

Chapter 14 **Street Trees in the Urban Forest Canopy** **189**
Acknowledgments 189
Summary 189
Results 201
References 204

Index **207**

List of Contributors

Darren S. Le Roux
The Fenner School of Environment and Society, the Australian National University, Canberra, Australia

Karen Ikin
The Fenner School of Environment and Society, the Australian National University, Canberra, Australia

David B. Lindenmayer
The Fenner School of Environment and Society, the Australian National University, Canberra, Australia

Adrian D. Manning
The Fenner School of Environment and Society, the Australian National University, Canberra, Australia

Philip Gibbons
The Fenner School of Environment and Society, the Australian National University, Canberra, Australia

Fardad Edalatkhah
Municipality of Tehran, Tehran, Iran

Mahdi Shafaghati
Payame Noor University, Taleghan, Iran

Akram Norouzi
Payame Noor University, Taleghan, Iran

Somaye Motaghi
Payame Noor University, Taleghan, Iran

Maryam Mirbahaei
Payame Noor University, Varamin, Iran

Dexter H. Locke
New York Citiy Urban Field Station,

J. Morgan Grove
USDA Forest Service,

Michael Galvin
SavaTree,

Jarlath P.M. O'Neil-Dunne
University of Vermont

Charles Murphy
TreeBaltimore

Matthew P. Wells
Central Forestry and Horticulture, New York City Department of Parks & Recreation, USA

Caiyun Zhang
Department of Geosciences, Florida Atlantic University, 777 Glades Road, Boca Raton, FL 33431, USA

Yuhong Zhou
Department of Geospatial Information Science , University of Texas at Dallas, 800 W Campbell Road, GR31, Richardson, TX 75080, USA

Fang Qiu
Department of Geospatial Information Science , University of Texas at Dallas, 800 W Campbell Road, GR31, Richardson, TX 75080, USA

Jessica M. Vogt
Center for the Study of Institutions, Population and Environmental Change, Indiana University, Bloomington

Burnell C. Fischer
The Vincent and Elinor Ostrom Workshop in Political Theory and Policy Analysis, and Center for the Study of Institutions, Population and Environmental Change, Indiana University, Bloomington

World Forestry Center
University of Nebraska - Lincoln

Mollie E. Freilicher
University of Massachusetts - Amherst

Camilo Ordóñez
School for Resource and Environmental Studies, Dalhousie University, 6100 University Avenue, Halifax, Nova Scotia, B3H 3J5, Canada

Peter N. Duinker
School for Resource and Environmental Studies, Dalhousie University, 6100 University Avenue, Halifax, Nova Scotia, B3H 3J5, Canada

Kirsten Schwarz
Department of Biology, Northern Kentucky University, Highland Heights, Kentucky, United States of America,

Michail Fragkias
Department of Economics, College of Business and Economics (COBE), Boise State University, Boise, Idaho, United States of America,

Christopher G. Boone
School of Sustainability, Arizona State University, Tempe, Arizona, United States of America,

Weiqi Zhou
Center for Eco-Environmental Sciences, Chinese Academy of Sciences, Haidian District, Beijing, China,

Melissa McHale
Department of Forestry and Environmental Resources, North Carolina State University, Raleigh, North Carolina, United States of America,

J. Morgan Grove
USDA Forest Service, Northern Research Station, Baltimore, Maryland, United States of America,

Jarlath O'Neil-Dunne
University of Vermont, Rubenstein School of Environment and Natural Resources and Spatial Analysis Lab, Burlington, Vermont, United States of America,

Joseph P. McFadden
Department of Geography, University of California Santa Barbara, Santa Barbara, California, United States of America,

Geoffrey L. Buckley
Department of Geography, Ohio University, Clippinger Laboratories 109, Athens, Ohio, United States of America,

Dan Childers
School of Sustainability, Arizona State University, Tempe, Arizona, United States of America,

Laura Ogden
Department of Global and Sociocultural Studies, Florida International University, FIU Modesto A. Maidique Campus, Miami, Florida, United States of America,

Stephanie Pincetl
Institute of the Environment and Sustainability, University of California Los Angeles, Los Angeles, California, United States of America,

Diane Pataki
Department of Biology, University of Utah, Salt Lake City, Utah, United States of America,

Ali Whitmer
Georgetown University, Washington D.C., United States of America,

Mary L. Cadenasso
Department of Plant Sciences, University of California Davis, Davis, California, United States of America

Brittain M. Briber
Department of Earth and Environment, Boston University, Boston, Massachusetts, United States of America,

Lucy R. Hutyra
Department of Earth and Environment, Boston University, Boston, Massachusetts, United States of America,

Andrew B. Reinmann
Department of Earth and Environment, Boston University, Boston, Massachusetts, United States of America,

Steve M. Raciti
Department of Earth and Environment, Boston University, Boston, Massachusetts, United States of America,
Department of Biology, Hofstra University, Hempstead, New York, United States of America,

Victoria K. Dearborn
Department of Earth and Environment, Boston University, Boston, Massachusetts, United States of America,

Christopher E. Holden
Department of Earth and Environment, Boston University, Boston, Massachusetts, United States of America,

Allison L. Dunn
Department of Earth, Environment, and Physics, Worcester State University, Worcester, Massachusetts, United States of America

Chang Zhao
Department of Geographical and Sustainability Sciences, University of Iowa, Iowa City, Iowa, United States of America

Heather A. Sander
Department of Geographical and Sustainability Sciences, University of Iowa, Iowa City, Iowa, United States of America

Jill L. Edmondson
Department of Animal and Plant Sciences, University of Sheffield, Sheffield, United Kingdom,

Odhran S. O'Sullivan
Department of Animal and Plant Sciences, University of Sheffield, Sheffield, United Kingdom,

Richard Inger
Environment and Sustainability Institute, University of Exeter, Penryn, Cornwall, United Kingdom

Jonathan Potter
Department of Animal and Plant Sciences, University of Sheffield, Sheffield, United Kingdom,

Nicola McHugh
Department of Animal and Plant Sciences, University of Sheffield, Sheffield, United Kingdom,

Kevin J. Gaston
Environment and Sustainability Institute, University of Exeter, Penryn, Cornwall, United Kingdom

Jonathan R. Leake
Department of Animal and Plant Sciences, University of Sheffield, Sheffield, United Kingdom,

Joseph Poracsky
Portland State University

David Banis
Portland State University

Preface

Urban Tree Management is the careful care and organizing tree populations in urban settings for the purpose of improving the urban environment. Urban tree management advocates the role of trees as a critical part of the urban infrastructure. Urban foresters plant and maintain trees, support appropriate tree and forest preservation, conduct research and promote the many benefits trees provide. Urban tree management is practiced by municipal and commercial arborists, municipal and utility foresters, environmental policymakers, city planners, consultants, educators, researchers and community activists. It contains tree selection, planting, care and protection and the overall management of trees as a collective resource. Urban Tree Management designates their benefits and describes their effects on quality of urban life and well-being facets that are increasingly important in these times of progressing urbanization. First chapter presents a simulation model to predict the future availability of native hollow-bearing trees in a rapidly expanding urban landscape. Second chapter explores on the role of trees in improving the urban landscape. The purpose of third chapter is to describe how diverse stakeholders can use urban tree canopy (UTC) Assessment and Prioritization tools to collaboratively achieve urban sustainability goals. Fourth chapter will look at the key research studies and how they have been used to justify and focus urban forestry programming. Fifth chapter presents an approach on urban forest inventory and we present in sixth chapter the Planted Tree Re-Inventory Protocol for citizen science-based monitoring of recently-planted urban trees. Seventh chapter explores on a technical guide to urban and community forestry. The purpose of the eighth chapter is to assess what it means to meet the four CARS measures and the purpose of ninth chapter is to further the thinking around urban-forest sustainability and to reveal alternative interpretations of the concept. In tenth chapter, we examine potential inequities associated with the distribution of urban tree cover in relationship to race/ethnicity and income. In eleventh chapter, we assessed the effect of land cover conversion from forest to a variety of urban land covers on vegetation dynamics. In twelfth chapter, we present an approach to assessing ecosystem service supply and demand in a spatially-explicit manner focusing on carbon storage and sequestration provided by urban trees. Thirteenth chapter highlights on urban tree effects on soil organic carbon and last chapter aims to evaluate the contribution of street trees to canopy.

Editor
Dr. Hazem Fouda, Ph.D.

INTRODUCTION

Forestry is the science of managing forests, plantations of trees, and related natural resources. Some people consider it to be as much of an art as it is a science. Urban forestry is the management of trees within urban spaces. Many cities have green spaces such as small parks or areas of trees that line streets, recreational areas, and monuments. Urban forestry is a kind of forestry that focuses on the location and maintenance of such trees. There are many types of people who participate in, support, or help to plan urban forestry. Such people include city planners, environmental policymakers, and arborists.

Because tree populations in urban landscapes are exposed to elements much different than those that are in the wild, it is important that they are carefully managed and tended to. Furthermore, people who focus on the importance of urban forestry are often concerned with how the trees improve the surrounding environment just as much as they are concerned with how the trees must be protected within the surrounding environment.

Trees can provide many benefits within urban areas. First of all, it is well known that trees and other plants are excellent at cleaning and refreshing the air. Air in urban areas, which is often tinged with vehicle and factory emissions to name just a few pollutants, can be greatly improved by the presence of trees. Furthermore, trees and plants help to naturally cool the air and the earth. This can be a great benefit. Cities are generally made mostly of steel, concrete, asphalt, and glass which reflect and conduct heat. Trees can help to make cities in hot climates and cities in climates with hot summers much more bearable during warm weather seasons.

Because cities can be such harsh environments, trees that are planted therein must be tended, otherwise they may wither and perish. One of the key elements in urban forestry is making sure that trees in urban spaces are planted in soil of acceptable quality and that they have enough space for their root systems to grow. Just as trees must have enough space for their roots to grow, they must also have enough space for their canopies. This means that trees must have enough overhead space so that their branches can extend and reach the sunlight that they need to thrive.

While those who practice urban forestry certainly have many challenged to face, it is clear that the challenges are worth the benefits. Trees and plants in urban spaces are wonderful environmental investments.

It is our goal to make your experience with Urban Tree Management significantly different than any other tree service. We are a service-focused business, meaning that we aim to meet the client's needs throughout the contracting experience. Growing and expanding is not our objective. We don't want to be the only tree service in Madison, but we want to be the best tree service in Madison. Our focus is never going to selling more work; pushing unnecessary services; or offering anything but the most informed, objective, and professional advice. We want you to be pleased with the trees on your property and see ourselves as facilitators to that end. Urban tree management is the key basis for greener cities of the future. It is a practical discipline which includes tree selection, planting, care and protection and the overall management of trees as a collective resource. Urban Tree Management aims to raise awareness for the positive impacts and benefits of city trees and for their importance to city dwellers. It describes their advantages and details their effects on quality of urban life and well-being aspects that are increasingly important in these times of progressing urbanization.

Urban trees and woodlands contribute greatly to the quality of life in towns and cities. They enhance the environmental, aesthetic and wildlife value of urban areas, and create a range of recreational and educational opportunities. Urban forestry was developed as an interdisciplinary approach to planning and managing forests and trees in and near urban areas.

The urban forest is a great resource for any town or city, and as such its management must be planned, systematic and integrated. For optimal management urban trees need to be listed, categorized and monitored, and objectives and aims need to be clearly defined. Due to the complexity of the manmade urban landscape, a wide range of expertise needs to be involved in developing and carrying out the strategies in order to achieve the aims and objectives. .

Urban tree management practices shall be adopted by developed countries so that effort in sustainable environment can be practically achieved. One of an example is preservation of urban forest. More than a natural gallery that holding valuable plants. Urban forest area also becoming a habitat for various kind of wildlife. Creating a biodiversity that make every living and non-living things related to each other. A place for creating breathing gases. Provide foods and natural resources furthermore balancing the condition of the environment.

This movement needs to be done in by all parties from government agencies. Private sectors and every individual. Each field of works has its own way to contribute at least an effort to help towards nature. Improvement and changes in all aspect of technology, education and also human behavior towards preservation of environment must done and strengthen up from time to time. Through this study it's an effort in educating and promoting importance of environment and why urban tree management plan is the important. This way it was hope to create new way of life and guidance for the future generations.

Trees become part of important contributor to the health and beauty of the urban environment. Their presence brings physical, psychological and spiritual well-being to a community and helps to create a sense of peace and harmony with nature. Trees moderate climate and air quality. They take carbon dioxide and other pollutants from the air, in exchange of oxygen in return. Their roots stabilize the soil and impede erosion.

Trees give life by softening the geometric rigidity and hard surfaces of streets and buildings within an urban environment. In creating diverse spatial configurations trees can be used with their infinite variety of form. Texture. Color, and mood. Why trees? Because trees bring life to a city. Trees inspire us and exemplify the continuity of time and space, and provide a sense of beauty for future generations to enjoy.

What is in an Urban Forest Management Plan (UFMP)?

An UFMP is a roadmap that creates a shared vision for the future of a tree canopy. It's a tailored plan that guides tree care professionals to proactively and effectively manage and provide for maximum, long-term benefits to the community. The plan provides recommendations based on the analysis of detailed inventories and includes additional components or documents, such as budgets, implementation schedules, policy and procedure manuals, standards and specifications, public education and monitoring plans, and existing ordinances. Annual work plans and budgets can then be developed based on the long-term UFMP.

In general, an effective UFMP plan will include:

- Vision for the urban forest
- Inventories and assessments of the current status of the urban forest
- Strategic Plan: goals, objectives, and actions based on the information analyzed and identified needs
- Implementation Plan with specific dates and assigned responsibilities
- Monitoring Plan with a system or matrix to check effectiveness and revise the UFMP as needed.

The Benefits of Urban Trees

Trees in towns bring with them both benefits and costs. Whilst many of the costs are well known to managers of urban trees, who watch the budgets and answer the phone to disgruntled residents, the benefits can be seen as nebulous and difficult to quantify or justify. Never the less, a considerable and expanding body of research exists on the benefits that urban trees bring. This briefing note is an attempt to summaries some of the benefits of urban trees. A number of papers relevant to the subject of the benefits of urban trees have, with the kind permission of their authors, been included in the appendices.

Urban Tree Policy

The Urban Tree Policy is a declaration of intent by the Public Works Department regarding urban trees. The main objective of the policy is to secure the vitality of Helsinki's urban trees, the values attached to them, and the ecosystem services provided by them through co-operation between the divisions within the Public Works Department. The Urban Tree Policy is a document by means of which the Public Works Department defines and publishes its principles and methods regarding urban trees for the information of those who work with the trees and city residents. The policy consists of eight strategies to fulfil the main objective. The vision for each of these strategies is defined in writing along with the methods to achieve this vision. The Urban Tree Policy is intended primarily for employees of the Public Works Department and decision makers, while some of its contents are intended also for residents. The Public Works Committee has approved the Urban Tree Policy to serve as guidelines on 13 May 2014. The Urban Tree Policy applies to trees planted in the built environment that are owned by the City of Helsinki and that are the responsibility of the Public Works Department, i.e. street trees and park trees. Trees growing in forest areas are not included in the policy. In this policy, urban trees are evaluated in terms of the monetary value of the tree assets, the cityscape, cultural heritage, the selection of tree species, the intangible benefits provided by the trees (ecosystem services) and natural diversity. The Public Works Department has primary responsibility for the city's tree assets, so the perspective of this policy is the development of the Public Works Department's activities in connection with urban trees

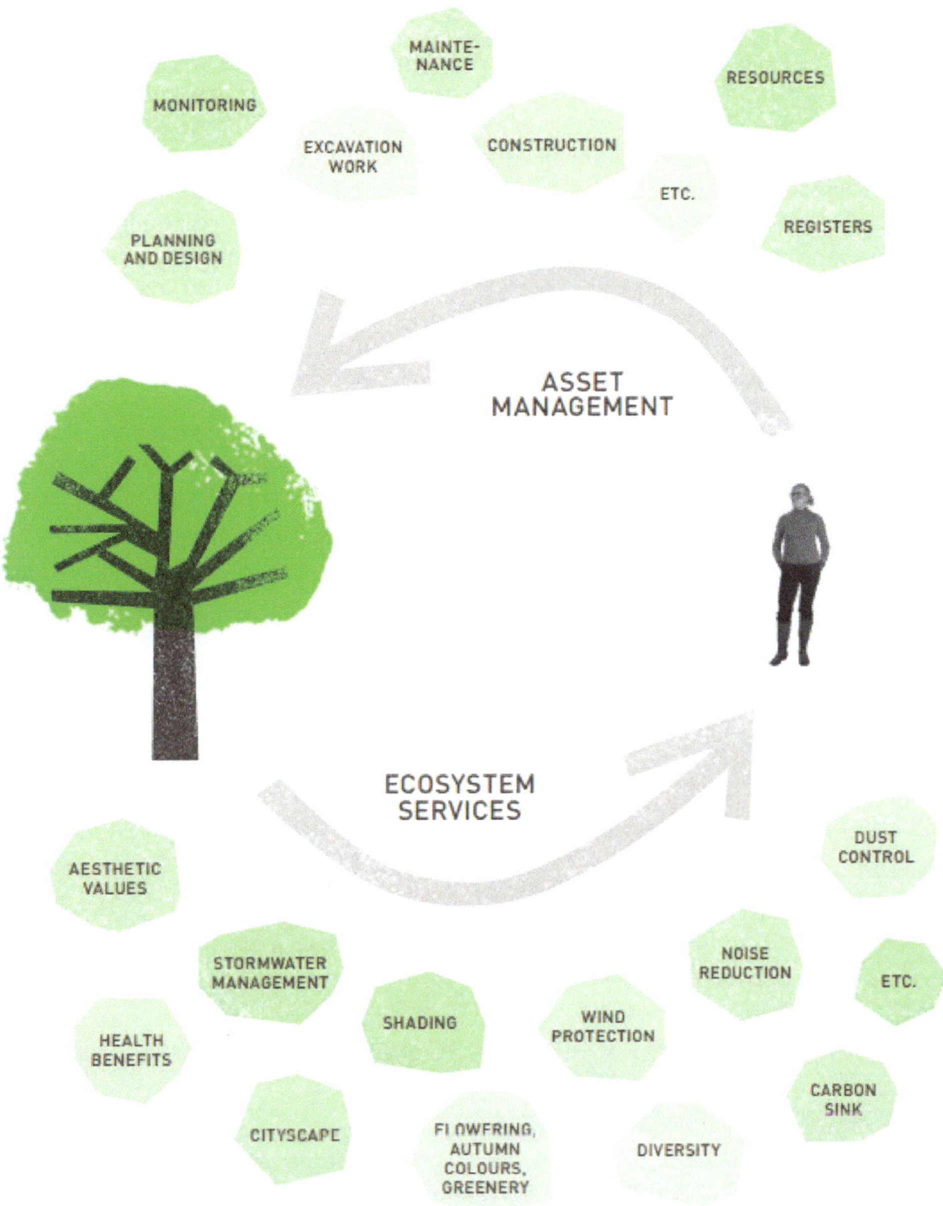

. This policy does not apply to urban trees in yards, gardens, cemeteries and plots owned or managed by other parties. Nevertheless, securing the vitality of trees belongs equally to all of the city's administrative bodies and actors within the city. The methods presented in this policy are applicable to other administrative bodies of the City of Helsinki, since the guidelines emphasize the necessity of co-operation and joint aims to secure the vitality of urban trees. The Urban Tree Policy is based on the assumption that thriving and healthy urban trees are in the interest of everyone.

Need for a Urban Tree Policy

The need for an Urban Tree Policy has arisen for many different reasons. A major inventory of Helsinki's street trees in 2010–2011 highlighted the need for better planning and forecasting for maintaining and regenerating the rows of street trees and tree-lined alleys in the city. Following the structural reorganization of the Public Works Department, the status of urban trees and the responsibilities for taking care of them were left unclear. With the increase in subcontracting, Helsinki's trees will most likely be tended to by subcontractors, who may change frequently. For such contractors, clear contract documents and guidelines for maintaining urban trees must be drawn up. In addition, the threats and opportunities of climate change should also be considered when selecting tree species. Trees are a living asset in the built urban environment that should be taken into consideration in all actions throughout the long life span of the trees. Efforts must also be made to help residents understand the needs of urban trees. In recent decades, the regeneration of certain tree-lined alleys in Helsinki that are important for both the cityscape and for residents has become topical. The regeneration of these alleys has created a lot of feedback for the Public Works Department, and these projects have taken a long time and a lot of work. Nevertheless, our generation must take responsibility for programming the regeneration of these alleys and create policies for such a major programmer. The Urban Tree Policy has been drawn up proactively so that the regeneration of Helsinki's oldest alleys can be properly initiated and so that the high quality of the cityscape can be retained also for future generations.

Development needs of the Public Works

Department regarding urban trees Helsinki possesses a lot of strengths and expertise related to urban trees. Retaining these strengths also in the future, however, requires a review of the operating methods of the Public Works Department regarding urban trees. The Public Works Department should have common targets regarding urban trees. Operations affecting urban trees should be clarified and responsibilities assigned, while the processes related to urban trees should also be identified. The further actions required to implement the Urban Tree Policy have been compiled in a draft action plan for internal use within the Public Works Department. The actions presented in this draft promote co-operation regarding urban trees and taking urban trees into consideration in decision making and processes. The actions propose, for example, improving processes, aspects of asset management, communications and expertise in order to secure the vitality of urban trees and increase awareness of the importance and value of urban trees.

Structure of the Urban Tree Policy

The main objective of Helsinki's Urban Tree Policy is to secure the vitality of Helsinki's urban trees, the values attached to them, and the ecosystem services provided by them through co-operation between the divisions within the Public Works Department. To achieve this objective, eight principles or strategies are presented in the Urban Tree Policy.

The administration of urban trees has been divided into eight sub-sections within this policy:

- Awareness and appreciation,
- Cultural heritage and cityscape,
- Diversity,
- The selection of tree species,
- Planning and design, construction and permits,
- Maintenance of the trees,
- Tree rows and alleys and
- Asset management.

A strategy has been defined for each of these sub-sections to help achieve the main objective. A long-term vision has been defined for each of the eight strategies. In connection with these visions, a catalogue of methods is listed by which, if systematically followed within the Public Works Department, the vision can be achieved or maintained.

Increase Awareness and Appreciation of Urban Trees

Vision

Urban trees are appreciated in Helsinki. There is a general awareness and appreciation of the numerous values of urban trees, such as social, economic, cityscape, cultural historic, architectonic, biological and ecological. Employees of the Public Works Department are aware of the ecosystem services provided by urban trees and their importance for public health and the aesthetic values of the city. The requirements for the vitality of urban trees are understood within the Public Works Department and the city's other administrative bodies, as well as among other actors. This appreciation is manifested as an investment in urban trees. Awareness of urban trees and their importance is also increasing among residents.

Methods

- Decision-making about urban trees is guided by legislation, the city's own strategies and guidelines, and expertise in trees.
- When making decisions about urban trees, the effects of the decisions on the longterm vitality of the urban trees and the ecosystem services they produce are taken into consideration.
- Communications about urban trees are transparent and based on expertise, and they are directed diversely at different target groups.
- Know-how about trees, their properties, their history and the conditions for their vitality is shared among different professional groups within the Public Works Department in order to increase expertise in trees.
- Seminars about urban trees are organised on a regular basis for the city's different administrative bodies and other actors.
- Park walks focusing on trees are included in the guided park walks for residents.

Urban Forest Management Plan

Trees on streets and on other publicly owned properties managed by public works agencies provide a multitude of aesthetic and environmental benefits to citizens, businesses and visitors alike. Beyond shade and beauty, trees also have practical benefits and a real monetary value that cities sometimes are unaware of—your urban forest provides valuable public services and could be worth over a million dollars. Unlike other public infrastructure components, properly planted and maintained trees increase in value over time. An urban forest management plan, based on recent tree inventory data and analysis of available staff, equipment, and budget resources, is an essential tool for protecting this valuable resource. An urban forest management plan is an action plan; it gives public works agencies detailed information, recommendations, and resources needed to effectively and proactively manage public trees. The purpose of having an urban forest management plan is to ensure that a community will enjoy the benefits of trees through proper arboricultural techniques and management practices. The goal of the plan is to state what is needed to manage the urban forest and to describe activities and services required to execute these responsibilities. If a management plan is based on analysis from an accurate tree inventory and developed with input from public works staff, arboricultural experts, and the citizens, then the public works agency responsible for the urban forest will realize many benefits:

Increased Public Safety

All public works agencies know that a large part of their primary mission is to assure safety and manage risk related to public infrastructure. A tree inventory and management plan will provide lists of trees requiring priority removal and pruning that a manager can carry out within the limits of budget and time. The inventory can be used subsequently to monitor trees for safety risks on a continual basis. By implementing recommendations made in the management plan, storm damage risks will also decline.

Increased Efficiency

Once an inventory has identified the work to be done and a management plan has prescribed a maintenance program, a manager can execute that work in a much more efficient manner than before. By scheduling all work in a given area to be done at the same time (rather than by reacting to single requests) the savings in travel and setup time are substantial, with historical examples showing about a 50 percent reduction in cost—especially when a system of rotational work and/or preventative maintenance is adopted. There is also increased efficiency in the office created by using an electronic inventory to locate and manipulate records and select and schedule work. The efficient response to citizen requests and questions also improves customer service.

Facilitate Short and Long-Term Planning

Planning can be made much easier by using the results of the tree inventory and the analysis of an urban forest management plan. Since maintenance and planting needs have been assessed, and other issues such as hardscape conflicts and right-of-way clearances, personnel levels and training, and even public relations are addressed in the plan, short and long-term planning for the forest is made easier.

Justify Budgets

An urban forest management plan provides the data and analysis needed to determine specific levels of funding for tree maintenance and tree planting projected over a multi-year period. With accurate data, a manager can establish, prioritize, and justify annual budget requests. The tasks and associated costs are clearly spelled out in the plan, and can be supported by detailed lists. Many public works managers have found that they have much greater success with budget requests that are based on the analysis of high-quality data. Also, a good inventory provides a solid basis for grant applications.

Management Plan Components

The components and variations of urban forest management plans are many, depending on the developmental stage of the urban forestry program within a public works agency. Generally, these elements are included or addressed in the plan:

1. Tree inventory data and analysis
2. Tree inventory and mapping data management software
3. Tree risk reduction/emergency storm response plan
4. Tree board or advisory council development
5. Public relations and education
6. Urban forest cost/benefit analysis

In the following sections, these six basic components of a plan will be discussed in more detail. They will be prioritized for the benefit of managers who are just beginning a program and for managers who have an established program and are looking to improve it.

Tree Inventories

What Is a Tree Inventory?

Public tree inventories are a statistically reliable survey of publicly owned and managed trees, used to determine the location and the exact or estimated measurements of quantity, quality, health, and trends of the urban forest, as well as a description of other urban forest attributes, such as potential planting sites, utilities present, and hardscape features.

Data commonly collected during an inventory includes:

- Location
- Species
- Diameter
- Condition
- Maintenance need and priority
- Proximity to utility lines, traffic signs and signals
- Sidewalk and other hardscape damage
- Insect and disease problems
- Potential planting sites

Inventories are generally completed by trained Certified Arborists or experienced inventory arborists. The tree attribute and location data are generally collected using handheld computers, geographic information systems (GIS) data, and/or geographic positioning systems (GPS) equipment. Types of Inventories - Depending on the size of your community and your resources, there are different types of inventories that can be accomplished to provide you with an accurate accounting of public trees.

CHAPTER 1

THE FUTURE OF LARGE OLD TREES IN URBAN LANDSCAPES

Darren S. Le Roux, Karen Ikin, David B. Lindenmayer, Adrian D. Manning, Philip Gibbons

INTRODUCTION

Large old trees have been defined as keystone ecological structures because, relative to their size, they are disproportionate providers of resources crucial to other species [1], [2]. As trees mature, they begin to form a set of unique physical attributes or structural elements, including large volumes of coarse woody debris and litter, peeling bark, dead branches and hollows [3],[4]. Habitat structures provided by large old trees take centuries to form and are typically not provided by younger trees [5]. For example, hollows in *Eucalyptus* typically begin to form in trees 120–220 years old [6]. Hollows alone provide critical nesting resources for a diverse range of taxa worldwide, including invertebrates [7], reptiles [8], birds [9], and mammals [10].

Once large old trees are removed, they can be extremely difficult to replace in the short term because of the prolonged time period needed for trees to mature. This time lag can have serious ecological and management implications, particularly in modified landscapes where the rate of large old tree removal exceeds the rate of tree replacement [11], [12]–[14]. Species that depend on large old trees for survival (e.g. hollow-dependent fauna) may face extinction in the short term without actions that reverse current patterns of tree decline [2].

Human activities such as land clearance, logging and livestock grazing are responsible for the decline of large old trees in a diverse range of ecosystems, including: conifer forests in Europe[15] and North America [16], tropical rainforest in South America [17], and agricultural land in Australia [18]. However, few studies have investigated the decline of large old trees in urban landscapes [19], [20]. This is a major concern given the unprecedented rate of global urbanisation, one of the most rapid and destructive forms of land-use change [21], [22]. Population growth and rising demand for urban living space invariably puts pressure on existing urban habitat that can be important for biodiversity [23], [24], [25]. However, a great deal of uncertainty remains about the future of habitat structures in urban landscapes, especially structures like large old trees that are known to limit some species [26], [27]. Large old trees are especially vulnerable to removal in urban landscapes worldwide due to the potential safety risks posed to the public and infrastructure from falling branches or trees [20], [28], [29]. Therefore, obtaining information about the future availability of large old trees in urban landscapes is of high priority, especially for practitioners who are challenged by balancing urban growth and maintaining critical habitat for biodiversity over the long term.

Citation: Le Roux DS, Ikin K, Lindenmayer DB, Manning AD, Gibbons P (2014) The Future of Large Old Trees in Urban Landscapes. PLoS ONE 9(6): e99403. doi:10.1371/journal.pone.0099403

Although there are parallels between urban landscapes and other modified environments (e.g. agricultural land), the management of trees in human-dominated urban settings poses a suite of unique and complex challenges. The key interacting drivers of tree loss in the urban matrix include: (1) urban sprawl and in-fill practices [30], (2) public safety policies that facilitate managed tree removal in existing greenspace to protect people and infrastructure [20], and (3) reduced tree regeneration [31]. Despite these challenges, urban environments also provide opportunities for innovative tree management, community engagement, people-led conservation strategies, and biodiversity offsets, which may include public tree planting initiatives and artificial nest box projects [32].

In this study, we used a simulation model to predict the future availability of native hollow-bearing trees in a rapidly expanding urban landscape. We used hollow-bearing trees as a surrogate for large old trees and other associated habitat structures such as coarse woody debris [4], [33], [34]. This is because it is well established that as trees age and their size increases so too does the probability of hollow occurrence [5], [35], [36]. Our four main study objectives were to: (1) compare future trajectories in hollow-bearing trees in urban greenspace with semi-natural nature reserves under existing land management practices; (2) identify which variables can be manipulated to increase the number of hollow-bearing trees occurring in urban greenspace over the long term; (3) test the efficacy of multiple alternative tree management strategies aimed at mitigating the decline of hollow-bearing trees; and (4) formulate recommendations that can be widely applied by practitioners to better maintain and perpetuate large old trees and their associated habitat structures in urban landscapes. Given the widespread nature of this issue in urban landscapes, we anticipate that our findings will be relevant to urban practitioners globally.

MATERIALS AND METHODS

Ethics statement

This research was conducted under ethical approval (protocol number A2012/37; The Australian National University Ethics Committee). Vegetation surveys undertaken on nature reserves and public greenspace were approved by permit from the ACT Government, Territory and Municipal Services in compliance with the Nature Conservation Act 1980. Field studies did not involve endangered or protected species.

Study area

We conducted our study in and around the city of Canberra, Australian Capital Territory (ACT), southeastern Australia (35° 17′35. 64″ S; 149° 07′27. 36″ E). Canberra is Australia's eighth largest city covering an area of 810 km^2. The city supports a population of 375,000 people, which is projected to double by 2056 [37]. Canberra is a highly planned city described as the "Bush Capital" because of the extensive suburban tree cover and 34 nature reserves flanking the urban boundary [38]. The city is situated in the ecologically diverse Southern Tablelands region west of the Great Dividing Range. Lowland box-gum *Eucalyptus* woodlands and grasslands once dominated the region [39]. Box-gum grassy woodlands are characterised by two dominant species, yellow box (*Eucalyptus melliodora*) and Blakely's red gum (*E. blakelyi*) that occur in association with other eucalypt species, including apple box (*E. bridgesiana*), red box (*E. polyanthemos*), red stringybark (*E. macrorhyncha*), and scribbly gum (*E. rossii*). Extensive land clearance for stock grazing and urban development has led to a near 95% decline in intact box-gum grassy woodlands, which is now listed as a critically endangered ecological community [40]. What vegetation remains exists in semi-natural nature reserves or as highly modified isolated remnant patches and scattered paddock and urban trees [41], [42].

Sampling design

We confined our sampling effort to a single vegetation type: the predicted pre-European (pre-1750) extent of box-gum grassy woodland. Within this vegetation type, we stratified sampling according to two dominant land-use types and five geographic zones, creating a total of 10 strata. Our land-use types were: (1) nature reserves, which are designated semi-natural areas managed for conservation; and (2) urban greenspace, made up of publicly accessible parklands (60%), roadside margins (24%), remnant vegetation (9%), and sports grounds (7%). Urban greenspace accounted for 11% of the total urban environment in our study area. We divided our study landscape into five geographic zones to capture variability and avoid biasing sampling effort to areas with specific local or historical attributes (e.g. fire history). An equal number of fixed area plots (50×20 m; 0.1 ha) were randomly allocated by land-use type (n=100) and geographic zone (n=40). This resulted in a total of 200 plots or 20 ha of sampled land from 28 reserves and 100 urban greenspaces. Plots were >250 m apart to minimise spatial dependence and allocated to greenspace ≥0.2 ha.

Data collection

We measured the diameter at breast height over bark (DBH; 1.3 m above ground) of every living and dead tree in each plot. We measured only the largest stem of multi-stemmed trees [43]. Trees with stems <1.3 m above the ground were measured at the base of the stem. The number of naturally regenerating and planted seedlings ≤10 cm (DBH) were counted in each plot and formed the first size class of our tree population. We identified all living trees to species level. Each tree was inspected for hollows from all angles on the ground using binoculars (10×25). One observer (DSL) completed this task to reduce multi-observer bias and maintain consistency in hollow identification [44]. Our objective was not to determine the absolute number of hollows but rather relative hollow occurrence per tree. We selected a minimum entrance size of 2 cm for hollows. This was because: (1) the full range of hollow-dependent vertebrate taxa, including marsupials, birds, and bats, would be accounted for; and (2) hollows smaller than 2 cm were difficult to reliably identify from the ground [45].

Simulation model

The simulation model described in [12], tracks the mean DBH of trees, including hollow-bearing trees, in separate size cohorts over time. The model has pre-defined rates of tree mortality and recruitment applied at each time step. For this study, we ran separate simulations for native tree populations occurring in nature reserves and urban greenspace. Exotic trees were recorded only in the urban greenspace and accounted for 30% of all recorded trees. We excluded exotic trees from our analyses because only native trees were recorded with hollows in our study area. Simulation models for both land-use types were parameterised with the following baseline data: the current number of native trees in existing stands sorted by DBH cohort; the predicted age and growth rate of trees; the frequency of regeneration events; the number of seedlings at each regeneration event; and the rate of tree mortality.

There were five principle steps in our modelling process (summarised in Fig. 1 and described further in Summary S1):

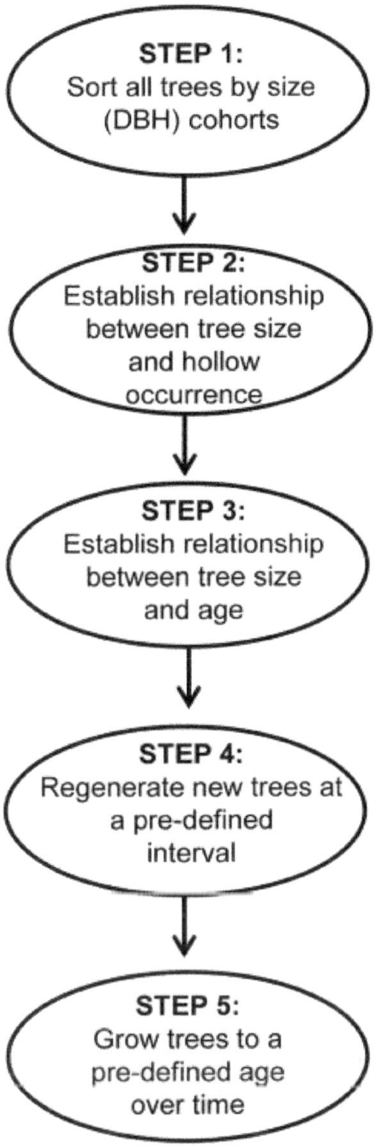

Figure 1: Simple schematic highlighting the five principle steps of our simulation model. http://dx.doi.org/10.1371/journal.pone.0099403.g001

(1) We calculated the mean number of trees in 10 cm DBH size cohorts (ranging from 0.1–10 cm to >100 cm) for each native tree species and dead trees, using data collected in each land-use type (Table S1).

(2) We used a generalised logistic regression model with a binomial distribution and logit link to establish a relationship between hollow occurrence (i.e. the presence of at least one hollow ≥2 cm; binary response) and tree size (i.e. DBH; explanatory variable). We also fitted tree species as an explanatory variable in our model. Based on correlations in hollow occurrence by DBH between individual species, we identified three distinct species groupings. Species group one included yellow box, apple box, brittle gum (*E. mannifera*), broad-leaved peppermint (*E. dives*), bundy (*E. goniocalyx*), mealy bundy (*E. nortonii*), brown barrel (*E. fastigata*), alpine ash (*E. delegatensis*), ribbon gum (*E. viminalis*), mountain gum (*E. dalrympleana*), candlebark (*E. rubida*) and ironbark (*E. sideroxylon*). Group two included Blakely's red gum, red box, red stringybark and scribbly gum. Group three was dead trees. We found that species groups differed significantly (Wald statistic =101.5; $P<0.001$) from each other (Table 1). The relationship between tree size and hollow presence was highly significant in our model (Wald statistic =388.1; $P<0.001$). The area under the receiver operating characteristic curve of our model was 0.92, indicating that the discriminating ability of our model was excellent [46]. For each species group, we derived separate model equations which took the form: *Logit (Pr. Hollows)*=−7.112+(0.086 x DBH) + (species group estimate).

Table 1: Generalised logistic regression model used to predict the proportion of hollow-bearing trees in each 10 cm DBH (diameter at breast height) cohort. http://dx.doi.org/10.1371/journal.pone.0099403.t001

Variables	Coefficient	Standard error	Lower 95% confidence interval	Upper 95% confidence interval	P-value
Intercept	−7.112	0.335	−7.769	−6.456	<0.001***
Species group 1	0.000	-	-	-	-
Species group 2	1.413	0.274	0.876	1.949	<0.001***
Species group 3	3.861	0.383	3.110	4.613	<0.001***
DBH	0.086	0.004	0.077	0.095	<0.001***

Coefficients, standard errors, 95% confidence intervals, and P-values are presented with species group one held as the reference level.
doi:10.1371/journal.pone.0099403.t001

(3) We established a relationship between DBH and tree age using the following equation: $Age = 0.02 \times \pi \times (DBH_{standardised}/2)^2$, where $DBH_{standardised}$ is the yellow box equivalent diameter for each tree. Yellow box is the only tree species for which data exist to establish a relationship between age and DBH [47]. We scaled all DBH values for each tree species relative to that of a yellow box equivalent using the method described in [18], [26]. To do this, we first calculated each DBH value as a proportion of the maximum DBH recorded for each tree species and then multiplied this value by the largest DBH recorded for yellow box in our study area (151 cm). Therefore, we assumed that all species had proportionally equal growth rates that were similar to that of yellow box. Although this approach is not ideal because it is unlikely to yield precise age estimates for each species, it currently is the most practicable solution available in the absence of age-DBH relationship data for other eucalypt species [26],[48]. Therefore, our model had a degree of uncertainty related to tree growth rates, as these data likely differ for each species. However, a previous study [12] found that long-term predictions for mature trees is not sensitive to uncertainty in this variable and suggests that the focus should instead be on testing the effects of uncertainty for other parameters in the model.

(4) We simulated tree regeneration in both land-use types to ensure that uncertainties associated with regeneration were reflected in our models. Tree regeneration is an event-driven process that can be sporadic and influenced by natural phenomena and/or anthropogenic factors such as climate, competition, and planting effort [31], [49]. At each regeneration event, viable seedlings may or may not establish and survive over time. To simulate these uncertainties, the number of seedlings ha^{-1} for each run of our model was drawn randomly from a Poisson distribution with the mean equal to the mean number of trees recorded in the 0–10 cm DBH cohort for each species group. For species group one and two in urban greenspace, the mean number of trees in the 0–10 cm DBH cohort was 11 and 13 seedlings ha^{-1}, respectively. For species group one and two in nature reserves, the mean number of trees in the 0–10 cm DBH cohort was 119 and 193 seedlings ha^{-1}, respectively. The time-step for each run of the model was equivalent to the average age of trees in the 0–10 cm DBH cohort for both land-use types, which was approximately 8 years.

(5) Annual tree mortality was modelled in a density-dependent manner to reflect declines in the number of trees over successive DBH cohorts or as trees age. Therefore, we assumed that tree densities would naturally thin out over time due to factors such as competition among conspecifics [50]. To simulate this process, we calculated annual mortality for each DBH cohort using the equation: $1 - s^{(1/y)}$, where s is the proportion of trees that survive from one cohort to the next, and y is the number of years it takes trees to progress from one cohort to the next by 10 cm DBH increments. However, in some urban greenspaces (e.g. roadside margins), density-dependent mortality may be less pronounced as tree survivability may instead be predominantly influenced by tree planting and protection efforts. Therefore, for urban greenspace, we also tested the mean annual mortality rate across all cohorts, which yielded similar model trajectories to density-dependent mortality. We decided to apply density-dependent mortality to both land-use types for consistency and because a majority of urban greenspace sampled constituted parklands and remnant vegetation where natural regeneration and density-dependent mortality may still occur. We set 500 years as the maximum age that living trees will remain standing in both land-use types. This is based on the only longevity estimate available for eucalypts in our study area [47]. It is reasonable to assume that for other eucalypt species this age would also be the upper limit of survivability. Therefore, model uncertainties pertaining to species longevity are likely to be over-estimated and based on a best-case longevity. We assumed that once trees died in urban greenspace, they no longer functioned as hollow-bearing trees into the next time step. This is based on local tree management policies that facilitate dead tree removal on public land [51]. However, for nature reserves, we conservatively estimated that dead trees could remain standing for at least 50 years after initial mortality (i.e. 550 years in total), based on observations of the standing life of dead trees in *Eucalyptus* forests [52], however, we acknowledge the paucity of available data to support this estimate.

The availability of hollow-bearing trees under existing management practices

We used our simulation model, parameterised with those data detailed above, to predict the mean number of hollow-bearing trees ha^{-1} occurring in nature reserves and urban greenspace over time under existing land management protocols. Simulations were undertaken over 300 years using a Monte Carlo simulation based on 300 runs of our model (i.e.

the number of iterations required for relatively well-defined distributions). This approach relies on random sampling over multiple simulations to generate probabilities in a heuristic manner [53]. Therefore, for each run of our model, input data for several variables were drawn randomly from defined distributions. The number of recruits was drawn from a Poisson distribution (step 4 above). Annual mortality was drawn from a normal distribution, where negative values were converted to zero. The maximum standing life of living trees was held at 500 years for nature reserves. However, for urban greenspace, values were drawn from a uniform distribution between 60 years (the estimated minimum standing life of trees in our study area) [54] and 500 years (the estimated maximum standing life of trees in our study area). This range of lifespans reflects variation in current tree management practices in different types of urban greenspace. Variables held constant in our model were the period between regeneration events (8 years) and coefficients for the age-DBH (0.019) and DBH-hollow (1.413) relationships.

Variables that can be manipulated to mitigate the decline of hollow-bearing trees

We performed a sensitivity analysis, as described in [55], to identify which variables can be manipulated in urban greenspace to mitigate the decline of hollow-bearing trees. For this analysis, we also used a Monte Carlo simulation based on 300 runs of our model. We repeatedly populated each run of the model with data drawn randomly from uniform distributions for each variable. Where applicable, values were drawn from a wider range than observed under existing management practices to more broadly test a range of alternative management strategies. Variables that can be manipulated by management included: (1) maximum standing life of trees (range: 60–500 years for species groups one and two, based on longevity estimates for urban trees in our study area); (2) number of seedlings ha^{-1} (range: 0–60 seedlings ha^{-1} for species groups one and two, testing various regeneration targets); (3) period between regeneration events (range: 1–50 years, testing various regeneration schedules); (4) rate of annual mortality (range: 0.03–0.1 model coefficients, testing various feasible survivability outcomes); and (5) rate of hollow formation (range: 1.5–3.7 model coefficients, testing a range of hollow acceleration strategies above an observed existing rate (i.e. 1.4) up to a rate observed for dead trees (i.e. 3.8) in our study area, which we assumed indicated a maximal hollow formation rate for living trees). We fixed the coefficient for the DBH-age relationship at 0.019 assuming that this could not be changed appreciably.

We used linear regression to test the relative sensitivity of our response variable (i.e. the mean number of hollow-bearing trees ha^{-1}) against the explanatory variables that are the parameters in our simulation model. We natural log-transformed (ln (x+1)) our response to satisfy assumptions of normality. There were no significant interactions between explanatory variables and interaction terms were dropped from the final additive model. We used stepwise regression to determine the model of best fit. Percentage variance accounted for by our final model was 40%. Due to the high number of replications used in simulation models, it is inappropriate to rely on conventional *P*-values to indicate statistical significance [56]. Instead, we used relative effect size, as indicated by variance ratios, to identify the most sensitive parameters in our model. Variance ratios were calculated as the mean square of each term change divided by the residual mean squares of the original maximal model (Table 2). Predictions are presented only for variables with the greatest relative effect sizes (i.e. most ecologically important), where all other explanatory variables are held at their mean model values.

Table 2: Linear regression model used to perform a sensitivity analysis of the mean number of hollow-bearing trees ha^{-1} (ln (x+1) transformed) perpetuated in urban greenspace over 300 years. http://dx.doi.org/10.1371/journal.pone.0099403.t002

Variables	Mean	Standard deviation	Coefficient	Standard error	Variance ratio
Intercept	-	-	0.602	0.204	-
Maximum standing life (years)	274.10	88.04	0.004	0.0003	138.61
Number of seedlings ha^{-1}	31.03	12.09	0.009	0.002	13.81
Rate of hollow formation (coefficient)	2.59	0.17	0.151	0.042	11.04
Rate of annual mortality (coefficient)	0.06	0.02	−1.290	1.450	0.31
Period between regeneration (years)	24.74	13.40	0.000	0.002	0.00

Means, standard deviations, coefficients, standard errors, and variance ratios, which indicate the relative importance or effect size of each model term, are presented for each explanatory variable used to parameterise our simulation model.
doi:10.1371/journal.pone.0099403.t002

The availability of hollow-bearing trees under alternative management strategies

We also simulated a series of alternative management strategies using our simulation model. We modelled the mean number of hollow-bearing trees ha^{-1} occurring in urban greenspace over 300 years. Scenarios were based on either: (1) a management strategy that manipulates only a single variable up to the maximum value defined in our regression model described above, or (2) a combined management strategy that manipulates all three variables for a set of values that we deemed most practicable for urban landscapes given other socio-economic constraints. Variables not manipulated were fixed at their mean values under existing management practices. In all simulated scenarios, management actions were assumed to take effect immediately. Statistical analyses were completed using GenStat (15th edition, VSN International Ltd, Hemel Hempstead, UK).

RESULTS

We recorded a total of 4,865 trees belonging to 16 eucalypt species. Of those trees, 85% (4,111 trees) were recorded in nature reserves and 15% (754 trees) in urban greenspace. The key difference between tree populations in nature reserves and urban greenspace was the number of seedlings recorded in the 0.1–10 cm DBH cohort (Fig. 2). In reserves, we recorded 315 seedlings ha^{-1}, which was 13 times the number recorded in urban greenspace, with 25 seedlings ha^{-1}.

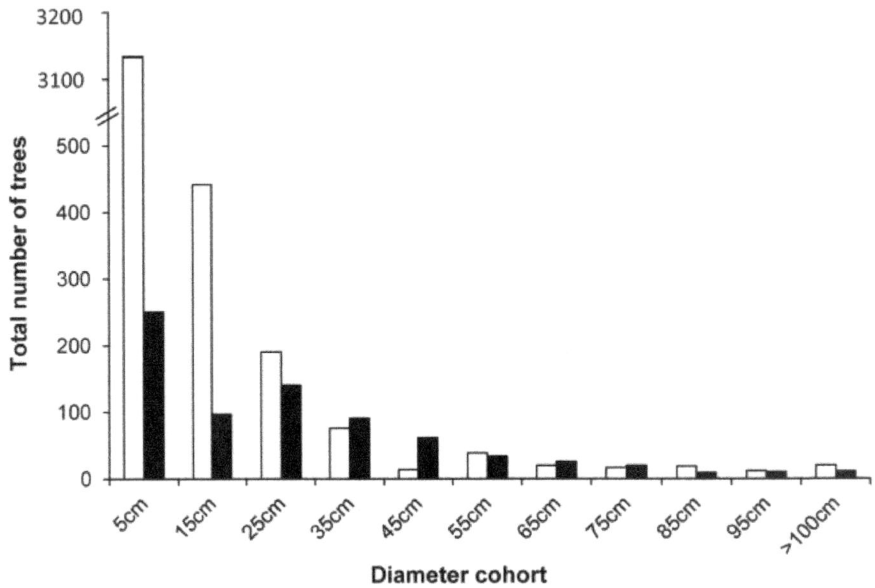

Figure 2: Frequency distribution of median tree diameter cohorts for tree stands (all species) in nature reserves (open bars) and urban greenspace (solid bars). http://dx.doi.org/10.1371/journal.pone.0099403.g002

The availability of hollow-bearing trees under existing management practices

In urban greenspace, we found that under existing management practices, the mean number of hollow-bearing trees ha^{-1} is predicted to decline by 87% over 300 years from an initial recorded stand density of 5.74 trees ha^{-1} to 0.76 trees ha^{-1} (Fig. 3). Conversely, in nature reserves, hollow-bearing tree densities fluctuate around a relatively stable mean density of 13.4 trees ha^{-1}. Prediction intervals for urban greenspace were more variable around the mean than for nature reserves. This is driven by highly variable standing lives that trees are permitted to reach in different urban greenspaces (i.e. 60–500 years old). Prediction intervals indicate that under a worst case scenario (i.e. lower 95% prediction interval) all hollow-bearing trees may be lost from urban greenspace within 115 years. Even under a best case scenario (i.e. upper 95% prediction interval) hollow-bearing trees steadily decline over time.

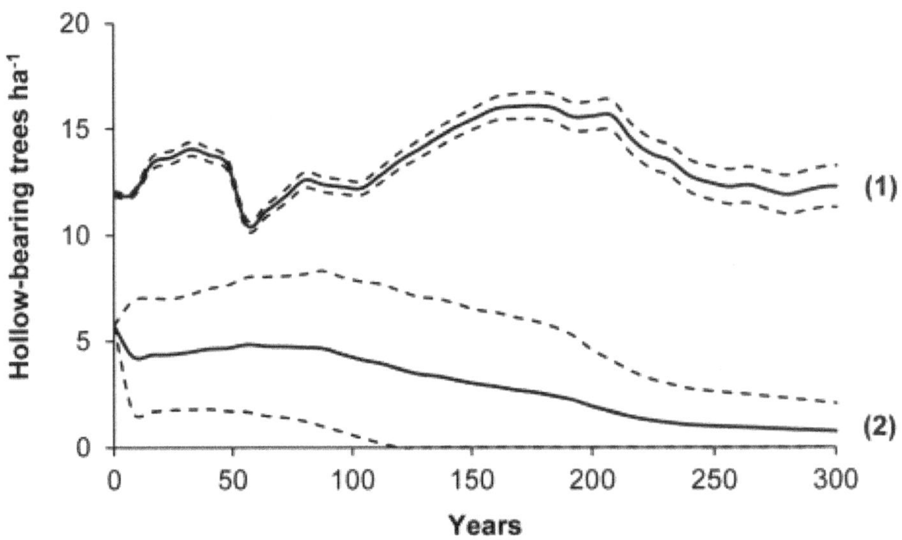

Figure 3: Simulations predicting the relative number of hollow-bearing trees ha^{-1}(mean **±95%** prediction interval) over 300 years under existing management practices in nature reserves (1) and urban greenspace (2). http://dx.doi.org/10.1371/journal.pone.0099403.g003

Variables that can be manipulated to mitigate the decline of hollow-bearing trees

Sensitivity analysis revealed that the mean number of hollow-bearing tree ha^{-1} was most sensitive to: (1) the maximum standing life of trees; (2) the number of seedlings ha^{-1}; and (3) the rate of hollow formation (Table 2). The mean number of hollow-bearing trees ha^{-1} was least sensitive to the period between regeneration events and annual mortality. We also did not identify meaningful interactions between maximum standing life and annual mortality, maximum standing life and the rate of hollow formation, and the number of seedlings ha^{-1} and the period between regeneration events.

Maximum standing life.

The number of hollow-bearing trees perpetuated in urban greenspace over the long term was most sensitive to the maximum standing life of trees (variance ratio =138.61). We predicted that hollow-bearing trees would increase in urban greenspace by approximately 0.8 trees ha^{-1} (22%) for each additional 50 years that trees are permitted to remain standing (Fig. 4A).

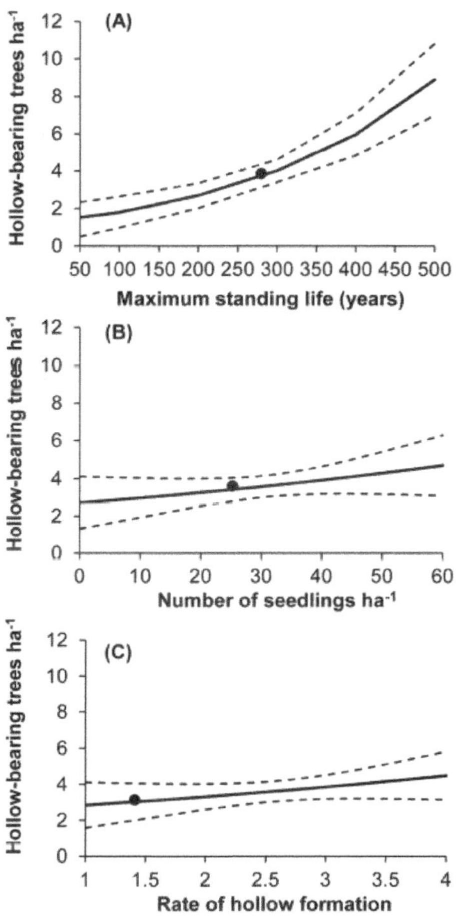

Figure 4: The predicted relative number of hollow-bearing trees ha^{-1} (mean ±95 prediction intervals) in urban greenspace over 300 years for a range of values for variables with the greatest relative effect sizes derived from a sensitivity analysis. Variables include: the maximum standing life of trees (A); the number of seedlings ha^{-1}(B); and the rate of hollow formation (represented by the coefficient for the probability of hollow occurrence; C). Predicted thresholds under existing management practices are provided for reference (solid circles). http://dx.doi.org/10.1371/journal.pone.0099403.g004

Number of seedlings.

The number of seedlings ha^{-1} also contributed to the number of hollow-bearing trees perpetuated in urban greenspace over the long term, although relative to maximum standing life this contribution was smaller (variance ratio =13.81). We predicted that for every 10 additional native seedlings ha^{-1}, the number of hollow-bearing trees would increase by 0.3 trees ha^{-1} (10%; Fig. 4B). However, we predicted that to perpetuate hollow-bearing trees even marginally above existing levels will require at least 30 seedlings ha^{-1} and all trees to remain standing for at least 200 years (Fig. 5A).

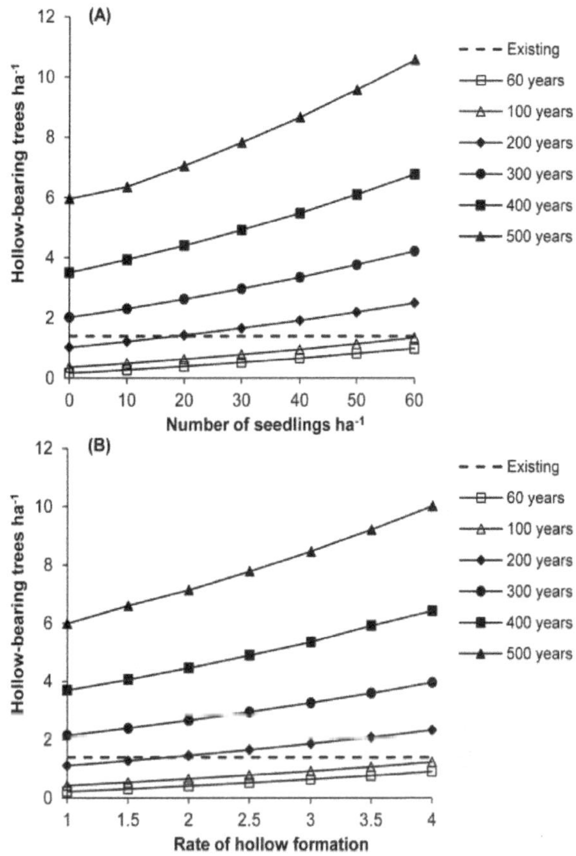

Figure 5: The predicted relative mean number of hollow-bearing trees ha^{-1} in urban greenspace over 300 years for a combination of values for: the maximum standing life of trees and the number of seedlings ha^{-1} (A); and the maximum standing life of trees and the rate of hollow formation (represented by the coefficient for the probability of hollow occurrence; B). http://dx.doi.org/10.1371/journal.pone.0099403.g005

Rate of hollow formation.

Similarly, the rate of hollow formation also contributed to the number of hollow-bearing trees perpetuated in urban greenspace over the long term, although relative to maximum standing life this contribution was smaller (variance ratio =11.04). We predicted that hollow-bearing trees would increase by 0.2 trees ha^{-1} (8%) for every 0.5 increase in the rate of hollow formation (Fig. 4C). However, we predicted that to perpetuate hollow-bearing trees even marginally above existing levels will require accelerating hollow formation to a rate of 2.5 (i.e. a 44% increase above the observed mean rate) and all trees to remain standing for at least 200 years (Fig. 5B).

The availability of hollow-bearing trees under alternative management strategies

Isolated management approach.

If tree standing life were maximised to 500 years and all other variables were unchanged (i.e. held at their mean values under existing management practices), then the mean number of hollow-bearing trees in urban greenspace is predicted to still decline by 64% over the long term, from an initial stand density of 5.74 trees ha^{-1} to 2.09 trees ha^{-1} (Fig. 6). If the number of seedlings ha^{-1} were increased only to 60 seedlings ha^{-1}, then the mean number of hollow-bearing trees in urban greenspace is predicted to still decline by 53% over the long term, from an initial stand density of 5.74 trees ha^{-1} to 2.68 trees ha^{-1}. If hollow formation were accelerated only to a rate of 3.7 (i.e. the maximum rate of hollow formation observed for living trees and a 62% increase above the observed mean rate), then the mean number of hollow-bearing trees in urban greenspace is predicted to initially increase to 9 trees ha^{-1} in the short term, but decline by 92% to 0.46 tree ha^{-1} over the long term.

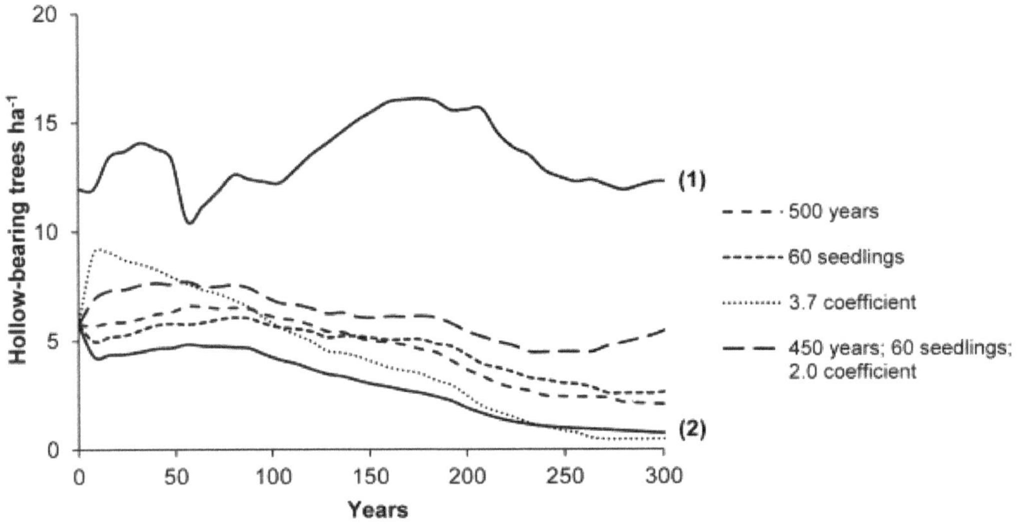

Figure 6: The predicted relative mean number of hollow-bearing trees ha^{-1} over 300 years under a series of alternative urban tree management scenarios (dashed lines). Simulated scenarios include: increasing the standing life of trees only up to 500 years; increasing the number of seedlings only up to 60 ha^{-1}; accelerating hollow formation only by 62% above the observed mean rate (as represented by the coefficient for the probability of hollow occurrence); and a combined management approach (i.e. our recommended management proposal), which manipulates all three variables simultaneously. Scenarios under existing management practices are provided for reference by solid black lines for nature reserves (1) and urban greenspace (2). http://dx.doi.org/10.1371/journal.pone.0099403.g006

Combined management approach.

In contrast, a combined management approach that manipulates all sensitive explanatory variables is predicted to increase the number of hollow-bearing trees ha^{-1} over the long term (Fig. 6). To achieve this will require at least: (1) increasing the standing life of trees to 450 years (approximately 40% longer average lifespans); (2) increasing the number of seedlings to 60 seedlings ha^{-1} (approximately 60% greater regeneration rate); and accelerating hollow formation up to a rate of 2.0 (approximately 30% greater hollow formation rate; see Table 3). Under this scenario, the density of hollow-bearing trees will initially need to be actively increased in the short term by accelerating hollow formation to achieve at least 7 hollow-bearing trees ha^{-1}. Over time, the density of hollow-bearing trees is predicted to first gradually decline before an increase occurs within 250 years.

Table 3: Summarised values for each variable used to parameterise our simulation model under existing management practices for nature reserves and urban greenspace. http://dx.doi.org/10.1371/journal.pone.0099403.t003

Variables	Nature reserves	Urban greenspace	Urban management recommendations
Maximum standing life (years)	500	60–500	450 (~40% increase)
Number of seedlings ha^{-1}(all species)	315	25	60 (~60% increase)
Rate of hollow formation	1.4	1.4	2.0 (~30% increase)
Rate of annual mortality	0.03	0.06	-
Period between regeneration (years)	8	8	-

Relative values are derived from raw vegetation data or, where applicable, published estimates. Urban management recommendations, derived from a series of simulated alternative management strategies, are indicated for variables identified as being the most ecologically important from a sensitivity analysis.
doi:10.1371/journal.pone.0099403.t003

DISCUSSION

Large old trees support unique habitat structures (e.g. hollows, coarse woody debris), which form over extensive time periods and cannot be provided by younger trees [5], [6]. The decline of large old trees in modified landscapes is a global conservation issue that has serious implications for biodiversity [11]. To date, few studies have addressed this problem in urban landscapes, which is a growing concern given the unprecedented rates of urbanisation in cities worldwide [25]. Using a simulation model, we investigated the decline of large old trees in an urban landscape over centuries. We predicted that hollow-bearing trees (a surrogate for large old trees) will decline by 87% over 300 years in urban greenspace under existing management practices. Under a worst case scenario, hollow-bearing trees may be entirely lost from urban greenspace within 115 years. Our analysis revealed that the decline of hollow-bearing trees in urban greenspace is most sensitive to: the maximum standing life of trees, the number of regenerating seedlings ha^{-1}, and the rate of hollow formation. To mitigate the decline of large old trees in urban greenspace over the long term, we recommend a management strategy that collectively: (1) maximises the standing life of trees, (2) increases tree regeneration rates, and (3) accelerates the formation of habitat structures provided by large old trees. These results, and the methods used, have important implications for ecologically sustainable urban development.

Existing management practices

Our results provide further evidence that urban landscapes face a concerning future of large old tree decline, which is comparable with other highly impacted landscapes, including agricultural land [12], [57] and production forests [15], [52]. We argue that predicted declines in hollow-bearing trees in urban greenspace (Fig. 3) will not only negatively impact hollow-dependent fauna (e.g. birds, bats, mammals and invertebrates), but also will impact a much wider range of plant and animal species that rely on large old trees and associated habitat structures (e.g. coarse woody debris, litter, peeling bark) for a range of purposes (e.g. foraging, spatial connectivity, epiphyte attachment). Ultimately, these species may face local extinction in urban landscapes. This is supported by recent research, which demonstrates that the removal of large old trees from existing urban habitats will likely impact animal populations and community assemblages [19], [20].

Predictions under existing management practices also highlight the important role that nature reserves play in bridging resource gaps across urban landscapes. In contrast to urban greenspace, we predicted that nature reserves adjacent to urban areas provide a relatively stable supply of hollow-bearing trees over time. Therefore, maintaining and establishing nature reserves in urban environments will likely provide important habitat refuge for species over the long term. However, nature reserves only represent a small proportion of the urban landscape and on their own are unlikely to achieve biodiversity conservation targets [58]. In addition, many species rely on networks of multiple habitat trees that extend over large areas of the landscape, including urban habitats [59]. For these reasons, we strongly encourage management strategies that focus on arresting large old tree decline within the 'working' urban matrix. This means that a re-evaluation of existing management practices in urban landscapes is needed to address the underlying drivers of tree decline.

Alternative management strategies

Large old trees are especially susceptible to removal in urban landscapes worldwide [20], [28],[29], [60]. With this in mind, we have formulated a set of targeted recommendations, based on results from our analyses, which we anticipate to be relevant to practitioners in a wide range of urban landscapes where trees are maintained.

Maximise tree standing life.

A major source of tree mortality in urban landscapes is due to managed tree removal [20]. This is facilitated by public safety policies and practices, which aim to minimise risk of injury to people and damage to property due to falling trees and branches. For example, in our study area it is estimated that by 2050, approximately 175,600 street trees (24% of all trees in urban greenspace) will have reached their safe standing life (ranging from 60 to 100 years old) and are likely to be removed [54]. Consequently, large old trees, hollow-bearing trees, dead trees and decaying branches are most susceptible to targeted removal prior to reaching their full potential in terms of forming and providing suitable habitat. We found that the number of hollow-bearing trees perpetuated in urban greenspace over the long term was most sensitive to the maximum standing life of trees (Table 2; Fig. 4A). Increasing the standing life of all trees by 50 years is predicted to increase the number of hollow-bearing trees ha^{-1} in urban greenspace by 22% over the long term.

Policymakers need to recognize the important habitat resources provided by large old trees and accordingly formulate or amend tree management protocols so that large old trees are afforded better protection. This may involve re-evaluating criteria used to guide tree felling decisions [29]. Practical strategies that maximise the safe standing life of trees should also be implemented. This may involve: (1) allowing trees to age more naturally in urban greenspace frequented less by members of the public and where risk to people and property is minimal (e.g. derelict land, areas along stormwater wetlands, and some parklands); (2) avoiding structural damage to trees (e.g. damage to roots due to road works); (3) creating safe zones or barriers that separate the public from potentially hazardous trees thereby minimising safety risks (e.g. landscaping around the base of the tree using shrubs); (4) physically re-enforcing the structural integrity of large, old trees (e.g. supporting frames, cables or poles); and (5) safely retaining dead trees wherever possible. However, our results indicated that management strategies that only maximise the standing life of trees will be insufficient at mitigating the decline of hollow-bearing trees over the long term (Fig. 6).

Increase tree regeneration.

We found that the rate of tree regeneration in urban greenspace (both natural and planted) was 13 times lower than in nature reserves (Fig. 2). A lack of young trees is a major contributing factor of large old tree decline in urban greenspace over the long term. Older trees that eventually die and are removed from any given landscape need to be replaced by younger trees, thereby perpetuating the formation of important habitat structures over multiple generations [18], [61]. We predicted that increasing tree regeneration by 10 native seedlings ha^{-1} would increase the number of hollow-bearing trees in urban greenspace by 10% over the long term (Fig. 4B).

Tree regeneration in urban habitats is typically achieved through planting initiatives and encouraging natural regeneration. Increasing the number of planted trees through government and community initiatives should increase the number of young trees persisting in urban habitats [62]. However, in some urban greenspace (e.g. roadside margins and residential areas), tree planting can be logistically challenging as practitioners need to balance multiple socio-economic and ecological factors when implementing planting strategies, including: site location, public safety, aesthetics, land ownership, and existing vegetation [63]. Furthermore, reducing seedling mortality in urban habitats is also an important consideration that may require additional protection measures (e.g. tree guards, supporting posts) and costs [64]. In some urban greenspace (e.g. parklands, wetlands) it may be more cost-effective over the long term to promote natural regeneration. Natural regeneration in urban habitats is predominantly limited because of: unfavourable seedbed conditions (e.g. impervious surfaces, pollution, and nutrient runoff), increased competition from invasive plants, and increased mortality due to mowing and pedestrian traffic [31], [65]. Strategies that promote natural regeneration could involve: fencing-off areas with existing re-growth, increasing public awareness of regenerating areas through signage, and enhancing local microclimates that favour seedling establishment and survival such as retaining litter and logs [31], [66]. However, our results indicated that management strategies based solely on increasing tree regeneration will be insufficient at mitigating the decline of hollow-bearing trees over the long term (Fig. 6).

Accelerate the formation of habitat structures provided by large old trees.

The formation of habitat structures such as hollows is a slow process more likely to occur in large old trees [35]. This is because trees with compromised structural integrity are more susceptible to wood decay resulting in the formation of hollows and other structures such as fallen logs and dead branches. Strategies promoting the formation of habitat structures by artificial means can bypass the time needed for these structures to form naturally. Our results indicate that the density of hollow-bearing trees could be increased in urban greenspace by accelerating hollow formation (Fig. 4C).

Accelerating hollow formation in urban areas is commonly achieved by replicating hollow structures, such as installing artificial nest boxes [32]. However, in urban areas, there are limitations with artificial habitat structures, including: occupancy by pest species, poor rates of target species occupancy, and rapid rates of attrition through collapse and decay of materials [67]. It may also not be feasible or practicable to install and maintain artificial habitat structures in large enough numbers across extensive areas over centuries. Therefore, strategies that accelerate the formation of habitat structures by other means should also be explored [68]. Methods previously proposed for hollows include: tree ringbarking or girdling[69], canopy topping [70], controlled fire burns [71], and injecting trees with herbicides [72]. These strategies are also likely to accelerate the formation of other important habitat structures provided by large old trees, including dead branches and coarse woody debris. In urban landscapes, sub-lethal methods of accelerating habitat structure formation are most preferable to also avoid compromising public safety. This may involve only partially injuring trees by carving out hollows on trunks and some branches [73] and using more invasive methods on trees with large diameters that are structurally robust in order to also maximise tree standing life [35]. More research is still needed to investigate methods aimed at accelerating habitat structure formation, especially in urban landscapes. Nevertheless, our results highlight that management strategies based solely on accelerating hollow formation can be effective at increasing the density of hollow-bearing trees in the short-term, but not over the long term (Fig. 6).

Our management proposal: A combined management approach is needed.

Our results emphasise that a combination of different management approaches, aimed at improving multiple aspects of tree management and maintenance, are needed to perpetuate hollow-bearing trees in urban greenspace over the long term (Fig. 5). We propose a management strategy based on simultaneously manipulating all three explanatory variables discussed above, which were identified as being the most sensitive model parameters in our analyses. Under this scenario (Fig. 6), we predicted that the decline of hollow-bearing trees in urban greenspace can be arrested within 250 years if: (1) trees remain standing for at least 450 years ensuring that they reach their maximum habitat potential; (2) at least 60 seedlings ha^{-1}are planted or naturally regenerated; and (3) hollow formation is accelerated to a rate of 2.0 in the short term by installing nest boxes and sub-lethally creating hollows by other methods. Our proposal considers the complexities associated with managing urban greenspaces for multiple purposes, including recreation and conservation. We recognize that it may not be possible to retain all trees to their maximum biological age due to public safety risks. It may also not be practical or feasible to accelerate the formation of habitat structures artificially on a large enough scale over prolonged time periods. Instead, we attempt to balance socio-economic and biodiversity benefits by combining multiple tree management and maintenance approaches in an achievable manner. Future research should also aim to investigate alternative management scenarios from a more financial perspective, which too would benefit practitioners (e.g. numbers of hollow-bearing trees gained per management dollar spent). However, even under our proposed management strategy, the density of hollow-bearing trees is predicted to first decline, or undergo a bottleneck, before increasing. This is because of an extinction debt or the time lag between implementing management actions and actually observing an increase in hollow-bearing trees. Delaying mitigation is anticipated to further exacerbate the effects of time lags and require more drastic measures at greater costs to reverse tree declines [26]. Immediate action will likely also reduce bottlenecks in urban plant and animal populations that depend on large old trees for survival.

CONCLUSION

We have quantified the decline of hollow-bearing trees in an urban landscape over centuries. We provided a novel assessment of the conservation implications associated with existing tree management practices and the efficacy of a range of alternative management strategies. It is evident from our results that existing urban tree management practices require urgent re-evaluation if hollow and tree-dependent biodiversity are to be maintained in urban landscapes. We recommend that: (1) large old trees are afforded better protection and remain standing over longer time periods; (2) tree regeneration is actively improved so that large old trees lost over time are replaced by younger trees; and (3) the formation of habitat structures provided by large old trees is accelerated to compensate for short term deficits in resource availability. Immediate implementation of these recommendations is needed to arrest the decline of large old trees, avoid lag effects, and avert long term risk to biodiversity in urban landscapes.

Table S1: List of recorded tree species and diameter size class distributions

A. Nature reserves	**Tree species**	**Diameter size class (cm)**										
		0.1-10	11-20	21-30	31-40	41-50	51-60	61-70	71-80	81-90	91-100	>100
	E. blakelyi (Blakely's red gum)	1,595	122	39	27	3	16	9	4	5	8	9
	E. bridgesiana (apple box)	244	58	49	9	1	6	2	2	3	0	2
	E. dalrympleana (mountain gum)	1	0	0	0	0	0	0	1	0	0	0
	E. dives (broad-leaved peppermint)	10	0	0	1	1	0	1	0	0	0	0
	E. goniocalyx (bundy)	0	0	1	0	0	0	0	0	0	0	0
	E. macrorhyncha (red stringybark)	60	39	17	7	2	0	2	0	0	0	1
	E. mannifera (brittle gum)	5	0	0	1	1	0	1	0	0	0	0
	E. melliodora (yellow box)	948	153	57	20	5	13	3	8	9	2	12
	E. nortonii (mealy bundy)	2	0	0	0	0	0	0	0	0	0	0
	E. polyanthemos (red box)	69	34	9	2	0	0	1	2	0	0	0
	E. rossii (scribbly gum)	216	32	21	10	1	4	2	0	0	1	1
	Dead trees	38	43	9	7	3	4	3	2	1	1	0
Sum		**3,188**	**481**	**202**	**84**	**17**	**43**	**24**	**19**	**18**	**12**	**25**
B. Urban greenspace												
	E. blakelyi (Blakely's red gum)	90	22	29	23	11	5	4	6	2	3	5
	E. bridgesiana (apple box)	14	4	12	3	0	1	0	0	0	0	0
	E. dalrympleana (mountain gum)	1	0	0	1	2	0	2	1	3	0	0
	E. delegatensis (alpine ash)	0	0	0	0	1	0	0	0	0	0	0
	E. dives (broad-leaved peppermint)	1	1	2	5	4	3	1	1	2	0	0
	E. fastigata (brown barrel)	0	0	3	0	2	0	0	0	0	0	0
	E. goniocalyx (bundy)	1	2	0	2	0	0	0	0	0	0	0
	E. macrorhyncha (red stringybark)	0	1	0	2	4	3	3	1	0	1	1
	E. mannifera (brittle gum)	20	21	37	26	11	7	3	2	1	0	0
	E. melliodora (yellow box)	68	26	29	10	11	10	3	4	1	4	6
	E. nortonii (mealy bundy)	2	1	1	1	1	0	0	0	0	0	0
	E. polyanthemos (red box)	1	8	5	3	1	0	1	0	0	0	0
	E. rossii (scribbly gum)	38	6	4	1	0	0	0	1	0	0	0
	E. rubida (candlebark)	8	1	2	4	4	0	0	1	0	0	0
	E. sideroxylon (ironbark)	1	5	7	2	2	3	4	1	0	0	0
	E. viminalis (ribbon gum)	4	3	8	9	5	3	2	2	0	1	1
	Dead trees	1	1	0	3	0	0	0	0	0	0	0
Sum		**249**	**102**	**139**	**95**	**62**	**35**	**23**	**20**	**9**	**9**	**13**

SUMMARY S1. DESCRIPTION OF THE SIMULATION MODEL USED FOR ANALYSES

The simulation model we used in our paper to predict the size-class distribution of trees in stands over time was developed using Visual Basic in Microsoft Excel and can be obtained from the authors on request. This simulation model is also described with respect to the management of scattered trees in agricultural landscapes [1], mature trees in wood production forests [2], and nest trees for a threatened species in [3].

This simulation model tracks the mean diameter at breast height (DBH) of trees in cohorts through time. For this exercise, we separately simulated trees in reserves and urban areas. We employed a Monte Carlo approach to simulation, whereby parameters in the model can be entered as random values within a range of observed or likely values so the predictions reflect the uncertainty of these parameters. The model is then run as many times as is specified (in this instance all predictions were based on 300 runs of the model), so predictions for any single scenario are a summary of values from multiple runs of the model. There are six key steps in the simulation.

Step 1 – Record the numbers of trees by size-class

The simulation commences with data on the mean number of trees in 10 cm diameter classes (i.e., 0-10 cm, 11-20 cm, etc.) for each tree species group (species distributions summarised in Table S1). Initial values at the commencement of the simulation (T=0 years) were based on data collected at 200 50 m x 20 m (0.1 ha) plots in urban and nature reserve sites within our study area. These data are expressed as a per ha basis for each simulation.

Step 2 – Estimate the numbers of trees with hollows

We predicted the proportion of trees that contained hollows separately for each tree species group and DBH cohort based on visual estimation of the presence/absence of hollows in trees across the study area using the equation: *Logit (Pr. Hollows)* = -7.112 + (0.086 x DBH) + (species group estimate). The area under the receiver operating characteristic curve of this statistical model was 0.92, indicating that its discriminating ability was excellent [4].

Step 3 – Grow trees for *t* years

To simulate tree-growth over time, we developed a relationship between tree age and DBH using the following equation developed by [5]: Age = $0.02 \times \pi \times (DBH_{standardized}/2)^2$, where $DBH_{standardised}$ is the yellow box (*Eucalyptus melliodora*) equivalent diameter for each tree as defined below. Data on the relationship between age and DBH only exist for one tree species (yellow box) in our study area [6]. To predict the ages of trees of other species, we followed the procedure outlined by [5] and calculated a yellow box equivalent diameter for each individual tree of the other eucalypt species. The procedure assumed that all eucalypt species in our study area follow an identical growth curve relative to their maximum attainable diameter, and have the same approximate life-span as yellow box. DBH values for all tree species were initially standardised as a proportion of the maximum attainable diameter for that species observed in the field. Those values were then multiplied by the maximum diameter observed for yellow box (151 cm) to obtain a yellow box equivalent diameter. We acknowledge that this procedure is unlikely to give precise age estimates, but it is a pragmatic solution given the paucity of data on tree ages available for trees in our study area. However, previous research [1,2] indicates that the number of mature trees perpetuated over time is not sensitive to this variable. The initial age of trees in each cohort is predicted by using the median DBH of each cohort in Equation 2, t years is added to this age (where t is the years between recruitment events) and then the inverse of Equation 2 is used, in turn, to predict the new DBH of the cohort after t years.

Step 4 – Recruit a specified number of trees at the beginning of each time-step

We recruited n new trees per ha into each landscape type (urban and reserve) every t years. For the *status quo* scenarios, n was the mean number of trees by tree species group in the smallest DBH class within each landscape and t was the age of this cohort (estimated using Equation 2). Recognising that tree recruitment can be highly variable from year to year, the value for n for each tree species group in each landscape type for the *status quo* scenarios was a random value drawn from a Poisson distribution with a mean taken from the smallest diameter cohort (0-10 cm DBH) for trees in each species group within each landscape type. For the alternative management scenarios in urban areas, we chose random values from a uniform distribution. Runs with negative values for recruitment were treated as zero recruitment.

Step 5 – apply tree mortality

We applied two sources of mortality during each time-step. We calculated tree mortality from data collected on changes in the mean numbers of trees in each DBH cohort, reflecting the density-dependent nature of tree mortality in natural stands. This was given as: $1 - s^{(1/y)}$, where s is the proportion of trees that survive from one cohort to the next, and y is the number of years between recruitment events.

For trees in reserves, we set 500 as the maximum number of years that trees will remain living, which is based on longevity estimates for yellow box reported by [6]. There were no other data from which this estimate could be derived and [1] reports that the number of scattered trees is not sensitive to this parameter in simulations of this type. For urban areas, the maximum number of years that trees remain living was selected randomly from a uniform distribution between 60-500 years, reflecting the existing policy of the government in our study area to remove trees as young as 60 years old for safety reasons. Once a tree had died, we allowed it to remain standing for 50 years in reserves, but in urban areas we removed the tree immediately in keeping with management practices in our study area.

At the completion of this step the number of surviving trees by DBH class and species group was tallied.

The predicted proportion of hollow-bearing tree was then multiplied by the mean numbers of surviving trees in each

DBH cohort to arrive at a predicted number of trees with hollows at the end of each time-step (*t*).

Steps 1-5 are repeated such that *t* (the period between regeneration events) is added to *T* (the total length of the simulation) until *T*=300 years. We reported the mean (±95% prediction interval) for all runs of the simulation model at each time-step between 0 and 300 years.

ACKNOWLEDGMENTS

We thank Dr. Peter Lane for his statistical support, the many private landowners and reserve managers who granted access to properties for vegetation surveys, and anonymous reviewers who improved earlier versions of this manuscript.

AUTHOR CONTRIBUTIONS

Conceived and designed the experiments: DSL KI DBL ADM PG. Performed the experiments: DSL PG. Analyzed the data: DSL PG. Contributed reagents/materials/analysis tools: DSL PG. wrote the paper: DSL KI DBL ADM PG.

REFERENCES

1. Manning AD, Fischer J, Lindenmayer DB (2006) Scattered trees are keystone structures - implications for conservation. Biological Conservation 132: 311–321. doi: 10.1016/j.biocon.2006.04.023
2. Lindenmayer DB, Laurance WF, Franklin JF, Likens GE, Banks SC, et al. (2013) New policies for old trees: averting a global crisis in a keystone ecological structure. Conservation Letters 00: 1–9. doi: 10.1111/conl.12013
3. Lindenmayer DB, Wood JT (2010) Long-term patterns in the decay, collapse, and abundance of trees with hollows in the mountain ash (Eucalyptus regnans) forests of Victoria, southeastern Australia. Canadian Journal of Forest Research 40: 48–54. doi: 10.1139/x09-185
4. Goodburn JM, Lorimer CG (1998) Cavity trees and course woody debris in old-growth and managed northern hardwood forests in Wisconsin and Michigan. Canadian Journal of Forest Research 28: 427–438. doi: 10.1139/cjfr-28-3-427
5. Ranius T, Niklasson M, Berg N (2009) Development of tree hollows in pedunculate oak (Quercus robur). Forest Ecology and Management 257: 303–310. doi: 10.1016/j.foreco.2008.09.007
6. Gibbons P, Lindenmayer DB (2002) Tree hollows and wildlife conservation in Australia. Victoria, Australia: CSIRO publishing.
7. Ranius T (2002) Influence of stand size and quality of tree hollows on saproxylic beetles in Sweden. Biological Conservation 103: 85–91. doi: 10.1016/s0006-3207(01)00124-0
8. Webb JK, Shine R (1997) Out on a limb: Conservation implications of tree-hollow use by a threatened snake species (Hoplocephalus bungaroides: Serpentes, Elapidae). Biological Conservation 81: 21–33. doi: 10.1016/s0006-3207(96)00160-7
9. Newton I (1994) The role of nest sites in limiting the numbers of hole-nesting birds: A review. Biological Conservation 70: 265–276. doi: 10.1016/0006-3207(94)90172-4
10. Lindenmayer DB, Cunningham RB, Tanton MT, Smith AP (1990) The conservation of arboreal marsupials in the montane ash forests of the Central Highlands of Victoria, south-east Australia: II. The loss of trees with hollows and its implications for the conservation of leadbeater's possum Gymnobelideus leadbeateri McCoy (marsupialia: petauridae). Biological Conservation 54: 133–145. doi: 10.1016/0006-3207(90)90138-f
11. Lindenmayer DB, Laurance WF, Franklin JF (2012) Global Decline in Large Old Trees. Science 338: 1305–1306. doi: 10.1126/science.1231070
12. Gibbons P, Lindenmayer DB, Fischer J (2008) The future of scattered trees in agricultural landscapes. Conservation Biology 22: 1309–1319. doi: 10.1111/j.1523-1739.2008.00997.x
13. Sodhi NS, Koh LP, Clements R, Wanger TC, Hill JK, et al. (2010) Conserving Southeast Asian forest biodiversity in human-modified landscapes. Biological Conservation 143: 2375–2384. doi: 10.1016/j.biocon.2009.12.029
14. Laurance WF, Laurance SG, Ferreira LV, Rankin-de Merona JM, Gascon C, et al. (1997) Biomass Collapse in Amazonian Forest Fragments. Science 278: 1117–1118. doi: 10.1126/science.278.5340.1117
15. Andersson R, Östlund L (2004) Spatial patterns, density changes and implications on biodiversity for old trees in the boreal landscape of northern Sweden. Biological Conservation 118: 443–453. doi: 10.1016/j.biocon.2003.09.020
16. Lutz JA, van Wagtendonk JW, Franklin JF (2009) Twentieth-century decline of large-diameter trees in Yosemite National Park, California, USA. Forest Ecology and Management 257: 2296–2307. doi: 10.1016/j.foreco.2009.03.009
17. Laurance WF, Delamonica P, Laurance SG, Vasconcelos HL, Lovejoy TE (2000) Rainforest fragmentation kills big trees. Nature 404: 836–836. doi: 10.1038/35009032
18. Fischer J, Zerger A, Gibbons P, Stott J, Law B (2010) Tree decline and the future of Australian farmland biodiversity. PNAS 107: 19597–19602. doi: 10.1073/pnas.1008476107
19. Stagoll K, Lindenmayer DB, Knight E, Fischer J, Manning AD (2012) Large trees are keystone structures in urban parks. Conservation Letters 0: 1–8. doi: 10.1111/j.1755-263x.2011.00216.x
20. Carpaneto G, Mazziotta A, Coletti G, Luiselli L, Audisio P (2010) Conflict between insect conservation and public safety: the case study of a saproxylic beetle (Osmoderma eremita) in urban parks. Journal of Insect Conservation 14: 555–565. doi: 10.1007/s10841-010-9283-5
21. Cohen JE (2003) Human Population: The Next Half Century. Science 302: 1172–1175. doi: 10.1126/science.1088665
22. Grimm NB, Faeth SH, Golubiewski NE, Redman CL, Wu J, et al. (2008) Global change and the ecology of cities. Science 319: 756–760. doi: 10.1126/science.1150195

23. Theobald DM, Miller JR, Hobbs NT (1997) Estimating the cumulative effects of development on wildlife habitat. Landscape and Urban Planning 39: 25–36. doi: 10.1016/s0169-2046(97)00041-8

24. Blewett CM, Marzluff JM (2005) Effects of urban sprawl on snags and the abundance and productivity of cavity-nesting birds. The Condor 107: 678–693. doi: 10.1650/0010-5422(2005)107[0678:eousos]2.0.co;2

25. McDonald RI, Kareiva P, Forman RTT (2008) The implications of current and future urbanization for global protected areas and biodiversity conservation. Biological Conservation 141: 1695–1703. doi: 10.1016/j.biocon.2008.04.025

26. Manning AD, Gibbons P, Fischer J, Oliver DL, Lindenmayer DB (2012) Hollow futures? Tree decline, lag effects and hollow-dependent species. Animal Conservation 16: 395–405. doi: 10.1111/acv.12006

27. Cockle KL, Martin K, Wesolowski T (2011) Woodpeckers, decay, and the future of cavity-nesting vertebrate communities worldwide. Frontiers in Ecology and the Environment 9: 377–382. doi: 10.1890/110013

28. Nagendra H, Gopal D (2010) Street trees in Bangalore: Density, diversity, composition and distribution. Urban Forestry & Urban Greening 9: 129–137. doi: 10.1016/j.ufug.2009.12.005

29. Terho M (2009) An assessment of decay among urban Tilia, Betula, and Acer trees felled as hazardous. Urban Forestry & Urban Greening 8: 77–85. doi: 10.1016/j.ufug.2009.02.004

30. Pauleit S, Ennos R, Golding Y (2005) Modeling the environmental impacts of urban land use and land cover change—a study in Merseyside, UK. Landscape and Urban Planning 71: 295–310. doi: 10.1016/s0169-2046(04)00083-0

31. Lehvävirta S, Rita H (2002) Natural regeneration of trees in urban woodlands. Journal of Vegetation Science 13: 57–66. doi: 10.1111/j.1654-1103.2002.tb02023.x

32. Harper MJ, McCarthy MA, van der Ree R (2005) The use of nest boxes in urban natural vegetation remnants by vertebrate fauna. Wildlife Research 32: 509–516. doi: 10.1071/wr04106

33. Thor G, Johansson P, Jönsson M (2010) Lichen diversity and red-listed lichen species relationships with tree species and diameter in wooded meadows. Biodiversity and Conservation 19: 2307–2328. doi: 10.1007/s10531-010-9843-8

34. Manning AD, Lindenmayer DB, Cunningham RB (2007) A study of coarse woody debris volumes in two box-gum grassy woodland reserves in the Australian Capital Territory. Ecological Management & Restoration 8: 221–224. doi: 10.1111/j.1442-8903.2007.00371.x

35. Gibbons P, Lindenmayer DB, Barry SC, Tanton MT (2000) Hollow formation in eucalypts from temperate forests in southeastern Australia. Pacific Conservation Biology 6: 218–228.

36. Harper MJ, McCarthy MA, van der Ree R (2005) The abundance of hollow-bearing trees in urban dry sclerophyll forest and the effect of wind on hollow development. Biological Conservation 122: 181–192. doi: 10.1016/j.biocon.2004.07.003

37. ACT Government (2011) Population and residential density in Canberra. Canberra: ACT Government.

38. Banks JCG, Brack CL (2003) Canberra›s Urban Forest: Evolution and planning for future landscapes. Urban Forestry and Urban Greening 1: 151–160. doi: 10.1078/1618-8667-00015

39. ACT Government (2004) Woodlands for Wildlife: ACT Lowland Woodland Conservation Strategy. In:Environment ACT, Urban Services, editor. Canberra: Environment ACT.

40. Department of Environment and Heritage (2006) EPBC policy statement 3.5 - white box - yelllow box - Blakely›s red gum grassy woodlands and derived native grasslands listing. Canberra, Australia.

41. Gibbons P, Boak M (2002) The value of paddock trees for regional conservation in an agricultural landscape. Ecological Management & Restoration 3: 205–210. doi: 10.1046/j.1442-8903.2002.00114.x

42. Manning AD, Lindenmayer DB, Barry SC (2004) The conservation implications of bird reprodcution in the agricultural «matrix»: A case study of the vulnerable superb parrot of south-eastern Australia. Biological Conservation 120: 363–374. doi: 10.1016/j.biocon.2004.03.008

43. Gibbons P, Briggs SV, Ayers DA, Doyle S, Seddon J, et al. (2008) Rapidly quantifying reference conditions in modified landscapes. Biological Conservation 141: 2483–2493. doi: 10.1016/j.biocon.2008.07.009

44. Rayner L, Ellis M, Taylor JE (2011) Double sampling to assess the accuracy of ground-based surveys of tree hollows in eucalypt woodlands. Austral Ecology 36: 252–260. doi: 10.1111/j.1442-9993.2010.02145.x

45. Gibbons P, Lindenmayer DB, Barry SC, Tanton MT (2002) Hollow selection by vertebrate fauna in forests of southeastern Australia and implications for forest management. Biological Conservation 103: 1–12. doi: 10.1016/s0006-3207(01)00109-4

46. Pearce J, Ferrier S (2000) Evaluating the predictive performance of habitat models developed using logistic regression. Ecological Modelling 133: 225–245. doi: 10.1016/s0304-3800(00)00322-7

47. Banks JCG (1997) Tree ages and ageing in yellow box. In: Dargavel J, editor. The coming of age Forest age and heritage values.Canberra, Australia: Environment Australia. pp. 17–28.

48. Fischer J, Stott J, Zerger A, Warren G, Sherren K, et al. (2009) Reversing a tree regeneration crisis in an endangered ecoregion. PNAS 106: 10386–10391. doi: 10.1073/pnas.0900110106

49. Wellington AB, Noble IR (1985) Post-Fire Recruitment and Mortality in a Population of the Mallee Eucalyptus Incrassata in Semi-Arid, South-Eastern Australia. Journal of Ecology 73: 645–656. doi: 10.2307/2260501

50. Schönau APG, Coetzee J (1989) Initial spacing, stand density and thinning in eucalypt plantations.

Forest Ecology and Management 29: 245–266. doi: 10.1016/0378-1127(89)90097-2

51. ACT Government (2013) Management of Trees on Public Land. In: Territory and Municipal Services, editor. Canberra, ACT.

52. Gibbons P, McElhinny C, Lindenmayer DB (2010) What strategies are effective for perpetuating structures provided by old trees in harvested forests? A case study on trees with hollows in south-eastern Australia. Forest Ecology and Management 260: 975–982. doi: 10.1016/j.foreco.2010.06.016

53. Mooney CZ (1997) Monte Carlo Simulation. California: Sage Publications, Inc.

54. ACT Government (2012) Canberra Tree Audit: Method Field Guidelines. In: Territory and Municipal Services, editor. Canberra: ACT Government.

55. McCarthy MA, Burgman MA, Ferson S (1995) Sensitivity analysis for models of population viability. Biological Conservation 73: 93–100. doi: 10.1016/0006-3207(95)90029-2

56. White JW, Rassweiler A, Samhouri JF, Stier AC, White C (2013) Ecologists should not use statistical significance tests to interpret simulation model results. Oikos 123: 385–388. doi: 10.1111/j.1600-0706.2013.01073.x

57. Pulido FJ, Díaz M, Hidalgo De Trucios SJ (2001) Size structure and regeneration of Spanish holm oak Quercus ilex forests and dehesas: Effects of agroforestry use on their long-term sustainability. Forest Ecology and Management 146: 1–13. doi: 10.1016/s0378-1127(00)00443-6

58. Rodrigues ASL, Andelman SJ, Bakarr MI, Boitani L, Brooks TM, et al. (2004) Effectiveness of the global protected area network in representing species diversity. Nature 428: 640–643. doi: 10.1038/nature02422

59. Rhodes M, Wardell-Johnson GW, Rhodes MP, Raymond BEN (2006) Applying Network Analysis to the Conservation of Habitat Trees in Urban Environments: a Case Study from Brisbane, Australia

60. Stagoll K, Lindenmayer DB, Knight E, Fischer J, Manning AD (2012) Large trees are keystone structures in urban parks. Conservation Letters.

61. Weinberg A, Gibbons P, Briggs SV, Bonser SP (2011) The extent and pattern of Eucalyptus regeneration in an agricultural landscape. Biological Conservation 144: 227–233. doi: 10.1016/j.biocon.2010.08.020

62. Pincetl S (2010) Implementing Municipal Tree Planting: Los Angeles Million-Tree Initiative. Environmental Management 45: 227–238. doi: 10.1007/s00267-009-9412-7

63. Wu C, Xiao Q, McPherson EG (2008) A method for locating potential tree-planting sites in urban areas: A case study of Los Angeles, USA. Urban Forestry & Urban Greening 7: 65–76. doi: 10.1016/j.ufug.2008.01.002

64. Nowak DJ, Kuroda M, Crane DE (2004) Tree mortality rates and tree population projections in Baltimore, Maryland, USA. Urban Forestry & Urban Greening 2: 139–147. doi: 10.1078/1618-8667-00030

65. Tonnesen AS, Ebersole JJ (1997) Human trampling effects on regeneration and age structures of Pinus edulis and Juniperus monosperma. Western North American Naturalist 57: 55–56.

66. Pauleit S (2003) Urban street tree plantings: identifying the key requirements. Proceedings of the ICE - Municipal Engineer. pp. 43–50.

67. Lindenmayer DB, Welsh A, Donnelly C, Crane M, Michael D, et al. (2009) Are nest boxes a viable alternative source of cavities for hollow-dependent animals? Long-term monitoring of nest box occupancy, pest use and attrition. Biological Conservation 142: 33–42. doi: 10.1016/j.biocon.2008.09.026

68. Bull EL, Partridge AD (1986) Methods of Killing Trees for Use by Cavity Nesters. Wildlife Society Bulletin 14: 142–146.

69. Shepherd KR (1957) Some aspects of ringbarking in Alpine ash forests. Australian Forestry 21: 70–75. doi: 10.1080/00049158.1957.10675343

70. Hane ME, Kroll AJ, Johnson JR, Rochelle M, Arnett EB (2012) Experimental effects of structural enrichment on avian nest survival. Forest Ecology and Management 282: 167–174. doi: 10.1016/j.foreco.2012.06.037

71. Adkins MF (2006) A burning issue: using fire to accelerate tree hollow formation in Eucalyptus species. Australian Forestry 69: 107–113. doi: 10.1080/00049158.2006.10676236

72. Whitford KR, Stoneman GL, Freeman IA, Reynolds MJ, Birmingham TC (1995) Mortality of Eucalyptus marginata (jarrah) and E. calophylla (marri) trees following stem injection: effects of herbicide, dose, season, and spacing of injections. Australian Forestry 58: 172–178. doi: 10.1080/00049158.1995.10674662

73. Carey AB, Sanderson HR (1981) Routing to Accelerate Tree-Cavity Formation. Wildlife Society Bulletin 9: 14–21.

CHAPTER 2

THE ROLE OF TREES IN IMPROVING THE URBAN LANDSCAPE

Fardad Edalatkhah, Mahdi Shafaghati, Akram Norouzi, Somaye Motaghi, Maryam Mirbahaei

Keywords: Urban landscape, tree species, Vali Asr Street Tehran, Urban environment

INTRODUCTION

Exactly as the establishment of developed urban areas is considered to be one of the greatest achievements of human civilization, city landscaping is can also be taken as a measure of the degree and nature of a civilization and the collective psyche of a nation. Although the view which regards cities as phenomena independent of the human will is not held as widely to date (3), there is still a deep-seated approach to the subject which has strong following and tries to interpret and analyze the city and her landscapes regardless of humankind as their simultaneous creator and perceiver (7).

CONCEPTS

The urban landscape is a concept that lingers in the human memory even after one has evacuated a landscaped spot and results from the interaction between man (the onlooker) and his living environment. (1)

The most decisive feature of such a concept is its being subjective-objective, human-concrete, social-spatial which can be comprehended solely via the human experience and his interaction with his surrounding environements.

There are three major approaches to the qualitative entity of the urban landscape:

a) The view that regards the urban landscape as a quintessential feature of the physical environment of a city, independent of the humankind as the onlooker and the perceiver;

b) The insight which regards the urban landscape an entirely subjective element created by the onlooker, independent of the structure and other characteristics of the physical environment;

c) The approach which considers the urban landscape to be an event emanating from the interaction between palpable physical characteristics of the surrounding environment and the existing paradigms, cultural symbols and intellectual competencies of the observer. In this phenomenological approach, the concept of the urban landscape can be interpreted in various ways. As compared to the previous two approaches, this is the most credible take on the concept of the urban landscape nowadays (7), (3).

Citation: Fardad Edalatkhah, Mahdi Shafaghati, Akram Norouzi, Somaye Motaghi, Maryam Mirbahaei, The Role of Trees in Improving the Urban Landscape, http://cdn.intechopen.com/pdfs-wm/39886.pdf

The case study on Vali Asr Avenue in Tehran

This is a major thoroughfare 18 kilometres long linking the northern and southern tips of the city along an almost straight north-south vertical line. Yet, perhaps the most important reason behind the outstanding role of this road in Tehran is the fact that it acts like a historical bridge connecting the junctures the city went through before and after the road was built. This road is, in effect, a showcase of Tehran's social, cultural, political, religious, etc developments, especially, over the past 100 years, and therefore, her historical backbone.

Figure 1: New picture from vali Asr street in Tehan city,Iran

THE PLANE TREES OF VALI ASR AVENUE

These have been lining the road ever since it was built over 80 years ago and now amount to a total of 11,000. The plane trees of the Avenue are long considered to be a unique symbol of Iran's capital city, a national treasure and a cultural heritage which for many residents of the older generation revive endless nostalgic memories of the city's rich and exciting history.

THE CURRENT SITUATION OF THE TREES

Unsettling reports have been published in ever-growing numbers on the immanent extinction of Tehran's historic plane trees. One of the best-researched among reports published very recently blames the following factors for the growing threat that has put the life of these priceless trees on the edge

a) The demolition of the surface- and ground-water runways of the city;

b) Mismanagement in irrigation;

c) `Thc rclcasc of sewage and other pollutants into the running water resources of Tehran;

d) The disposal of a wide range of toxic chemicals such as detergents, industrial chemicals, cement and other construction materials, industrial oils and gasoline in the city's soil and water resources;

e) Physical damage incurred on the root system of the trees as a result of constant diggings, curb lining and concrete injection projects;

f) Changes in top soil levels exposing the trees' root systems to ambient, erosive climatic factors;

g) Physical mistreatment of the trees by humans;

h) High-rise buildings which block solar radiation reaching the canopy of the trees;

i) Highly polluted air all the year round (6), (2)

Figure 2: Old picture from vali Asr street in Tehan city,Iran

SPOTTING OPTIMAL LOCATIONS FOR LANDSCAPING

Landscaped plots must be centrally positioned, that is, they must be located at the heart of a neighborhoods, zone or district of the city. Moreover, such plots should be consistent with the physical structure of the environments that surrounds them. Another crucial criterion is accessibility. City landscapes must be linked with the traffic and transport grid in their four directions to become easily accessible for the greatest number of people and to become safer places for the public as enforcing public security measures in such locations is optimized thus.

THE ROLE AND EFFECTS OF TREES IN URBAN LANDSCAPES

City landscaping is the art of visually and structurally integrating the complex of buildings, roads and elements present in them, as well as all those spaces which give shape to urban environments. These include natural spaces inside city precincts which play a decisive role in creating urban landscapes with a decisively different appearance. Among natural elements influencing urban landscapes, trees are of paramount importance owing to their particular structure and spatial function. Like buildings, trees have structures which help create plant-based architectural styles. In lieu with overall changes that a city experiences in the course of its subsequent historical stages, the selection criteria for the varieties of trees to be planted in the city change, too. Meanwhile, almost all cities in the world today host a large proportion of open space with distinctive physical characteristics each. These include streets, squares, and spaces adjacent to car parks. Tree selection criteria for each of these spatial categories vary accordingly as determined by professional city landscaping architects.

RECOMMENDATIONS

Any successful management of urban green space requires the serious enforcement of the following measures and/or principles:

1. Granting a greater role to the public as the direct benefactors: This can be realized through raising public awareness and strengthening public responsibility among others;
2. Boosting the ecological productivity of urban green space: this can be realized through the protection and management of urban green space, the informed selection of the right plant varieties, integrated management systems in planting, growing and training the selected plant varieties, urban woodland management, intelligent plot spotting for the expansion of urban green space, evaluating the feasibility of extending urban green space to abandoned plots and mountainous ecosystems;
3. Raising the social productivity of urban green space: this can happen via ensuring safety and security, the provision of basic facilities and amenities, ease of access, consistency and cohesion with the greater physical environment, proactive management, balanced geographical distribution, etc.

CONCLUSION

Both from an architectural-aesthetic point of view and in order to raise the quality of the essential elements in urban livelihoods, paying attention in urban management processes to the image of urban areas overall and urban landscapes in particular is of paramount importance. However, the scientific study of these elements and their contribution to urban development in view of their economic as well as budgetary repercussions is no less significant. In other words, imageand landscape-related projects should not be taken into account barely from an architectural and aesthetic but also from an economic standpoint. As regards important traffic axes, strategic decisions must be made in accordance with the prominent position of spatial planning and landscaping projects in urban areas. In other words, the simple fact that budgetary provisions are available to municipal managers must not automatically lead to the illusion that such resources can be allocated to imagined beautification projects arbitrarily to boost development efforts in urban areas in turn.

REFERENCES

1. Soltani,B.2010,view to green area of Tehran city, journal of environmental science,environment organization press.
2. Bijan zad,M,2010,evaluation of green area of Tehran city, jihad daneshgahi ,press, Tehran.
3. Satari,J,2006,Myth of Tehran city, office of cultural research, environment organization press.
4. Rasouli, A, 2009, buttonwood, holy tree in Iran, cultural heritage press, Tehran
5. Riyazi, B, 2009, Development of green space in Tehran, proceeding of green space seminar, Tehran city parks and green space organization, Tehran
6. Majnoniyan, H, 2010, Debates about parks, green space and promenade, Tehran city parks and green space organization, Tehran
7. http://www.darioush-shahbazi.com
8. McPherson, E.G & Nowak,D.G, 2011,The Chicago urban forest climate project.

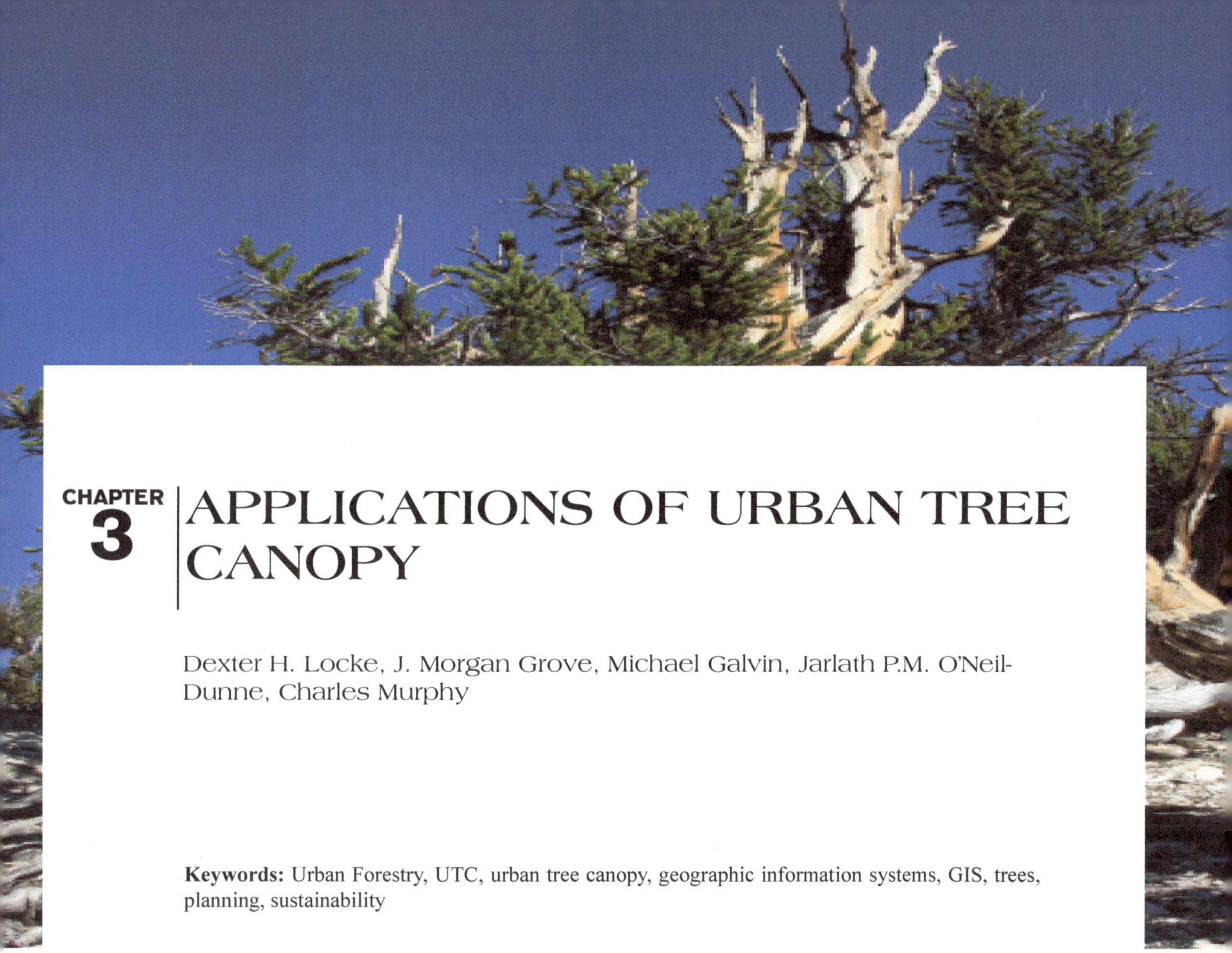

CHAPTER 3 APPLICATIONS OF URBAN TREE CANOPY

Dexter H. Locke, J. Morgan Grove, Michael Galvin, Jarlath P.M. O'Neil-Dunne, Charles Murphy

Keywords: Urban Forestry, UTC, urban tree canopy, geographic information systems, GIS, trees, planning, sustainability

INTRODUCTION

The purpose of this article is to describe how diverse stakeholders can use urban tree canopy (UTC) Assessment and Prioritization tools to collaboratively achieve urban sustainability goals. The motivations for this article are practical, methodological, and theoretical. From a practical perspective, there is a growing need for diverse stakeholders to work collaboratively on interrelated strategies to achieve comprehensive, urban sustainability goals. Methodologically, these same stakeholders need tools that integrate social, economic, and ecological data. These tools need to be relevant to stakeholders' management goals, transparent, time efficient, and interoperable with their data systems. Theoretically, the application of these tools creates new research questions for the scientific community to address. We use our experience working in Baltimore, Maryland to describe this process in order to share the methods we have developed and the lessons we have learned.

There is an emerging shift in urban land management practice from a focus primarily on street trees and parks to an "All Lands, All People" approach (Grove 2009). This "All Lands, All People" approach is an inclusive framework based on all urban land use types; multi-sector coordination and collaboration; and integration of social and ecological knowledge and goals. In order to operationalize this approach, it is necessary to characterize land cover at the unit of individual property parcels, and integrate social and ecological data corresponding to diverse sustainability goals so that all neighborhoods and parcels can be prioritized for tree planting (Grove, 2009; Locke et al., 2010).

Urban Tree Canopy (UTC) tools provide a set of methods to operationalize the "All Lands, All People" approach. Currently there are two types of UTC methods. UTC Assessment tools are used to assess existing and possible UTC at any spatial scale from the property parcel level to a regional extent. The second set of tools are the UTC Prioritization protocols which are used to identify where to increase UTC at watershed, neighborhood, and parcel levels based upon diverse stakeholders' goals and preferences among goals. These goals may be ecological, social, or economic in nature – or a mix of the three.

Significant progress can be seen in the development and application of UTC tools with our team completing 68 UTC Assessments since 2006 . These assessments cover 8,780 sq. miles, 837 communities, include over 28,000,000 people, and range in geographic size from smaller urban areas like Scranton and Lancaster, PA, to counties such as Jefferson WV, and

Citation: Locke, Dexter H.; Grove, J. Morgan; Galvin, Michael; O'Neil-Dunne, Jarlath P.M.; and Murphy, Charles, Applications of Urban Tree Canopy, http://digitalcommons.lmu.edu/cate/vol6/iss1/7

Montgomery, MD. A sample of UTC reports can be found here: http://www.nrs.fs.fed.us/urban/utc/pubs/ and http://letters-sal.blogspot.com/2011/04/tree-canopyassessment-reports.html. This achievement suggests both demand for UTC analyses and the ability of UTC teams to develop cost-effective approaches for analyzing large amounts of data and produce operational information to decision makers in a timely manner. In the case of New York City, the Department of Parks and Recreation used its UTC Assessment and Prioritization analysis to set a goal and prioritize its $400 million tree program, the Million Trees NYC initiative, over a ten year period .

METHOD

The case study we describe here for Baltimore uses UTC Prioritization tools described by Locke and others (2010). Baltimore may be similar to many other cities in the United States in that there is a large and diverse number of "tree" stakeholders from public, NGO, neighborhood, and private sectors. In Baltimore, we identified 25 stakeholder organizations from these sectors. There is insufficient space on public lands to achieve the City's UTC Goal of 40% of city land area. High resolution estimates of tree canopy revealed that if every opportunity for street tree's canopy to grow in the right of way were realized, and if all parks reached 100% tree canopy, the city will would achieve only ~10% of its total 40% canopy goal. The reaming 30% of the tree canopy goal have to be established on other lands in the City (Galvin et al., 2006; O'Neil-Dunne, 2009). Completely canopied parkland may not be desirable, either. An "All Lands, All People" approach that includes public, private, community, and abandoned lands is needed. Independent action is inadequate: no agency, organization, single landowner or business has sufficient funds or land to achieve a city's UTC goal. Coordination and collaboration are needed and depend upon identifying common or complementary interests, categories of programs, or areas for action.

Based on these needs the UTC Prioritization framework was built. Given the known and studied benefits of trees as described by Locke and others (2010) and catalogued in depth elsewhere spatial data are gathered to identify where those benefits of trees are lacking. Next organizations whose mission or mandate aligns with that particular benefit, function, or property of trees are identified. For example, reducing impervious surfaces and planting trees to intercept, filter, slow, and uptake storm water may reduce infrastructure costs associated with treatment while reducing surface water pollution (Beattie et al. 2000; Nowak et al. 2007). Therefore, an NGO interested in water quality, or a government agency charged with meeting water quality standards may consider using trees as part of an overall water quality strategy. Data on impervious surfaces, citizen reported floods, and/or proximity to surface waterways could be used to identify where to plant trees for the specific water quality management objectives. UTC Prioritization works by matching known benefits of trees, to places lacking those benefits, and then matches those locations to organizations positioned to manage those issues that trees help ameliorate.

We began our work in Baltimore in partnership with the City's TreeBaltimore Program, which is part of the Division of Forestry in the Department of Recreation & Parks. We had developed the methods and piloted the first UTC assessment in 2006 using Baltimore as our case study (Galvin et al., 2006; Raciti et al., 2006) and later refined our land cover mapping techniques (O'Neil-Dunne, 2009). Subsequently, the City established a UTC goal of increasing the City's tree canopy to 40% in 30 years . This goal was set by staff in the Department of Recreation and Park's Forestry Division and the Director of Recreation and Parks made a recommendation to the Mayor. Staff from the Parks & People Foundation and Blue Water Baltimore participated in the evaluation of the UTC Assessment data and agreed to the goal. All three subsequent Mayors have also endorsed the goal. Staff from the USDA Forest Service and the Maryland Department of Natural Resources provided the technical support, data analysis, and the report. As the Division of Forestry worked to achieve the City's UTC goal, it became obvious to Forestry staff that they had insufficient resources to achieve the City's UTC Goal through a combination of planting, canopy maintenance and loss reduction efforts. TreeBaltimore staff began to search for additional partnerships and resources to leverage. While the initial premise was that the Forestry Division would work with non-profit and private partnerships to achieve the City's goal, it was increasingly clear through informal conversations that other city agencies had regulatory requirements or programmatic interests that involved tree planting.

Common interests and little coordination or sharing of resources among City agencies' and local NGOs highlighted the high degree of fragmentation in the activities used to achieve the City's urban tree canopy goal. These conditions are not new to resource management, and have been described by Yaffee's (1997) as "recurring nightmares:"

(1) a process in which short-term interests out-compete long-term visions and concerns;

(2) conditions in which competition supplants cooperation because of the conflicts that emerge in management issues;

(3) the fragmentation of interest and values;

(4) the fragmentation of responsibilities and authorities (sometimes called "functional silos" or "stove pipes"); and

(5) the fragmentation of information and knowledge, which leads to inferior solutions.

To address these "recurring nightmares," we worked with Division of Forestry staff to develop a two phase processes for stakeholder involvement to prioritize planting. In the first phase we met with staff from different city agencies on an individual basis to explain prioritization and elicit feedback on criteria. In the second phase we met with stakeholder organizations from public, non-profit, community, and private sectors in a workshop setting to provide specific input for prioritization in the form of votes. All of these stakeholder organizations had expressed interest previously in tree issues in the City. Ultimately, the goal of this process was to develop interest and support from other city agencies to increase tree canopy and to involve public, NGO, and private stakeholders to identify tree canopy priorities to promoting biodiversity.

Phase I: Public Agencies

Working with TreeBaltimore staff, we identified and interviewed key representatives from five other City agencies that had an interest in tree canopy. These agencies included the Departments of Planning, Housing & Community Development, Transportation, Public Works, Public Health, and Education. We met individually with each agency. First, we described the UTC Assessment and Prioritization tools using our Assessments in Baltimore and both Assessments and Prioritization for New York City so that they had a sense of data needs, analytical capabilities, and possible products. After our presentation, we discussed their agency's regulatory requirements and programmatic interests in tree canopy and which variables they would use, if they had a prioritization tool, to prioritize where to plant trees to meet their requirements or interests. We revised our presentation to incorporate each agency's suggestions for prioritization criteria and variables to use, (Table 1). Then we asked them if they would provide the data they identified. Each meeting took approximately two hours. GIS staff from each agency provided the data they had listed so they could be incorporated in to subsequent prioritizations for their organization or others.

The interests among public agencies were varied. For instance, the Departments of Transportation and Public Works sought to reduce the amount of impervious surfaces in the City. Housing and Community Development considered greening to be a strategy for community stabilization and re-development. Public Health understood trees to be important for reducing cases of heat-related stress and asthma. Education sought to create greener and more attractive school campuses, while Planning considered trees to be important to a variety of sustainability goals from reducing energy consumption to improving water quality and reducing the severity of flood events.

Phase II: NGOs, Community Groups, Private Businesses

In Phase II we worked with NGOs, community groups, businesses, and government agencies. We repeated the process from Phase 1 with several modifications. We made our presentation in a large group setting to representatives of 25 organizations (Appendix 1), we conducted a paper survey (Appendix 2) using a refined version of Table 1 (Appendix 2, Question 2), and the entire meeting lasted 2 hours.

The survey had three parts. First, participants were presented with more than two dozen criteria variables for prioritizing tree planting and asked to allot ten votes among the variables representing their preferences. This voting approach assumed that people voted in ways that are reflective of their organization's management objectives. Variables could be voted for multiple times. Variables are indicators of where the benefits of trees are lacking. Trees and access to open space (frequently tree dominated landscapes) are commonly associated with improved health outcomes (Bell et al. 2008; Jackson, 2003; Lovasi et al. 2008; Mitchell and Popham, 2008; Takano et al. 2002), therefore planting in areas with poor public health may be desirable. A "write-in" option was also provided, but not all write-in options could be accommodated because the associated data did not exist or were not otherwise available. Based upon "write-in" requests, two additional variables were subsequently added: the percent of each block group's area that is a historic district and the percent of the block group area beneath the canopy of trees 50 feet or taller.

Table 1: Summary of Baltimore City public agencies' criteria and variables related to trees and their benefits, as well as the linkages to goals in the City's sustainability plan. City Agencies column: Agencies that self-identified with a major criteria, or were identified by another agency. Sustainability Plan column: linkages of increasing tree canopy to other goals, as shown in Table 2.

Major Criteria	Criteria Variables	City Agencies	Sustainability Plan
Public Health & Safety	Life expectancy; rates including mortality, mortality from heart disease; mortality from strokes, mortality from respiratory disease, mortality from diabetes, infant mortality rate	Health, Planning	Pollution Prevention Goals 2, 4 & 5
	Asthma by zip code	Health, Planning	Pollution Prevention Goals 2, 4 & 5
	Dependency Ratio	Health, Planning	
	Urban Heat Island: surface temperature and solar exposure	Health, Recreation and Parks, Planning, Transportation	Pollution Prevention Goal 2
	Crime: Personal, Property and Total	Police	
	Transportation Connections	Health, Planning, Transportation, Education	Transportation Goals 1, 2, 3 and 4
Environmental Justice	Toxic Releases Inventory	Health, Planning	Pollution Prevention Goals 2 & 4
	Brownfields	Planning, Public Works	Cleanliness 3, Pollution Prevention Goal 4
	Poverty	Planning	
	Race	Planning	
	Home ownership	Planning	
	Percent Parks	Recreation and Parks	Greening Goal 3
Water Quality	Percent Impervious Surface	Public Works, Transportation, Planning	Pollution Prevention Goal 3
	Watershed H20 quality assessments	Public Works	Pollution Prevention Goal 3
	Stream corridors	Public Works, Recreation and Parks	Greening Goal 4
	Flood Plains	Public Works, Recreation and Parks, Planning	
	Critical Area	Recreation and Parks, Planning	Greening Goal 4
	Greenstreets	Public Works, Transportation, Recreation and Parks, Planning	Cleanliness Goal 1, Pollution Prevention Goal 3, Greening Goal 1
	Blue alleys	Public Works, Transportation, Recreation and Parks, Planning	Cleanliness Goal 1, Pollution Prevention Goal 3, Greening Goal 1
	Flooding	Public Works, Planning	
Air Quality & Noise Pollution	(Major) Road Density	Transportation	Pollution Prevention Goal 2
Critical Places	Schools, hospitals, libraries, recreation centers, and elderly care facilities	General Services, Recreation and Parks, Planning, Housing and Community Development, Education	Education & Awareness Goals 1, 2, 3 &4
	Population density (per square mile)	Planning, Public Works	
Community Presence	Potential stewardship	Recreation and Parks, Planning, Housing and Community Development	Education & Awareness Goals 2, 3 & 4

Table 2: Key to Sustainability Goals listed in Table 1.

Sustainability Plan Key:	
Cleanliness Goal 1	Eliminate litter throughout the City
Cleanliness Goal 3	Transform vacant lots from liabilities to assets that provide social and economic benefits
Education & Awareness Goal 1	Turn every school in Baltimore City into a green school
Education & Awareness Goal 2	Ensure that all city youth have access to environmental stewardship programs and information
Education & Awareness Goal 3	Raise the environmental awareness of the Baltimore community
Education & Awareness Goal 4	Expand access to information on sustainability
Greening Goal 1	Double Baltimore's Tree Canopy by 2037
Greening Goal 3	Provide safe, well-maintained public recreational space within ¼ mile of all residents
Greening Goal 4	Protect Baltimore's ecology and biodiversity
Pollution Prevention Goal 2	Improve Baltimore's air quality and eliminate Code Red days
Pollution Prevention Goal 3	Ensure that Baltimore water bodies are fishable and swimmable
Pollution Prevention Goal 4	Reduce risks from hazardous materials
Pollution Prevention Goal 5	Improve the health of indoor environments
Transportation Goal 1	Improve public transit services
Transportation Goal 2	Make Baltimore bicycle and pedestrian friendly
Transportation Goal 3	Facilitate shared-vehicle usage
Transportation Goal 4	Measure and improve the equity of transportation
Source: Baltimore Sustainability Plan http://www.baltimoresustainability.org/sites/baltimoresustainability.org/files/Baltimore%20Sustainability%20Plan%20FINAL.pdf	

Most criteria were included because of the known benefits of urban trees. For example, heat may be lethal in the summer months in urban areas (Son et al. 2012). Trees reduce temperatures by intercepting the sun's rays and casting shadows, the evapotranspiration process cools air, and planting may be accompanied by a reduction in impervious surfaces which store and emit heat (Akbari et al. 2001; Akbari and Konopacki 2005; Nowak, 2002; Nowak et al. 2007; Rosenfeld et al. 1998; Streiling and Matzarakis 2003). Therefore trees may help prevent heat-induced injuries and mortality. High surface temperatures and percentages of impervious surface can be combined with measures of population vulnerability (dependency ratio) to identify places to plant trees for improved health outcomes.

A couple of criteria were identified by participants for practical motivations unrelated to the City's sustainability plan. First, historic landmarks or districts may have maintenance budgets that can be used for beautifying a site with trees and retaining the "historic" appearance of an area. Second, neighborhoods with existing canopies of "big trees" may be concerned that these trees will be lost because of their height and susceptibility to windstorms, or because they fear the trees are getting old and are at risk to disease or senescence. Thus, neighborhoods with big trees may want to plant the next generation of trees to ensure a sustained succession of canopy over the long.

A second part of the survey asked participants about the types of sites where they work, or Categories of Interest (termed "COI"). The suggested categories were based upon A Typology of Forestry in an Urban Ecosystem, first described by Grove and others (2005) and later refined by Raciti and others (2006). Categories of Interest include street trees, abandoned lots, private residential lands, parks, schools, stream valleys-riparian areas, shoreline areas or other.

In the third part, representatives of participating organizations were also asked, "Where do you work" and asked to choose from four Areas of Interest (termed "AOI's"). Answer choices were Neighborhood, Watershed, Council Districts, City or Other. When taken together, combinations of COI's and AOI's may form the basis for different strategies for reaching tree canopy goals. A future goal is to include a further developed typology that would enable improved targeting of neighborhoods based on a blend of site type characteristics to make outreach efforts more locally applicable. See Appendix 2 for the complete survey instrument. In return for completing the survey, we promised each organization that we would provide them with their own prioritization map based upon their prioritization votes as well as a summary map of all the organizations votes combined. We promised to provide these maps in both paper and digital formats (jpeg and pdf) within one month. Finally, we asked to schedule a meeting in one month so that we could present and discuss the results and decide on next steps.

After the survey response data were collected and tabulated, descriptive statistics were calculated to characterize the survey responses. Hierarchical cluster analyses were performed using the R Programming language (R Development Core Team 2012) on the prioritization criteria, COI and AOI votes. A Euclidean distance matrix was first defined for each set of questions, and then Ward's method of agglomeration (Ward 1963) applied using the hclust() function in the statistical package R (See Supplemental Material for the R scripts created for this paper). Used in this way, cluster analysis acts as an exploratory and visualization tool for identifying relative similarities simultaneously across both rows and columns (or organizations and their votes in this case).

FINDINGS

Using our interviews in Phase I and surveys in Phase II, we summarize our results by several dimensions of coordination and collaboration among stakeholders: 1) overlap among public agencies' programs; 2) stakeholder preferences for different UTC criteria; and 3) stakeholder programs by a) categories of interest and b) areas of interest.

Overlap among Public Agencies' Programs.

Significant opportunities exist for coordination and collaboration among government agencies in the City of Baltimore. Nine agencies were identified with programs that were directly or indirectly related to trees. Several agencies were named as having an interest even though they were not interviewed. The opportunities for coordination and collaboration were uncovered through these meetings, the workshop, and our surveys. The agencies included Department of General Services, Education and the Police Department. For the major prioritization criteria, there were mutual interests among public agencies for 5 of the 6 criteria, as indicated in Table 3 and the maps in Appendix 3 indicate. In several cases, the number of agencies with mutual interest for a Major Criteria ranged from 4 to 5 agencies. Some agencies were "landowners", such as Recreation and Parks, and Transportation. These lands were often physically adjoining, which represents opportunities for coordination and collaboration. In other cases, agencies were "landless" such as Planning, Public Works, and Health. This fact represents an opportunity for collaboration among "landless" agencies, technical assistance and resources to meet their programmatic requirements, and landowner agencies who are in deep need for resources and additional expertise to better meet the City's diverse sustainability goals

Results from our interviews of public agencies show that increasing UTC was more than just a "greening" goal and was linked to many of the City's Sustainability goals, as shown in Table 2. While it might be expected that UTC was related to other Greening Goals, it was unexpectedly related to other Sustainability Goals that included Cleanliness, Education and Awareness, Pollution Prevention, and Transportation (Table 2, Appendix 3). Thus, tree planting had a multi-functional role that addressed numerous sustainability goals at the same time.

Stakeholder Preferences for Different UTC Criteria

Survey responses from our stakeholder interviews indicated a strong interest in diverse environmental, social, health, and cultural priorities. These included percent impervious surfaces (24.4%), potential stewardship (11.6%), urban heat island (10.0%), stream corridors (7.2%) and "critical places" - places such as schools, hospitals, and recreation centers (6.8%). Impervious surfaces was by far the most popular criteria variable, receiving 61 votes (24.4 %) which was more than double the second most popular criteria variable, potential stewardship, which gathered 29 votes (11.6%). The fact that the five most popular criteria include environmental, social, health, and cultural priorities suggests that these stakeholder groups are thinking holistically, more collectively and beyond professional "silos" (Table 4). The substantial support for reducing impervious surfaces and improving water quality is likely attributable, in part, to the cultural and historic significance the Chesapeake Bay as well as the existing regulatory emphasis on water (See, for example Chesapeake Executive Council 1983, and Chesapeake Bay Program 2004). The City's sustainability plan states clearly in its rationale for doubling its tree canopy: trees "reduce the amount of stormwater from running into the harbor" (Baltimore Sustainability Plan 2009).

Some organizations were more alike in their criteria preferences, and the affinity among organizations was not based upon whether the organization was a public agency, NGO, community group, or business (Figure 4).

STAKEHOLDER PROGRAMS A

Categories of Interest: The Types of Sites Where Organizations Work

The stakeholder survey asked questions about the urban forest categories on which groups worked. For instance, does the organization work on street trees, at schools, or in riparian areas? Street trees were the most popular category of interest (receiving 21.7 % of votes) followed by schools (15.9% of votes), with stream valleys - riparian buffers ranking third (13.0% of votes) (Figure 5). There was a disproportionate focus on street trees over residential lands. Yet, the City is predominantly private residential land area, most tree canopy is found on private residential, and most opportunities for tree planting are found on private residential lands (O'Neil-Dunne 2009).

Table 3: Stakeholder organizations and their preferences for different tree planting prioritization criteria. Numbers indicate the number of votes each criteria received by each organization. Each organization had 10 votes and were asked to distributed them across criteria to reflect their priorities. Acronyms are spelled out in Appendix 1.

	Percent Impervious Surface	Potential stewardship (positive)	Urban Heat Island using surface temperature	Stream corridors (1 ft buffer of each water element, sum area)	Schools, hospitals, libraries, recreation centers, and elderly care facilities	Life expectancy (Inverse)	Crime: Robbery, Burglary, Theft	Percent White (Inverse)	Road Density	Historic Districts	Transportation Connections	Critical Area	Population density (per square mile)	Percent of tree canopy that is over 5%	Percent Parks	Flood Plains (2 most risky)	Dependency Ratio	Toxic Releases Inventory
BCRP - Capital Development	1		1	1	1					2		1		1	1	1		
Office of Public Health Preparedness and Response (OPHPR)	1		2			1		2					1			1	2	
Baltimore Tree Trust	2	2	1	1	2	1		1										
Parks & People Foundation	2	1	1	1	1	1						1		1	1			
Department of General Services	2	1	1		1	1		1	1		1						1	
Bartlett Tree Experts	2	2	1	1	2				2									
Baltimore Medical System (E Baltimore)	2	1	2		3	2												
Department of Planning	2	3	3									1	1					
Baltimore Medical System (City)	1	3	3			2	1											
Mahan Rykiel		3		1	1					4	1							
The Friends of Wyman Park Dell	2	2				1				4					1			
CPHA							4	2			2		2					
Park Heights Renaissance					3		3	2			2							
Baltimore Office of Sustainability	3	1	2						1			1	1					1
SouthEast CDC	3	1	2						1		1		2					
Brooklyn and Curtis Bay Coalition, Inc.	3	1	1	1	1		2						1					
Patterson Park Neighborhood Association	5	2	2								1							
RK&K	5		1	3								1						
DOT	5		1	1							1	1				1		
Roland Park Civic League Greater Roland Park Master Plan Implementation (City)	4	1			1	1								2	1			
DPW	4	1		1	1	1						1				1		
Roland Park Civic League Greater Roland Park Master Plan Implementation (Roland Park)	3	3				1								3				
Wyndhurst Neighborhood	3	1	1				1	1	3									
Midtown Community Benefits District - Midtown Green	3						1	2	2			1			1			
BWB	3			7														
Criteria Totals	61	29	25	18	17	12	12	11	10	10	9	8	8	7	5	4	3	1

Table 4: The expressed preferences for each prioritization criteria summed across all workshop participants. Major criteria are summary categories that represent clusters or groupings of variables.

Major Criteria	Criteria Variables	Total Votes	Percentage of Total
Public Health &Safety	Life expectancy (Inverse[8])	12	4.8%
	Dependency Ratio	3	1.2%
	Urban Heat Island using surface temperature	25	10.0%
	Crime: Robbery, Burglary, Theft	12	4.8%
	Transportation Connections	9	3.6%
Environmental Justice	Toxic Releases Inventory	1	0.4%
	Percent White (Inverse[8])	11	4.4%
	Percent Parks	5	2.0%
Water Quality	Percent Impervious Surface	61	24.4%
	Stream corridors (100 ft buffer of each water element, sum area)	18	7.2%
	Flood Plains (2 most risky)	4	1.6%
	Critical Area	8	3.2%
Air Quality & Noise Pollution	Road Density	10	4.0%
Critical Places	Schools, hospitals, libraries, recreation centers, and elderly care facilities	17	6.8%
	Population density (per square mile)	8	3.2%
Community Presence	Potential stewardship (positive)	29	11.6%
Aesthetic	Restore Historical Sites	6	2.4%
Design	Historic Districts	4	1.6%
Replacement	Percent of tree canopy that is over 50%	7	2.8%

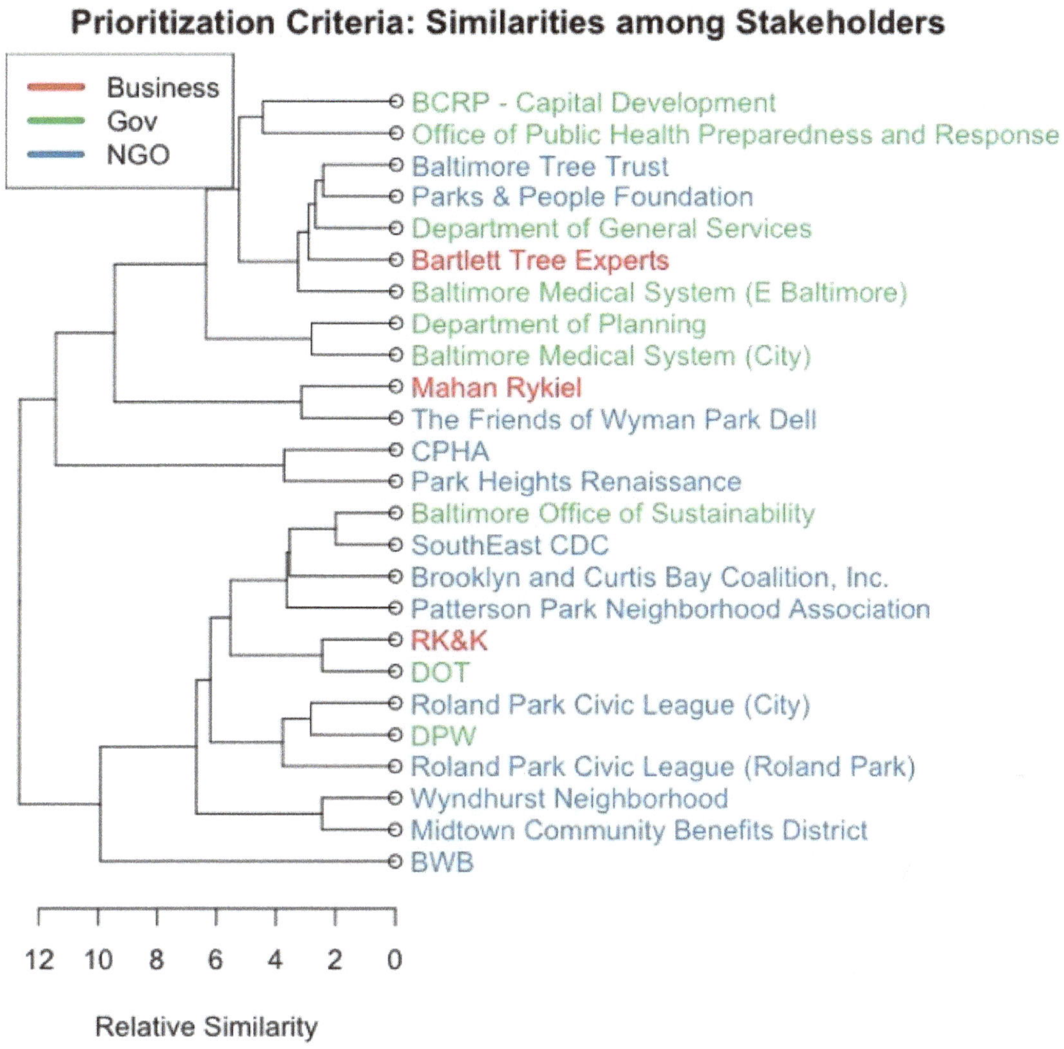

Figure 4: Hierarchical clustering helps highlight the relative similarities and differences across all prioritization criteria. Sectors are indicated by color. Data for Figure 4 can be accessed here.

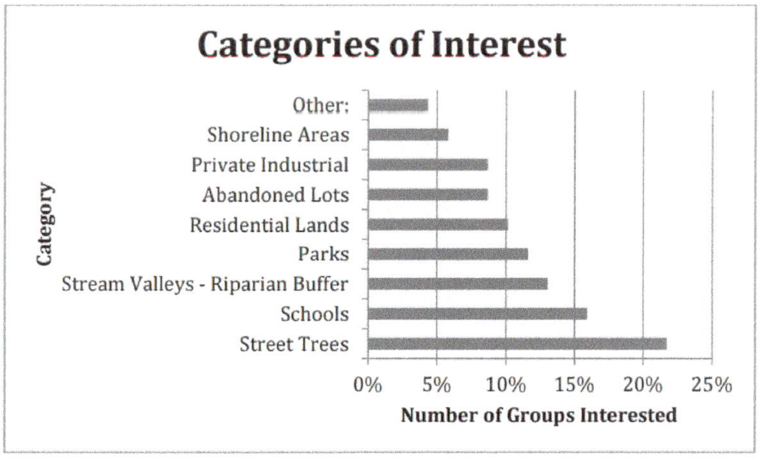

Figure 5: Categories of Interest (COI) as expressed by workshop attendees. Attendees could vote for items multiple times if they work across multiple COI's.

Survey results can be examined for similarities among organizations in terms of Categories of Interest (Figure 5). As was the case with Prioritization Criteria, similar types of organizations had diverse types of urban forest categories on which they worked, and different types of organizations worked on similar categories. For instance, the private consulting firm RK&K and Department of Public works worked on similar forest types, while the Departments of Public Works and General Services worked on different types. Figure 6 illustrates the specific similarities and differences. Both public agencies are involved in street tree planting, but the Department of Public Works also works on stream buffers and shorelines, while the Department of General Services concentrates on private industrial and residential lands.

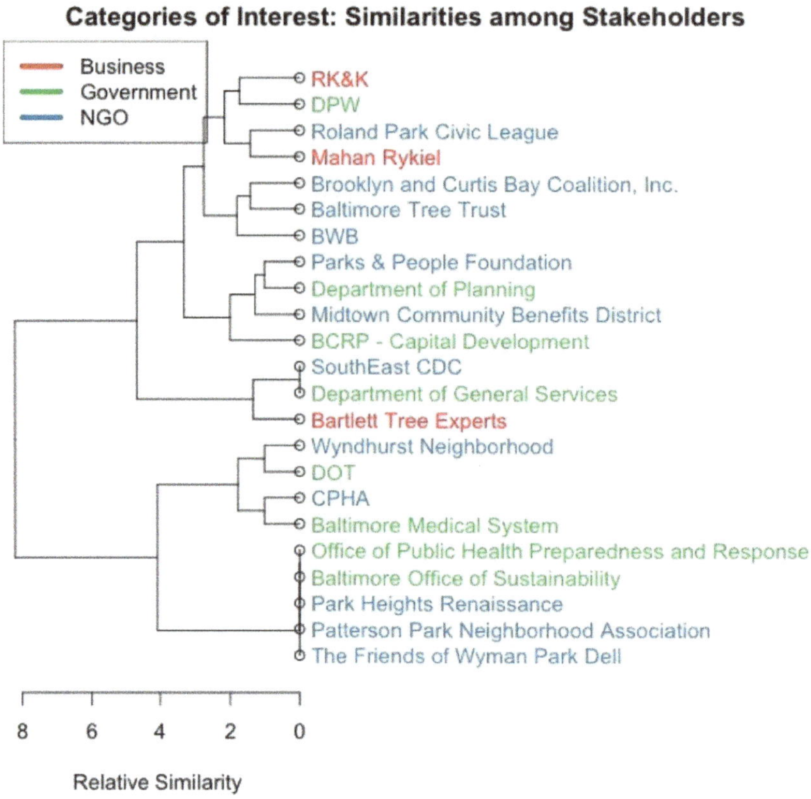

Figure 6: Hierarchical clustering helps highlight the relative similarities and differences across all Categories of Interest. Sectors are indicated by color, not all workshop participants completed the entire survey. Data for Figure 6 can be accessed here.

Areas of Interest: The Scale and Scope of Where Groups Work

Neighborhoods were the most popular type of area (receiving 29.4% of votes), followed by stakeholders who worked on a city-wide basis (26.5% of votes)(Figure 7). It is somewhat surprising that relatively few groups answered "watershed" as an area of focus given the high stated preferences for using trees to achieve water quality goals (Table 4) and the need for stormwater compliance. One manager offered an alternative explanation: managers know the entire city drains into the Bay, so city or neighborhood might be synonymous with watershed because every place is part of a watershed.

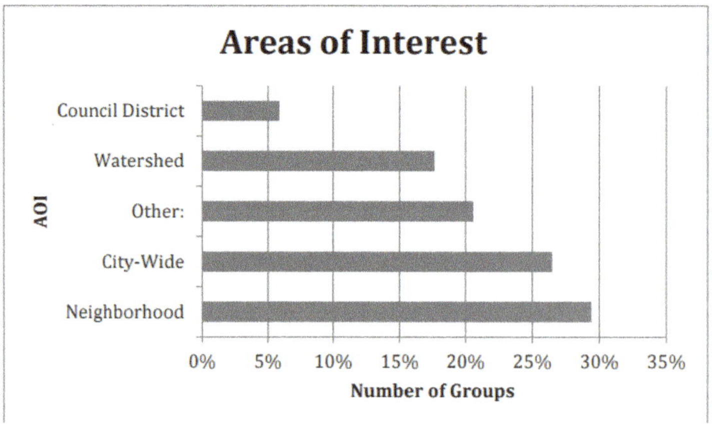

Figure 7: Areas of Interest (AOI) as expressed by workshop attendees. Attendees could vote for items multiple times if they work across multiple AOI's.

Stakeholders can be grouped based upon their Areas of Interest. The resulting groups are different from prioritization preferences or categories of interest. For instance, the two major, non-profit organizations—the Parks & People Foundation and BlueWater Baltimore—were very dissimilar in terms of prioritization preferences. In the case of Areas of Interest, however, they are nearly identical and form a group with the Department of Public Works. Unsurprisingly, almost all of the neighborhood organizations form a major branch of the tree diagram. Figure 8 permits examination of specific factors that cause organizations to be different or similar. Within the "neighborhood" branch, some neighborhood organizations focus on a specific neighborhood while others incorporate a collection or region of neighborhoods and have an associated watershed focus. At the bottom of the tree diagram are organizations that have a watershed and city-wide focus.

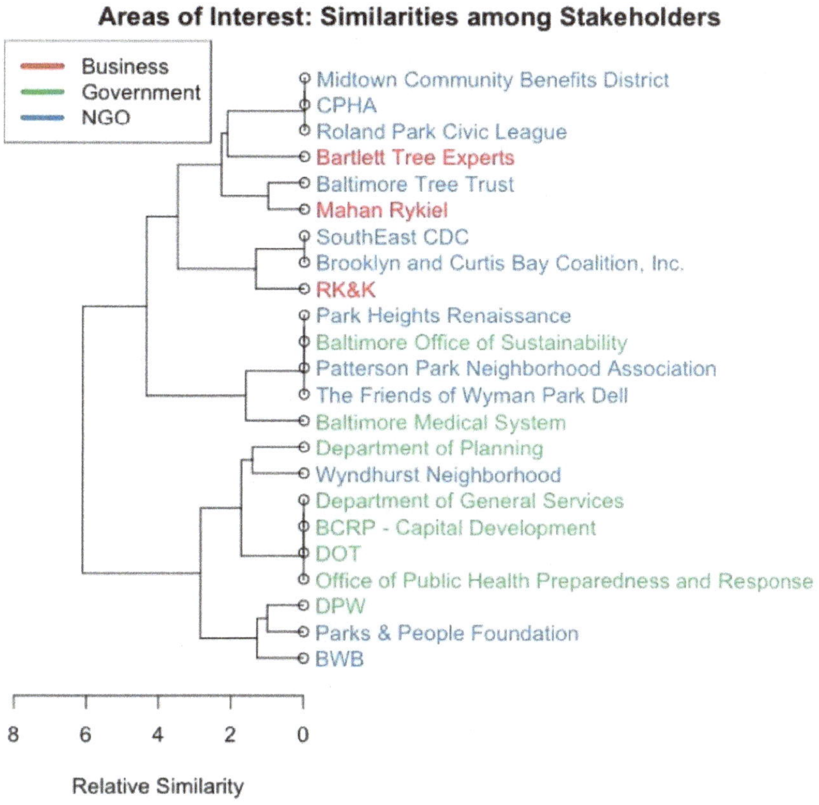

Figure 8: Hierarchical clustering helps highlight the relative similarities and differences across all Areas of Interest. Sectors are indicated by color, not all workshop participants completed the entire survey. Data for Figure 8 can be accessed here.

DISCUSSION

Opportunities for Coordination and Collaboration.

Opportunities for coordination and collaboration among stakeholders can be identified by analyzing the survey results in terms of groups. For instance, one can use the combination of prioritization mapping results, priority preferences, categories of interest, and areas of interest to ask, "which groups make a natural coalition to work in the high priority area of the Brooklyn and Curtis Bay region (mapping), focused on water issues (preference), addressing a combination of shoreline, street trees, schools, and residential areas (categories), and work on a neighborhood and watershed basis (areas)? Another example could be that a local energy company is interested to make school areas healthier by reducing extremely high local temperatures. The question could then be "which groups are most likely to work together on urban heat island (preference), focusing on schools (categories) in high priority UHI areas (mapping and areas)?"

Survey results can also be used to identify gaps in capacities and need for coalitions. For instance, few organizations work on private lands even though private property—as a type of owner—are the areas with the most land available for planting trees (O'Neil-Dunne, 2009).

These results can be used to identify high priority areas with significant residential tree planting opportunities (mapping) and to recruit groups who work on residential lands (categories).

Production of Maps and Databases

We learned several lessons about the production of maps and databases through our UTC Prioritization process. First, each organization was eager to receive a map of their prioritization preferences. It was important that the processing time for our UTC Prioritization tools were not time-intensive and that we could produce maps that were cartographically simple and selfexplanatory (Figures 9 and 10 are shown as examples, Appendix 3 contains 26 maps, one for each organization plus a map created by summing all 25 participants' ten votes). Second, each organization wanted their map, the summary map of how all organizations' preferences overlapped in space regardless of individual organizations' preferences, and the individual maps from other organizations. The motivation for accessing all of these maps was so that each organization could see, spatially, how their preferences produced priority areas that might be similar or different from the overall preferences of the stakeholder group and other individual organizations. Third, organizations wanted to use their maps for reports and proposals. Thus, maps had to have a resolution appropriate for publication, yet small enough in terms of file size so that they could be manipulated in word processing programs.

Organizations wanted to be able to integrate the UTC prioritization data with their information systems. Our fourth lesson was that we produce our data at units of analysis and in formats that organizations could use. For example, both GIS data and Excel spreadsheets were provided in addition to maps. We created an FTP site so that maps and data could be downloaded as needed. Finally, organizations can now use these maps to communicate internally and externally. Managers seeking additional funding and other resources within their organization may use a priority map to communicate an implementation strategy, while government officials can offer maps when explaining where they are planting and why. Sometimes areas within Baltimore were identified as high priority by different organizations for different reasons. These maps then allow organizations to see common cause based on location. We have discussed the differences among organizations in terms of priorities, Categories, and Areas of Interest. Yet, one of the common observations in follow-up stakeholder meetings is that the groups tended to agree on the same locations, particularly for high priority areas (Appendix 3). This suggests that UTC prioritization tool can be used not only as a tool for identifying priority areas, but also as a tool for building consensus among stakeholder groups.

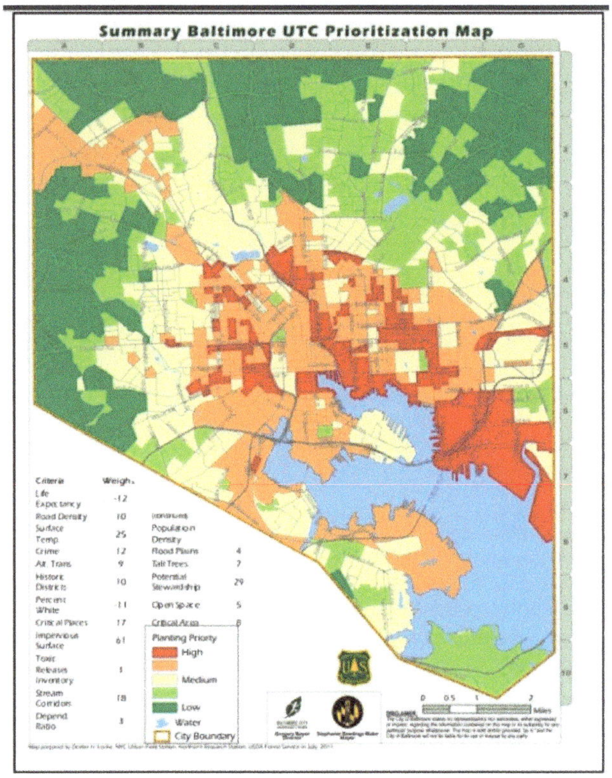

Figure 9: A sample map created by weighting Urban Tree Canopy Prioritization (UTC-P) criteria variables by one stakeholder's votes. See Appendix 3 for maps of all 25 stakeholder groups.

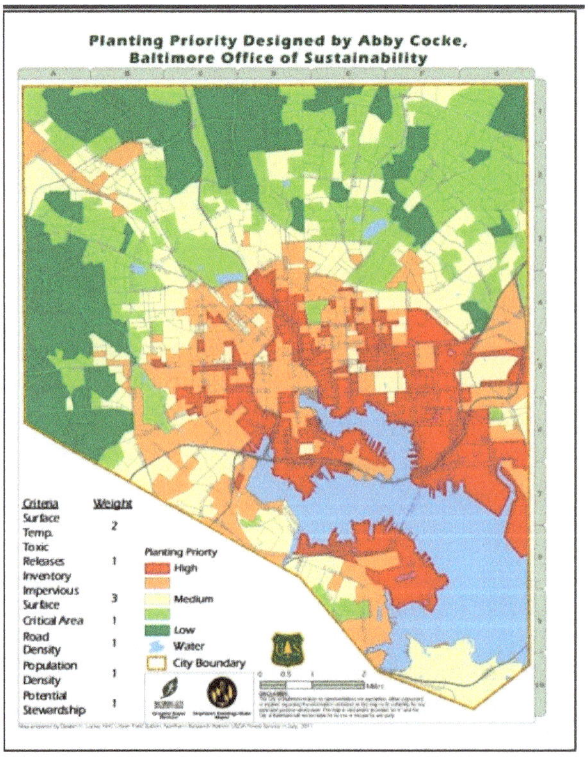

Figure 10: This map was created by summing up all 25 stakeholder's stated preferences for each criteria.

Linking Prioritization Analysis to other City Policy and Planning Initiatives

It may be important to link prioritization analyses to other policy and planning initiatives in a City. We created UTC prioritization maps for the City using Census Block Groups / Neighborhoods as the unit of analysis. This was effective for showing spatial variation with a familiar and useful social unit that is the basis for policy and planning.

The Department of Public Works is interested to explore how UTC prioritization can be used to implement its municipal stormwater utility. Specifically, UTC Prioritization maps can be used to target where to locate green infrastructure projects using stormwater utility revenue and to communicate how those stormwater utility fees are being strategically spent. Further, some landowners, particularly industrial landowners, may be interested to fund stormwater mitigation projects off-site in order to reduce their assessment. UTC Prioritization maps can also be used to identify potential sites and provide an opportunity for the industry to take "good citizen" credit for other UTC benefits, such as heat island mitigation, associated with the site.

The Department of Planning was interested to know how UTC priorities varied by its Housing Market Typology. The City's Housing Market Typology is the result of a cluster analysis used to define categories of housing markets9 and serves as a tool to help the City determine how it prioritizes investments and services. The five classes are Regional Choice, Middle Market Choice, Middle Market, Middle Market Stressed, and Distressed. We learned three things by linking the UTC prioritization with the City's Market Typology. First, even the City's highest quality neighborhoods (Regional Choice) had neighborhoods that were rated high priority for increasing UTC in the summary map with all groups votes combined (Figure 11). Each housing market type contains the range of UTC priorities when using the map containing the sum of all stakeholder's preferences. This combined UTC Prioritization and housing market analysis makes clear that areas which are high priority for UTC are not synonymous with poor or undesirable neighborhoods and that increasing tree canopy in high priority areas is not identical to addressing issues of environmental equity.

The second thing we learned by combining UTC and the City's Housing Market Typology is that UTC prioritization—its implementation and the coalitions who support it—can be used to further the City's larger agenda of urban stabilization and revitalization. Third, the motivations, preferences, and capacities for increasing tree canopy may vary among the housing markets that are also areas of high priority for increasing canopy. The following are two hypothetical examples. Households in Regional Choice neighborhoods may be motivated by property values, aesthetics, and stormwater mitigation. They may prefer small, flowering ornamental trees nearby, large shade trees and extensive lawns. They may plant trees on their own or work with a private landscape service. In contrast, households living in Distressed neighborhoods may be on a fixed income because they are retired or are on public assistance. They may not want to increase their property values because it may either lead to higher annual taxes, or because they do not own their home. But these residents may be motivated by aesthetics, job programs for youth, public health, and reducing trash and crime. They may prefer mid-size shade trees and planters since space might be more limited. They may be more likely to work with local churches, community associations, and non-profit organizations.

Additional Research Needs

Two major types of research needs were identified through the UTC Prioritization Process. The first type is methodological. Stakeholders asked if there was a way to identify trees that are in poor health or at risk of mortality because of insects or blights. Particularly, could we identify all the Ash trees, which are at risk of dying due to the Emerald Ash Borer? This is currently only possible using either field surveys, which are limited to public lands or private lands with the owner's permission, or through hyperspectral remote sensing techniques. Hyperspectral approaches are particularly attractive because they could be used to create a complete census of the landscape and, when combined with LiDAR, could quantify canopy cover by species, height, and condition. Hyperspectral approaches have been developed for rural forest areas. However, these areas tend to be relatively homogenous in terms of forest pattern and species diversity. Urban environments tend to be much more heterogeneous in spatial pattern and species diversity. Research is needed to determine the feasibility of hyperspectral sensors to map tree species and their attributes in urban environments and for these systems to be cost-effective.

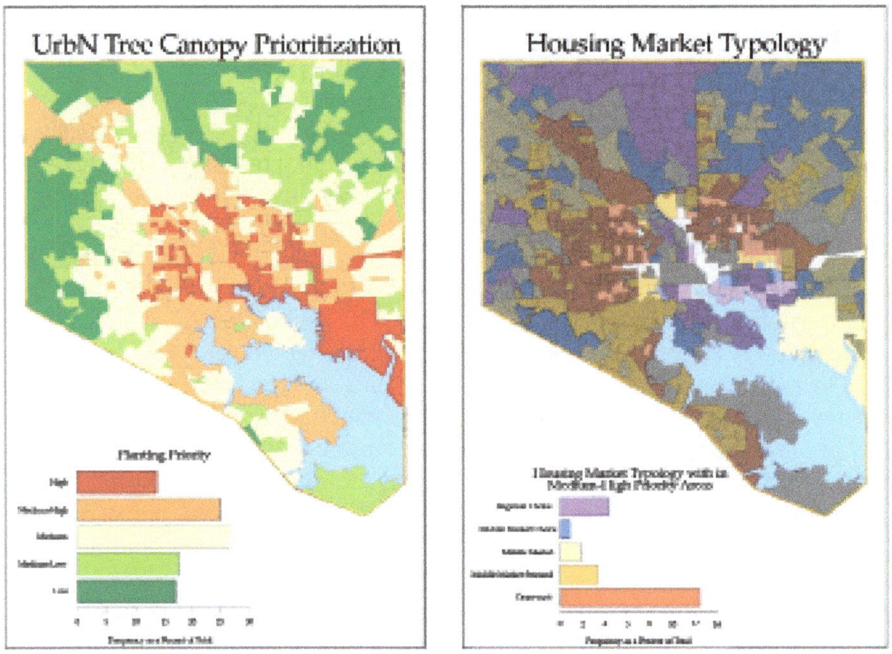

Figure 11: When only the high priority categories of UTC Prioritization map are shown, all five of Baltimore City's Housing Market Typology classes are found (right). High planting priority does not co-vary with the housing market. Different urban forestry strategies may be more effective in Regional Choice markets than in Distressed housing markets.

The second major type of research is theoretical: what social-ecological factors affect household motivations, preferences, and capacities to steward urban trees? Here, we use the term "steward" in reference to planting new trees or trees that have been either planted or grown through natural regeneration. Acts of stewardship involve a variety of behaviors that maximize the growth and minimize the loss of the urban canopy (Luley and Bond 2002). In the case of tree planting, novel approaches may be needed to increase tree canopy in high priority areas on community and private lands. Historically, public agencies have had limited jurisdiction for tree planting on private lands, and have focused their attention on public rights-of-way. To reach private landowners on private lands, public agencies have used tree programs that depended upon reactive approaches such as "first come, first served" tree giveaway programs. Who participates in these programs and why is an area of active research in Baltimore as well as Washington D.C. UTC high priority areas may not be areas with a high interest in tree planting on private land. Thus, market analysis research may need to be developed to better understand potential social, economic, and environmental motivations, preferences, and capacities of the community groups and residents associated with high priority UTC areas and low participation in tree giveaway programs (Grove et al. 2006b).

Market analysis research could benefit by examining household and neighborhood behaviors in the context of stewardship networks. Research on stewardship organizations and their networks is growing: factors affecting volunteerism in public parks (Fisher et al. 2010, Fisher et al. 2011), the organizational characteristics and degree of professionalization of civic environmental groups (Fisher et al. 2012), history of environmental stewardship organizations (Svendsen 2010; Connolly et al 2013), and their networks (Connolly et al 2013). Similar studies on the connections and collaborations among stewardship groups in Chicago (Belaire et al. 2011), Baltimore (Romolini under review), and Seattle (Wolf et al. 2011, Romolini under review) further extend this active area of research. This growing research has permitted cross-site analyses of stewardship networks and canopy cover in Seattle and Baltimore (Romolini, under review).

CONCLUSION

In this paper we have described the practical use of UTC Prioritizations tools in situations where there are a diverse set of stakeholders', whose interests spanned social, economic, and environmental issues. A transparent process was important to build confidence and trust among the stakeholders who participated, and to work toward a collaborative "All Lands, All People" vision that encompasses not just publicly funded activities on public land, but greening opportunities throughout the urban forest. An important step in the stakeholder process was a stakeholder survey that asked about what UTC services and benefits were important to them and their programs, the categories of forestry on which they focused, and the areas in which they were interested. Results from the survey were important for learning about the diverse community of stakeholders and identifying opportunities for coordination and collaboration and potential gaps in capacity. Results from the mapping process created prioritization areas, which was important for building consensus and coordination of limited resources among the stakeholders who participated.

The planning process described in this Baltimore example can be used in other cities, with local data and criteria adapted to meet the needs of urban forest managers, planners and sustainability goals. This process can used to advance the shift from planning for a Sanitary City to planning for a Sustainable City (Grove 2009, Pincetl 2009). Ultimately, UTC Prioritization can help groups transition from "tree people" to "sustainability people."

APPENDIX 1: List of Organizations who attended UTC Prioritization Workshop

Individual on behalf of	**Organization**
Amy Gilder-Busatti	Baltimore Office of Sustainability/Department of Planning
Duncan Stuart	Department of Public Works
Abby Cocke	Baltimore Office of Sustainability
Justin Reel	RK&K
William Andersen	Baltimore City Recreation and Parks - Capital Development
Tom McGilloway	Mahan Rykiel
Tom McGilloway	The Friends of Wyman Park Dell
Anne Draddy	Department of General Services
Barbara Shea	Baltimore Tree Trust
Robbyn Lewis	Patterson Park Neighborhood Association
John Bishop	Wyndhurst Neighborhood
Fernando Guerra	Midtown Community Benefits District - Midtown Green
Mel Freeman	Citizens Planning and Housing Association, Inc.
Chad Vrany	Bartlett Tree Experts
Al Copp	Roland Park Civic League - Greater Roland Park Master Plan Implementation
Guy Hager	Parks & People Foundation
Robert I. Catlin, II	Brooklyn and Curtis Bay Coalition, Inc.
Rebecca Ruggles	Baltimore Medical System
Darin Crew	Blue Water Baltimore
Odessa Neale	Park Heights Renaissance
Jonathan Gross, MPH	Office of Public Health Preparedness and Response (OPHPR), Baltimore City Health Department
Kari Snyder	SouthEast Community Development Corporation
Valorie Lacore	Department Of Transportation
UTC Team	
Jarlath O'Neil-Dunne	University of Vermont, Spatial Analysis Laboratory
J. Morgan Grove	USDA Forest Service, Northern Research Station and Baltimore Ecosystem Study
Mike Galvin	SavATree Consulting Group
Dexter H. Locke	Graduate School of Geography, Clark University (formerly USDA Forest Service, Northern Research Station).

APPENDIX 2: Survey Instrument

Baltimore City TreeBaltimore Meeting:

UTC Prioritization Survey

Thursday, June 16th

1. Contact information: Name, Organization, Phone, Email

2. UTC Prioritization

You get 10 "Xs" to weight your priorities. You can vote more than once for your priorities.

Major Criteria	Criteria Variables	Priority Weightings
Public Health & Saftey	Life expectancy, mortality rate, mortality rate from heart disease, mortality rate from strokes, mortality rate from respiratory disease, mortality rate from diabetes, infant mortality rate	
	Asthma by zipcode	
	Dependency Ratio (< 18 + > 65) / sum >18 - <65	
	Urban Heat Island: surface temperature and solar exposure	
	Crime: Personal, Property and Total	
	Transportation Connections	
Environmental Justice	Toxic Releases Inventory	
	Brownfields	
	Poverty, Race, Home ownership	
	Percent Parks	
Water Quality	Percent Impervious Surface	
	Watershed H20 quality	
	Stream corridors	
	Flood Plains	
	Critical Area	
	Greenstreets	
	Blue alleys	
	Flooding	
Air Quality & Noise Pollution	(Major) Road Density	
Critical Places	Schools, hospitals, libraries, recreation centers, and elderly care facilities	
	Population density (per square mile)	
Community Presence	Potential stewardship	

3. Where do you work? Categories of Interest, please circle:

- Street Trees
- Abandoned Lots
- Private Residential Lands
- Private Industrial
- Parks
- Schools
- Stream Valleys – Riparian Areas
- Shoreline Areas
- Other

4. Where do you work? Areas of Interest, please specify name:

- Neighborhood
- Watershed
- Council Districts
- City
- Other

ACKNOWLEDGEMENTS

We thank the anonymous reviewers and Lynne Westphal for their helpful comments. We are also grateful for the workshop participants. Without them this work would not be possible, and Baltimore City would not have the prioritization materials. Funding was provided by the DC-BC ULTRA-Ex NSF-DEB-0948947.

REFERENCES

1. Akbari, H. and S. Konopacki. 2005. Calculating energy-saving potentials of heat-island reduction strategies. Energy Policy 33(6):721-756.
2. Akbari, H., M. Pomerantz, and H. Taha. 2001. Cool surfaces and shade trees to reduce energy use and improve air quality in urban areas. Solar Energy 70(3):295-310.
3. Baltimore Sustainability Plan. 2009. The Baltimore Sustainability Plan. http://www.baltimoresustainability.org/uploads/files/Sustainability_Plan.pdf Accessed 10/31/12
4. Beattie, J., C. Kollin, and G. Moll. 2000. Trees tackle clean water regs. American Forests 106(2):18.
5. Belaire, J. A., Drilbin, A. K., Johnston, D. P., Lynch, D. J., & Minor, E. S. (2011). Mapping stewardship networks in urban ecosystems. Conservation Letters, 4, 464-473. doi:10.1111/j.1755-263X.2011.00200.x
6. Bell, J. F., J.S. Wilson, and G.C. Liu. 2008. Neighborhood greenness and 2-Year changes in body mass index of children and youth. American Journal of Preventive Medicine 35(6):547-553.
7. Chesapeake Bay Program. 2004. Guidelines for Implementing the Chesapeake Bay Program's Urban and Community Tree Canopy goals. 11 pp. http://www.urbanforestrysouth.org/resources/library/guidelines-for-implementing-thechesapeake-bay-program2019s-urban-and-community-tree-canopy-goals/file Accessed 5/24/12.

8. Chesapeake Executive Council. 1983. 1983 Chesapeake Bay Agreement. http://www.chesapeakebay.net/content/publications/cbp_12512.pdf Accessed 5/24/12.

9. Connolly, J., Svendsen, E.S., Fisher, D.R., Campbell, L.K. (2013). Organizing Urban Ecosystem Services through Environmental Stewardship Governance in New York City. Landscape and Urban Planning.

10. Fisher, D. R., Campbell, L. K., & Svendsen, E. S. (2012). The organisational structure of urban environmental stewardship. Environmental Politics, (January): 37-41.

11. Fisher, D. R., Connolly, J. J., Svendsen, E. S., & Campbell, L. K. (2010). Who Volunteers to Steward the Urban Forest in New York City? An analysis of participants in MillionTreesNYC planting events. Environmental Stewardship Project White Paper #1. 33 p. Accessed from: http://nrs.fs.fed.us/nyc/localresources/downloads/2010_WhoAreVolunteerStewards_ESPWhitepaper1.pdf October 9, 2012.

12. Fisher, D. R., Connolly, J. J., Svendsen, E. S., & Campbell, L. K. (2011). DIGGING TOGETHER: Why people volunteer to help plant one million trees in New York City.

13. Environmental Stewardship Project at the Center for Society and Environment of the University of Maryland White Paper #1. 36 p. Accessed from: http://nrs.fs.fed.us/nyc/local resources/downloads/Digging_Together_White_Paper.pdf October 9, 2012.

14. Galvin, M. F., Grove, J. M., & O'Neil-Dunne, J. P. M. (2006). A Report on Baltimore's Present and Potential Urban Tree Canopy. Scenario. Retrieved from http://www.fs.fed.us/nrs/utc/reports/UTC_Report_BACI_2007.pdf

15. Gillespie, T. W., Pincetl, S., Brossard, S., Smith, J., Saatchi, S., Pataki, D., & Saphores, J.-D. (2011). A time series of urban forestry in Los Angeles. Urban Ecosystems, 15(1), 233– 246. doi:10.1007/s11252-011-0183-6

16. Grove, J. M. (2009). Cities: Managing Densely Settled Social-Ecological Systems. In C. Folke, G. P. Kofinas, & F. S. Chapin, III (Eds.), Principles of Ecosystem Stewardship (pp. 281– 295). New York, NY: Springer New York. doi:10.1007/978-0-387-73033-2

17. Grove, J. M., Burch, W. R., & Pickett, S. T. A. (2005). Social Mosaics and Urban Community Forestry in Baltimore, Maryland Introduction: Rationale for Urban Community Forestry Continuities From Rural to Urban Community Forestry. In R. G. Lee & Field D. R. (Eds.), Communities and Forests: Where People Meet the Land (pp. 248-273). Corvalis: Oregon State University Press. Retrieved from http://web.mac.com/jmorgangrove/Morgan/Products_files/Grove_et_al_Social Mosaics.pdf

18. Grove, J.M., J. O'Neil-Dunne, K. Pelletier, D. Nowak, & J. Walton. (2006a). A report on New York City's present and possible urban tree canopy: Prepared for Fiona Watt, Chief of the Division of Forestry and Horticulture. New York Department of Parks and Recreation, USDA Forest Service, Northern Research Station. 28 pp. http://nrs.fs.fed.us/nyc/localresources/downloads/Grove_UTC_NYC_FINAL.pdf (accessed 01/26/2011).

19. Grove, J. M., Troy, A. R., O'Neil-Dunne, J. P. M., Burch, W. R., Cadenasso, M. L., Pickett, S. T. A., & O'Neil-Dunne, J. P. M. (2006b). Characterization of Households and its Implications for the Vegetation of Urban Ecosystems. Ecosystems, 9(4), 578–597. doi:10.1007/s10021-006-0116-z

20. Jackson, L.E. 2003. The relationship of urban design to human health and condition. Landscape and Urban Planning, 64(4): 191-200.

21. King, K.L., Locke, D. H., Svendsen, E.S., Campbell, L.K., Small, C., Falxa-Raymond, N., Fisher, D.R. & Lu, J.W.T. (In Review). Urban environmental stewardship and changes in vegetative cover in New York City neighborhoods (2000-2010).

22. Locke, D. H., Grove, J. M., Lu, J. W. T., Troy, A. R., O'Neil-Dunne, J. P. M., & Beck, B. D. (2010). Prioritizing Preferable Locations for Increasing Urban Tree Canopy in New York City. Cities and the Environment (CATE), 3(1), 1–18. Retrieved from http://digitalcommons.lmu.edu/cgi/viewcontent.cgi?article=1066&context=cate

23. Lulely, C.J. & J. Bond. 2002. A Report to the Northeast State Foresters Association. A Report to Integrate Management of Urban Trees into Air Quality Planning. Davey Resources Group. Naples, NY. Pp. 70

24. Lovasi, G.S., J.W. Quinn, K.M. Neckerman, M.S. Perzanowski, and A. Rundle. 2008. Children living in areas with more street trees have lower asthma prevalence. Journal of Epidemiology and Community Health 62, 647-649.

25. MacFaden S.W., O'Neil-Dunne J.P.M., Royar A.R., Lu J.W.T., & Rundle A.G. (2012) Highresolution tree canopy mapping for New York City using LiDAR and object-based image analysis. Journal of Applied Remote Sensing 6:063567

26. McGee, J., Day, S., Wynne, R., & White, B. (2012). Using Geospatial Tools to Assess the Urban Tree Canopy: Decision Support for Local Governments. Journal of Forestry, (August).

27. Retrieved from http://www.ingentaconnect.com/content/saf/jof/2012/00000110/00000005/art00006

28. Mitchell, R. and F. Popham. 2008. Effect of exposure to natural environment on health inequalities: An observational population study. Lancet 372:1655-60.

29. Nowak, D.J. 2002. The effects of urban trees on air quality. Syracuse, NY: USDA Forest Service. 5 pp. http://nrs.fs.fed.us/units/urban/localresources/downloads/Tree_Air_Qual.pdf (accessed 01/26/2011).

30. Nowak, D.J., R.E. Hoehn, D.E. Crane, J.C. Stevens, and J.T. Walton. 2007. Assessing urban forest effects and values: New York City's urban forest. Resour. Bull. NRS-9.

31. Newtown Square, PA: U.S. Department of Agriculture, Forest Service, Northern Research Station. 22 pp. http://www.nrs.fs.fed.us/pubs/rb/rb_nrs009.pdf (accessed 01/26/2011).

32. Nowak, D. J. (2012). Contrasting natural regeneration and tree planting in fourteen North American cities. Urban Forestry & Urban Greening, 11(4), 374–382. doi:10.1016/j.ufug.2012.02.005

33. Nowak, D. J., & Greenfield, E. J. (2012). Tree and impervious cover change in U.S. cities. Urban Forestry & Urban Greening, 11(1), 21–30. doi:10.1016/j.ufug.2011.11.005

34. O'Neil-Dunne, J.P.M. (2009). A Report on the City of Baltimore's Existing and Possible Tree Canopy. The Spatial Analysis Lab at the University of Vermont's Rubenstein School of the Environment and Natural Resources. (p.5). http://www.fs.fed.us/nrs/utc/reports/UTC_Report_BACI_2007.pdf

35. O'Neil-Dunne, J. P. M., MacFaden, S. W., Royar, A. R., & Pelletier, K. C. (2012). An object based system for LiDAR data fusion and feature extraction. Geocarto International, (June 2012), 1–16. doi:10.1080/10106049.2012.689015

36. Pincetl, S.S., (2010). From the Sanitary City to the Sustainable City: Challenges to Institutionalizing Biogenic (Nature's Services) Infrastructure, Local Environment 15(1): 43-58.

37. R Development Core Team. 2012. R: A language and environment for statistical computing. R Foundation for Statistical Computing, Vienna, Austria. ISBN 3-900051-07-0, http://www.R-project.org/

38. Raciti, S., M. F. Galvin, J. M. Grove, J. P. M. O'Neil-Dunne, A. Todd & S. Clagett (2006). Urban Tree Canopy Goal Setting: A Guide for Chesapeake Bay Communities, United States Department of Agriculture, Forest Service, Northeastern State & Private Forestry, Chesapeake Bay Program Office, Annapolis, Md.

39. Rosenfeld, A.H., H. Akbari, J.J. Romm, and M. Pomerantz. 1998. Cool communities: Strategies for heat island mitigation and smog reduction. Energy and Buildings 28(1): 51-62.

40. Romolini, M., J.M. Grove, & Locke, D. H. (In review). Assessing and comparing relationships between urban environmental stewardship networks and land cover in Baltimore and Seattle.

41. Son J-Y, Lee J-T, Anderson GB, Bell ML 2012. The Impact of Heat Waves on Mortality in Seven Major Cities in Korea. Environ Health Perspect 120:566-571. http://dx.doi.org/10.1289/ehp.1103759

42. Streiling, S., and A. Matzarakis. 2003. Influence of single and small clusters of trees on the bioclimate of a city: A case study. Journal of Arboriculture 29(6): 309-316.

43. Svendsen, E. (2010) Civic Environmental Stewardship as a form of Governance in New York City. Doctoral Dissertation. Columbia University, Departement of Urban Planning.

44. Takano, T., K. Nakamura, and M. Watanabe. 2002. Urban residential environments and senior citizens' longevity in megacity areas: The importance of walkable green spaces. Journal of Epidemiology and Community Health 56(12): 913-918.

45. Ward, J.H. (1963). Hierarchical grouping to optimize an objective function. Journal of the American Statistical Association. 58:236–244.

46. Wolf, K. L., Blahna, D. J., Brinkley, W., & Romolini, M. (2011). Environmental stewardship footprint research: linking human agency and ecosystem health in the Puget Sound region. Urban Ecosystems. doi: 10.1007/s11252-011-0175-6.

47. Yaffee, S. L. 1997. Why Environmental Policy Nightmares Recur. Conservation Biology 11:338 - 337.

CHAPTER 4 TREES, PEOPLE AND THE BUILT ENVIRONMENT PROCEEDINGS OF THE URBAN TREES RESEARCH

Matthew P. Wells

INTRODUCTION

Urban forestry managers have continually strived to find the precarious equilibrium between the needs of trees and the needs of people. Often the pressures of liability and limited resources have forced these managers to focus solely on tree maintenance and tree removals. There has been some excellent research completed in the fields of tree mechanics and hazard tree evaluation. This research has been coupled with numerous studies on the social and psychological benefits of humans interacting with their natural environment. However, this arboricultural and social research has a limited use for urban forest managers battling to holistically manage a diverse resource at a city or town level. Only recently have urban forest managers had more to help them secure funding and guide urban forest programming. The US Forest Service has recently released a number of free useful tools for urban forest managers. These tools allow urban forest managers to quantify the annual environmental benefits provided to their town or city by their urban forest. These quantified environmental benefits have allowed policy makers to understand and appreciate the urban forest. These tools have very much put trees on the policy map. The New York City Department of Parks and Recreation (NYC DPR) has used two of these tools to analyse the city's urban forest. The Urban Forest Effects Model (UFORE) calculated the environmental benefits of the entire urban forest, while the Street Tree Resource Analysis Tool for Urban Forest Managers (STRATUM) focused solely on the street tree population. NYC DPR coupled the results of these tools with other pertinent research to justify the inclusion of trees into Mayor Bloomberg's sustainability plan for New York City (NYC) called PlaNYC. In PlaNYC, trees play a major role in greening the landscape and are also being actively deployed in helping to capture stormwater and cleaning the air. Their inclusion was only possible through NYC DPR being able to prove and quantify the annual environmental benefits provided by them. However, the research did not only justify why additional resources should be allocated into the urban forest. This research also provided key information that allowed proper attainable urban forest goals, policies and strategies to be established to maximize the benefits of New York's urban forest.

This paper will look at the key research studies and how they have been used to justify and focus urban forestry programming in NYC. Alongside this central theme will be the importance and power of in-house collection of administrative data and its analysis. NYC DPR has very successfully used in-house resources, volunteers and interns to help perform vital research.

Citation: Matthew Wells, Trees, people and the built environment Proceedings of the Urban Trees Research, ISBN 978-0-85538-849-2

THE SOCIAL VALUE OF THE URBAN FOREST AND URBAN TREES

The social value of the urban forest has been well researched, although these studies have not been able to quantify this value in dollars. It is understood that views of trees and nature are known to help improve mental wellbeing (Kaplan and Kaplan, 1989) and also help with recovery from illness (Ulrich, 1984). It has been shown that humans derive pleasure from trees (Lewis, 1996). Other research has also shown that outdoor spaces with trees facilitate greater interactions among local residents, which improves neighbourhood socializing (Kou et al., 1998). This research is fascinating and very valuable and reinforces what many of us have always instinctively believed about trees and the urban forest. However, these social values alone do not provide the strongest justification or argument for urban foresters trying to preserve existing trees or find resources to plant new ones, especially if liability is also a concern.

Only when more recent research emerged that started to quantify the environmental benefits and the associated financial value provided by the urban forest did trees become an essential element in a city rather than just a feel-good luxury item. A great deal of this research has been done by the US Forest Service (McPherson et al., 2007; Nowak et al., 2007; Peper et al., 2007). They provide a number of free tools for urban forest managers via their i-Tree software suite. Two of these tools, the Urban Forest Effects Model (UFORE) and the Street Tree Resource Analysis Tool for Urban Forest Managers (STRATUM), have been invaluable to urban foresters in NYC, especially when combined with other relevant research.

RESEARCH ON THE ENTIRE URBAN FOREST IN NEW YORK CITY

New York City (NYC) is America's largest metropolis and home to an estimated 8.2 million people (US Census Bureau, 2006). NYC is extremely urban in its environment and even though it is home to one of the most famous parks in the world, Central Park, it is not otherwise known for its trees and open spaces.

The Urban Forest Effects Model (UFORE)

The U.S. Forest Service completed a UFORE (now called iTree Eco) survey and analysis of NYC's entire urban forest in 1996, and estimated that it contained 5.2 million trees (Nowak et al., 2007). This was somewhat of a surprise. Furthermore, the UFORE study put the structural value of NYC's urban forest at $5.2 billion and estimated that 50% of the urban forest fell under the jurisdiction of the New York City Department of Parks and Recreation (NYC DPR). UFORE also estimated that NYC had a 20.9% tree cover, with 42.7% of the trees being over 6 inches (15.25 cm) in diameter. But perhaps the most interesting findings were the environmental benefits the urban forest was delivering to New Yorkers. The urban forest worked to remove 1998 tonnes of air pollution each year at an annual value of $10.6 million and stored 1.22 million tonnes of carbon at an estimated value of $24.9 million. Finally, the urban forest was sequestrating 38 374 tonnes of carbon annually at an annual value of $779 000. It should be noted that despite all this impressive data, the UFORE study acknowledged that additional social and environmental benefits were not included. These key figures about NYC's urban forest immediately provided NYC DPR with a reason to request additional resources for forestry. Ultimately, a federal agency had proved that NYC's urban forest was providing substantial and valuable environmental benefits to the city.

The UFORE report was more that just a report on environmental benefits. It also provided essential data to aid in the correct management of the urban forest. It identified the most common species as being tree of heaven (Ailanthus altissima) at 9.0%, black cherry (Prunus serotina) at 8.1% and sweetgum (Liquidambar styraciflua) at 7.9% (Nowak et al., 2007). It also confirmed what many already assumed, that large-canopied trees, provide the greatest benefits, with ironically the London plane (Platanus x hispanica) having the greatest importance in NYC based on total leaf area and abundance. UFORE also helped us understand the potential threat of the invasive Asian longhorned beetle (ALB) to NYC. ALB was discovered in the NYC borough of Brooklyn in 1996 and this was actually the first time it had been discovered on the US mainland. ALB is a beetle that destroys certain species of trees through boring damage. UFORE concluded that 43.1% of the urban forest was potentially at risk from ALB. This knowledge made federal, state and city agencies very aware of the implications of ALB for NYC as $2.25 billion worth of trees were potentially at risk.

Urban tree canopy coverage

In April 2006, NYC DPR commissioned the US Forest Service and the University of Vermont's Spatial Analysis Laboratory to conduct an analysis of urban tree canopy (UTC) coverage in the city. NYC DPR wanted to understand if achieving an UTC goal of 30% by 2030 was possible. The completed research established that 24% (17 972 hectares) of NYC's total land area was already covered by UTC (Grove et al., 2006). The study also calculated that 42% (32 052 hectares) of the city's total land area had the potential to be covered by UTC because no roads or buildings were present. The report concluded that a goal of 30% UTC by 2030 was achievable if 4856 hectares were added. The report also recommended that progress towards attaining this UTC goal should be monitored by using remote sensing at five-year intervals.

RESEARCH ON STREET TREES IN NEW YORK CITY

Street trees are perhaps the most visible and easily defined component of any urban forest. They are the trees outside people's homes and places of work that touch their lives on a day-to-day basis. Street trees therefore usually require the most intensive management by urban foresters and their location tends to make them the ones that people are most interested in for either positive or negative reasons. They are the public face of trees.

The 2005–2006 street tree census

Every decade the NYC DPR undertakes a census of the street tree population. The last census undertaken in 2005–2006 was called 'Trees Count'. The census was conducted with the help of more than 1100 volunteers logging over 30 000 hours (New

York City Department of Parks & Recreation, 2007). This level of participation represented a 57% increase from the previous census in 1995–1996 where only 700 volunteers participated. Volunteers were required to attend a three-hour training session and collected 42% of the census data. The remainder was completed by in-house staff and by an urban forestry consultant. The census collected over 15 million pieces of data across the five boroughs. To facilitate the data collection, the city was divided into 1649 survey zones that were assigned to the individuals taking part in the census. For each tree counted, the surveyor recorded information such as location, species, diameter at breast height (dbh), condition, tree pit type, soil level, sidewalk condition, presence of overhead wires and infrastructure conflicts. Survey results were reported back to NYC DPR using an interactive census website application or on paper

The published results of the tree census identified 592 130 street trees in NYC; this represented a 19% increase from the census a decade earlier (New York City Department of Parks & Recreation, 2007). London plane was the most prominent species making up 15.3% of the population with Norway maple (Acer platanoides) not far behind at 14.1%. Other important species were Callery pear (Pyrus calleryana) at 10.9%, honey locust (Gleditsia triacanthos) at 8.9% and pin oak (Quercus palustris) at 7.5%. This data immediately highlighted that NYC needed greater species diversification and no one species should really exceed 10% of the total population (Peper et al., 2007).

Table 1 shows the tree condition results of the census and Table 2 shows the size of the trees. The census data provided a good snapshot of the entire street tree population within a relatively small time band. This is not achieved when surveying a portion of the street tree population on an annual basis over multiple years.

Table 1: Tree condition results of the 2005–2006 tree census (New York City Department of Parks & Recreation, 2007).

Tree condition	Percentage of the population
Excellent	23.9%
Good	66.4%
Poor	8.3%
Dead	1.4%

Table 2: Tree size results of the 2005–2006 tree census (New York City Department of Parks & Recreation, 2007).

Tree size	Percentage of the population
Small (0-15 cm)	25%
Medium (15-46 cm)	50%
Large (46-76 cm)	20%
Extra large (over 76 cm)	5%

Other interesting information that came out of the census was that 15% of the tree population suffered from trunk wounds and 5.3% had a cavity of some type. Finally, the census highlighted some of the key conflicts that NYC's tree population has with infrastructure (see Table 3).

Table 3: Trees with urban conflict results of the 2005–2006 tree census (New York City Department of Parks & Recreation, 2007).

Urban conflict	Number of trees	Percentage of the population
Overhead wires	209 171	35.8%
Raised sidewalks	100 829	17.3%
Cracked sidewalks	65 299	11.2%
Close paving	43 409	7.4%
Choking wires	13 865	2.4%
Choking guard/grate	3918	0.7%
Tree lights	3918	0.4%

The number of trees impacted by urban conflicts in NYC is considerable (Table 3). Therefore, mitigating these street tree conflicts with infrastructure, as far as reasonably possible, is a key challenge for NYC DPR. The census recorded that nearly 36% of the population was under wires and could be subjected to utility clearance pruning. The census also identified that 17.3% of the trees surveyed had raised adjacent sidewalk and 11.2% of the population had cracked adjacent sidewalk. In NYC property owners are responsible for the maintenance of the sidewalk adjacent to their land (New York City Department of Transportation, 2008). Damaged sidewalks and the disturbance of utility wires are often cited as a reason for requesting removal of a tree or protesting against the planting of a new one. The authors of recent research analysed complaints to NYC DPR about the placement of new tree planting. A total of 33% of these complainants objected because of the potential of the tree to cause utility service disturbance and 14% objected because of the potential of future sidewalk damage (Rae et al., 2010). These are obviously both significant factors when considering urban forestry programming and the concerns of property owners.

The tree census data allowed NYC DPR to consider their street tree inventory at a borough level and the change in that inventory since the census in 1995–1996 (Table 4). The census clearly showed that certain boroughs had considerably more trees than others, as detailed in Table 4. It can be seen that Staten Island had the greatest rise in its street tree population since 1995–1996 with a 33% increase. Manhattan had the least with just a 9% increase and Queens was not far behind at only a 10% increase. The census data also identified that London plane was the most common species citywide, but is only the dominant species in Brooklyn (24%) compared to honey locust in the Bronx (13%) and Manhattan (23%), Callery pear in Staten Island (25%) and Norway maple in Queens (18%) (New York City Department of Parks & Recreation, 2007).

Table 4: Number of trees recorded per borough in the 2005–2006 tree census versus 1995–1996 (New York City Department of Parks & Recreation, 2007)

Borough	1995-1996 census	2005-2006 census	% increase
Bronx	47 995	60 004	25%
Brooklyn	112 400	142 747	27%
Manhattan	45 793	49 858	9%
Queens	217 111	239 882	10%
Staten Island	75 171	99 639	33%
Totals	**498 470**	**592 130**	**19%**

The tree census also identified other borough trends in tree health and infrastructure conflicts. The Bronx's tree population was in the worst condition, with 12% of trees falling into the dead or poor condition categories, followed closely by Manhattan at 11.3% and Queens at 10% (New York City Department of Parks & Recreation, 2007). Staten Island's trees were recorded as being in the best condition with only 6% of trees falling outside the good and excellent tree condition categories. As stated previously, 36% of the total citywide tree street population was recorded as being under utility wires. However, when we look at this percentage at a borough level, it rises significantly to 48% in Queens but falls back to 23% in Staten Island and is lower still in the Bronx at 12%. In summary, management policies should account for the distinct differences in the urban forest even within a single city or town.

STREET TREE RESOURCE ANALYSIS TOOL FOR URBAN FOREST MANAGERS (STRATUM)

STRATUM (Street Tree Assessment Tool for Urban Forest Managers) is now known as i-Tree Streets and is another application available from the US Forest Service. STRATUM uses street tree inventory data to calculate the annual environmental and aesthetic benefits generated. It is distinctly different from UFORE because it does not consider the urban forest in its entirety. The STRATUM model is more accurate in its results compared to UFORE because the size, species and condition of each and every tree is known. It is possible to perform a STRATUM analysis using just a sample of the street tree population (Kling, 2008), although this was not done in NYC. The quantified benefits calculated by STRATUM include energy conservation, air quality improvement, carbon dioxide reduction, and stormwater catchment. The model also looks at the aesthetic contribution of street trees in terms of increasing property value.

STRATUM analysis for a city could cost more than $100 000 to survey and analyse growth data for 800 trees (Kling, 2008). So that this cost would not be prohibitive, the US Forest Service split the USA mainland into 16 climatic zones. Within each zone, an in-depth analysis has taken place at a single reference city. The reference city research involves detailed data collection on 30–60 trees for each of the predominant 20 species. NYC is the reference city for the Northeast region. The concept is that any city or town within a particular zone can then feed their street tree inventory data into the model to produce a fairly accurate calculation of the aesthetic and environmental benefits of their tree stock without the associated cost of having their own individual analysis done by the US Forest Service (Kling, 2008).

In 2007, the US Forest Service's Center for Urban Forest Research produced a STRATUM report for NYC DPR's Commissioner Adrian Benepe (Peper et al., 2007). This STRATUM analysis calculated that the street tree population of NYC, identified in the 2005–2006 tree census, provided an estimated $121.9 million in annual benefits. This translates to $209 per tree. These benefits are broken down in Table 5 below:

Table 5: Annual benefits provided by New York City's street tree population as estimated by STRATUM (Peper et al., 2007)

Annual benefits	Total value ($)	Value ($) per tree
Energy	$27 818 220	$47.63
Air quality	$5 269 572	$9.02
Stormwater catchment	$35 628 224	$61.00
Carbon dioxide reduction	$754 947	$1.29
Aesthetic/other	$52 492 384	$89.88
Total	**$121 963 347**	**$208.82**

At the time of the report NYC DPR estimated that it spent $21.8 million annually on planting new trees and maintaining existing street trees (Peper et al., 2007). Therefore, the street tree population provides $100.2 million or $172 per tree in net annual benefits to the city. It can also therefore be deduced that for every $1 spent on tree care operations, the city receives $5.60 in benefits. Aside from these benefits, STRATUM also estimated the replacement costs of the NYC street tree population at $2.3 billion or $3938 per tree (Peper et al., 2007)

JUSTIFYING GREATER RESOURCES THROUGH RESEARCH

A greater appreciation of the value and functions of an urban forest can be used to justify increased support and resources for its correct management (McPhearson et al., 2010). In NYC the quantified figures for environmental benefits produced by UFORE and STRATUM have been invaluable and very influential. NYC DPR's Commissioner Benepe said of STRATUM, 'It was probably the single most important sales tool we used to convince policy makers to put money into trees' (McIntyre, 2008). Putting dollars figures on trees perhaps does not sit well with all parties, but, just as with proper tree valuation, it is essential. David Nowak said on this subject 'the monetizing (of trees) is a necessary evil. We know trees have great value but they're intrinsically underrated. You have to talk the language of the people who make decisions' (Jonnes, 2011). In essence the establishing of the benefits of an urban forest will become a vital, if not mandatory, duty of any manager trying to convince policy makers to invest in trees.

Mayor Bloomberg invests in trees through PlaNYC

The knock-on effects of UFORE and STRATUM were dramatic in NYC. On Earth Day 2007, Major Bloomberg launched a comprehensive sustainable development plan for greener, greater NYC called PlaNYC (City of New York, 2007). PlaNYC lays out initiatives for the city to strive towards in five key dimensions of the urban environment. Trees play a significant role in 60% of those areas: namely land, water and air. The role of trees in this plan can be directly attributed to policy makers now understanding the vast potential that trees offer in combating many of the most worrying urban environmental challenges. UFORE data is actually quoted in PlaNYC as justification for the inclusion of trees in the initiatives. Furthermore, trees are relatively inexpensive, easy to access and return far more than is needed to be invested in them. Table 6 is a breakdown of the PlaNYC initiatives involving trees.

Table 6: PlaNYC initiatives involving trees (City of New York, 2007).

Dimension of the environment	Initiative	Goal
Land	Fill every street tree opportunity in NYC to achieve 100% stocking	Plant 22 000 street trees annually to fill the estimated 220 000 open planting opportunities by 2017
Water	Plant trees with improved pit designs	Maximize the ability of tree pits to capture stormwater
Air	Reforest 809 hectares of parkland	Complete reforestation project by 2017
Air	Partner with stakeholders to help plant one million trees	Plant one million trees in the city on both private and public property by 2017

To achieve the PlaNYC initiatives involving trees, Mayor Bloomberg massively increased NYC DPR's annual urban forestry budget. $118 million was listed in the Capital budget (FY 2008–2017) for the 809 hectares of new forest and $247 million for the estimated 220 000 street trees needed to obtain 100% stocking level (City of New York, 2007). Prior to PlaNYC, NYC DPR was annually planting around 6000 trees; with PlaNYC, this figure sky-rocketed to 22 000 trees. It should be noted that the 220 000 street trees and those planted through the reforestation initiative will make up the majority of the city's 60% commitment to the million tree goal. The remaining 40% (400 000 trees) will be planted by private and community organizations and homeowners (MillionTreesNYC, 2007a, 2007b).

In conclusion, Mayor Bloomberg planned to invest $365 million alone in tree planting over a decade because science and research had shown they play such a key role in producing a healthier and more sustainable environment for New Yorkers.

USING RESEARCH TO DIRECT URBAN FORESTRY PROGRAMMES

In addition to research being used to justify and secure resources for trees, it also should play a vital role in determining how those resources are used, or else the potential benefits of those additional resources may be squandered or lost. Research can be used to help set up programmes and monitor the progress of these programmes once operational. It can also be used to give insight into the outcomes of certain management decisions. Overall, research should be used to establish achievable goals and to formulate the most effective and efficient urban forestry programmes to reach them. Urban foresters should endeavour to run research driven programmes to guarantee success. The 2006 report by the US Forest Service and the University of Vermont's Spatial Analysis Laboratory on the present and possible urban tree canopy (UTC) in NYC was clearly a key reference for Mayor Bloomberg's staff when formulating realistic initiatives and goals for PlaNYC. As stated before, the research established that NYC's UTC could be increased from 24% to as high as 42% (Grove et al., 2006). The report identified numerous opportunities where this UTC increase could be realized based on land use type. For example, UTC on the Public Right of Way could be increased from 6% (4317 hectares) to 9% (6497 hectares). Therefore these figures reinforce the management decision in PlaNYC to plant an additional 220 000 street trees to reach a 100% stocking to take full advantage of this potential 3% UTC. In terms of other land uses, the report established that there was around 2000 hectares of car parks in NYC, approximately 1% of the NYC land area, and these were covered by 76 hectares of UTC. The report estimated that this land use had the potential to contain as much as 478 hectares of UTC, so this represented another significant opportunity to add around 402 hectares of UTC. PlaNYC included an initiative for changing planning regulations mandating perimeter landscaping and adjacent street tree planting for commercial and community run parking lots over 557 square metres (City of New York, 2007). In addition, for parking lots over 1115 square metres, a specific number of canopy trees would be required inside those lots in planting islands.

UFORE made recommendations relating directly to air quality because the study had shown that the urban forest was taking in 38 374 tonnes of carbon each year and also removing 1998 tonnes of pollutants (Nowak et al., 2007). The UFORE report for NYC included a tree planting index map that used census data and tree stocking data to identify areas of high population with low tree stocking densities. UFORE recommended that these areas should be prioritized for planting first. This management concept has been taken forward and evolved in PlaNYC. In PlaNYC it states that the planting of the 220000 street trees by NYC DPR will prioritize neighborhoods with the lowest UTC levels and the highest air quality concerns (City of New York, 2007). In practice NYC DPR has identified six neighbourhoods with lower than average tree stocking but higher than average asthma rates among young people (MillionTreesNYC, 2007a, 2007b). These geographical areas are called Trees for Public Health (TPH) neighbourhoods and they are being prioritized first for tree planting.

IN-HOUSE URBAN FORESTRY RESEARCH

NYC DPR also has a rich history of performing its own research and analysis. The tree census is a great example of a relatively simple research project using predominantly volunteers and in-house staff to produce a vast wealth of invaluable information about the street tree inventory. This information was not only used to run the STRATUM analysis but is also used on a regular basis to help guide urban forestry programming. A clear understanding of every aspect of a resource can only aid in its successful management.

Young tree mortality study

Perhaps some of the most impressive research undertaken by NYC DPR is a young street tree mortality study using inhouse staff and interns. This study randomly selected and surveyed 13 405 street trees that had been in the ground between three and nine years (Lu et al., 2010). The survey was completed in the summers of 2006 and 2007 and examined how biological, social and urban design factors affected young street tree mortality. The results showed that 74.3% of the trees surveyed were alive, with the rest either dead or missing. This percentage was raised to 82.7% for trees planted in one and two-family residential areas and dropped to 60.3% for trees in areas with heavy traffic. This number dropped even further to a 53.1% survival rate for trees located in central street medians. The research also highlighted some other very interesting data on the impacts of species, tree guards and the tree pit type on mortality rates. Alarmingly, the London plane tree had the lowest survival rate when compared to 19 other species, especially when STRATUM identified it as the most important tree in the urban forest in terms of environmental benefits delivered (Peper et al., 2007). Surprisingly, this study also concluded that tree pit size had little impact on survival rates and that the presence of animal waste was actually associated with a higher survival rate. This in-house research is obviously an invaluable resource in helping guide NYC DPR in reaching 100% stocking of live trees in its streets.

September 2010 tornado

Another example of the use of in-house research is perhaps less obvious and occurred when a tornado passed through NYC on 16 September 2010. After any storm event, gaining situational awareness of the type and location of damage is vital. This information is usually not available until qualified staffers have completed comprehensive field inspections, which could take several days if not weeks. Within two hours after the tornado, NYC DPR had received around 1000 calls reporting storm damage and had incorporated this into their forestry management system, ForMS. Using the addresses of these calls, NYC DPR was able to produce an initial map of the areas in the city that had suffered the brunt of the tree damage. This allowed for NYC DPR to provide key situational awareness data to the Mayor's Office and also the city's Office of Emergency Management. Valid situational awareness is essential in tempering an appropriate response to a tornado both in terms of requesting help and also in activating emergency debris clearance contracts. Eventually, just under 10 000 calls had been made to NYC DPR to report storm damage. NYC DPR used 15 years of previous storm data to explain to decision makers how severe this event was compared to previous storms and hurricanes. This provided the justification for a vast increase in the resources available for cleaning up the damage and for the help that was asked from other entities including the Federal Emergency Management Agency. NYC DPR also used previous storm data to extrapolate from the confirmed number of uprooted trees how many of those had potentially caused sidewalk damaged when they fell. This data was then provided directly to the New York City Department of Design and Construction (NYC DDC) who were tasked with repairing these damaged sidewalks. This allowed DDC to start the process of bidding out emergency contracts without having to wait for all the field inspections to be completed. Essentially, NYC DPR used research and analysis to give rapid situational awareness of the storm damage. This allowed for a far quicker gathering and deployment of appropriate resources needed to perform the clean-up operation and also communicating the severity of the damage to policy makers.

CONCLUSIONS AND FUTURE RESEARCH

This paper has endeavoured to illustrate the vital role of research in shaping the NYC urban forest and the programs of NYC

DPR. Urban forestry research has placed trees into the toolbox of urban planners battling to mitigate the negative impacts of city life and also take a responsible stance on the wider issue of climate change. Research should be an essential component of any urban forestry programme. Even in-house research of existing programmes can provide vital data and guidance for maximizing the benefits generated by those efforts. Research is a compass to guide urban forestry efforts as well as to help justify additional resources. NYC DPR has recently opened an urban field station in partnership with the US Forest Service at Fort Totten in Queens. This facility supports research by providing a fully equipped base for researchers to carry out their studies within NYC's urban forest. NYC DPR intends to use this resource to continually identify, pursue and undertake urban forestry research that assists the agency in its goal of providing the highest quality, hardest working and most sustainable urban forest to New Yorkers it possibly can.

REFERENCES

1. CITY OF NEW YORK (2007). PlaNYC: A Greener, Greater, New York. City of New York, Mayor Michael R. Bloomberg.
2. GROVE, J.M., O'NEIL-DUNNE, J., PELLETIER, K., NOWAK, D. AND WALTON, J. (2006). A report on New York City's present and possible urban tree canopy. Prepared for Fiona Watt, Chief of the Division of Forestry and Horticulture. New York City Department of Parks & Recreation, US Department of Agriculture Forest Service, Northern Research Station. [Online] Available at: http://nrs.fs.fed .us/nyc/localresources/downloads/Grove_UTC_NYC_FIN AL.pdf [Accessed 15/01/11].
3. JONNES, J. (2011). What is a tree worth? The Wilson Quarterly, Winter 2011. [Online] Available at: http:// www. wilsonquarterly.com/article.cfm?aid=1772 [Accessed 23/02/11].
4. KAPLAN, R. AND KAPLAN, S. (1989). The Experience of Nature: A Psychological Perspective. Cambridge University Press, New York.
5. KLING, J. (2008). Greener cities: US Forest Service software package helps cities manage their urban treescape. Science Perspectives Fall 2008. US Department of Agriculture Forest Service, Pacific Southwest Research Station.
6. KOU, F.E., BACAICOA, M. AND SULLIVAN, W.C. (1998). Transforming inner city landscapes: trees, sense of safety, and preference. Environment and Behavior 30(1), 28–59.
7. LEWIS, C.A. (1996). Green Nature/Human Nature: The Meaning of Plants in Our Lives. University of Illinois Press, Chicago.
8. LU, J.W.T., SVENDSEN, E.S., CAMPBELL, L.K., GREENFELD, J., BRADEN, J., KING, K.L. AND FALXA-RAYMOND, N. (2010). Biological, social, and urban design factors affecting young street tree mortality in New York City. Cities and the Environment 3(1), article 5. [Online] Available at: http://escholarship.bc.edu/cate/vol3/iss1/5 [Accessed 20/02/11].
9. MCINTYRE, L. (2008). Treeconomics. Landscape Architecture, February 2008. [Online] Available at: http://ccuh.ucdavis .edu/projects/qti/Treeconomics.pdf [Accessed 20/02/11].
10. MCPHEARSON, P.T., FELLER, M., FELSON, A., KARTY, R., LU, J.W.T., PALMER, M.I. AND WENSKUS, T. (2010). Assessing the Effects of the Urban Forest Restoration Effort of MillionTreesNYC on the Structure and Functioning of the New York City Ecosystems. Cities and the Environment 3(1), article 7. [Online] Available at: http://escholarship .bc.edu/cate/vol3/iss1/7 [Accessed 20/02/11]
11. MCPHERSON, E.G., SIMPSON, J.R., PEPER, P.J., GARDNER, S.L., VARGAS, K.E. AND XIAO, Q. (2007). Northeast Community Tree Guide: Benefits, Costs and Strategic Planting. General Technical Report, PSW GTR-202. US Department of Agriculture, Forest Service, Pacific Southwest Research Station, Albany, California.
12. MILLIONTREESNYC (2007A). About a Million Trees. MillionTreesNYC. [Online] Available at: http:// www. milliontreesnyc.org/html/about/about.shtml [Accessed 30/01/11].
13. MILLIONTREESNYC (2007B). Getting to a Million Trees, Target Neighborhoods. MillionTreesNYC. [Online] Available at: http://www.milliontreesnyc.org/html/ million_trees/neighborhoods.shtml [Accessed 30/01/11].
14. NEW YORK CITY DEPARTMENT OF PARKS & RECREATION (2007). Tree Count!. New York City Department of Parks & Recreation. [Online] Available at: http://www.nycgovparks .org/sub_your_park/trees_greenstreets/treescount/ treecount_documents.php [Accessed 30/01/11].
15. NEW YORK CITY DEPARTMENT OF TRANSPORTATION (2008). Sidewalks, the New York City Guide for Property Owners [leaflet]. New York City Department of Transportation, Office of Sidewalk Management.
16. NOWAK, D.J., HOEHN, R., CRANE, D.E., STEVENS, J.C. AND WALTON, J.T. (2007).Assessing urban forest effects and values: New York City's urban forest. USDA Forest Service, Northern Resource Bulletin NRS-9. Newtown Square, Pennsylvania. 24 pp.
17. PEPER, P.J., MCPHERSON, E.G., SIMPSON, J.R., GARDNER, S.L., VARGAS, K.E. AND XIAO, Q. (2007). New York City, New York, Municipal Forest Resource Analysis. Technical Report to Adrian Benepe, Commissioner, Department of Parks & Recreation, New York City, New York. Center for Urban Forest Research, USDA Forest Service, Pacific Southwest Research Station.
18. RAE, R.A., SIMON, G. AND BRADEN, J. (2010). Public reactions to new street tree planting. Cities and the Environment 3(1), article 10. [Online] Available at: http:// escholarship.bc.edu/cate/vol3/iss1/10 [Accessed 20/01/11].
19. ULRICH, R.S. (1984). View through a window may influence recovery from surgery. Science 224 (4647), 420–421.

20. US CENSUS BUREAU (2006). State and County QuickFacts: New York City, New York. US Census Bureau. [Online] Available at: http://quickfacts.census.gov/qfd/states/36/ 3651000.html [Accessed 15/01/11].

21. US FOREST SERVICE (2011). i-Tree Urban Forest Assessment Applications. US Department of Agriculture, Forest Service. [Online] Available at: http://www.itreetools.org/ applications.php [Accessed 15/01/11].

CHAPTER 5 URBAN FOREST INVENTORY

Caiyun Zhang, Yuhong Zhou and Fang Qiu

Keywords: LiDAR; individual tree extraction; tree metrics estimation

INTRODUCTION

An urban forest is defined as a forest or a collection of trees that grow within a city, town, or suburb. Urban forests provide a number of benefits such as saving energy and reducing air pollution. To maximize these benefits, an urban forest inventory is often needed for planning and management purposes. However, urban forestry databases are rare, incomplete, and infrequently updated because traditional ground surveys are time-consuming, costly, and labor-intensive. Thus, relatively little is known about urban forest resources. With the emergence of Light Detection and Ranging (LiDAR) techniques, it has been anticipated that the urban forest inventory can be automated.

LiDAR is facilitated with the ability to "see" the ground through openings in canopies and to detect the 3-D structure of trees. Research into the application of LiDAR in forests started in 1980s. Early work mainly focused on the quantification of a forest using profiling LiDAR systems at the stand level [1,2]. Stand level measurements are more important than the individual tree level for natural forest management. However, an urban forest is a mosaic of many different species and ages, and commonly has a higher degree of spatial heterogeneity. Thus, more detailed information is usually needed by community managers.

Modern scanning LiDAR technology has allowed for the estimation of inventory elements at the individual tree level due to the great increase in sampling rate. A range of methods have been explored to detect single trees, many of which are based on a LiDAR-derived canopy height model (CHM), a raster surface interpolated from LiDAR points hitting on the tree canopy surface (e.g., [3,4,5,6,7,8,9,10,11,12,13,14,15,16,17]). These algorithms are favored in many commercial and business environments mainly because of the speed of processing and the accessibility to software that commonly uses regularly spaced data (*i.e.*, rasters). There are several problems with using these methods. First, the derivation of CHM introduces errors and uncertainties due to both the interpolation methods and the grid spacing chosen [18]. This will ultimately affect the subsequent estimation of tree metrics [19]. Second, the Gaussian smoothing step for alleviating the rough surface of the canopy may lead to under- or overestimates of tree height [20].

Several fundamental tree parameters such as height, base height, crown depth, and crown diameter can actually

Citation: Caiyun Zhang, Yuhong Zhou and Fang Qiu, Urban Forest Inventory, doi:10.3390/rs70607892

be estimated from the LiDAR point clouds, as illustrated in Figure 1. Tree height is the length along the main axis from the treetop to the base of a tree. Crown depth is the length along the main axis from the treetop to the base of the crown. Base height is the length from the base of the crown to the base of the tree. Crown diameter, also known as crown width, is the span of the crown of a tree. Several researchers have developed automated methods for extracting individual trees from the LiDAR point clouds leading to varying degree of success. Mosdorf *et al.*(2004) [21] applied the *k-means* clustering algorithm to detect individual trees. Their clustering method depends on seed points extracted from the LiDAR-derived CHM, thus this method is not an approach relying on point clouds only. Reiberger *et al.* (2009) [22] used a normalized-cut segmentation approach that subdivides the tree area into a voxel space and then extracted single trees from identified graphs. Lee *et al.* (2010) [23] developed an adaptive clustering method to segment individual tree crowns. Tittmann *et al.* (2011) [24] developed a RANdom SAmple Consensus (RANSAC)-based approach using a geometric model fitting strategy to identify individual tree crowns. Li *et al.* (2012) [25] proposed a spacing-based tree segmentation algorithm for a mixed conifer forest. Sima and Nüchter (2012) [26] used a graph-based segmentation approach to LiDAR point clouds. These methods have been tested in natural forests dominated by conifers. Several researchers reported the applicability of LiDAR data in urban forests. Haala and Brenner (1999) [27] extracted buildings and trees in urban environments by classifying rasterized LiDAR data and multispectral imagery. Wu *et al.* (2013) [28] developed a voxel-based method to identify individual street trees and estimate morphological parameters from mobile laser scanning data. Liu *et al.* (2013) [29] applied a surface growing algorithm to segment individual tree crowns from LiDAR point cloud data. Holopainen *et al.* (2013) [30] evaluated the accuracy and efficiency of airborne, terrestrial, and mobile laser-scanning methods for urban tree mapping. None of these methods were designed to estimate tree metrics for urban forest inventory purposes. Most recently, Saarinen *et al.* (2014) [31] attempted to update urban tree attributes using multisource single tree inventory.

The challenges for urban forest inventory using LiDAR mainly come from three sources: the complexity of urban areas, the spatial heterogeneity of urban forests, and the diverse structure and shape of urban trees. Urban regions are a mosaic of buildings, roads, rivers, trees, and other ground surface features, which makes the LiDAR filtering process (separation of ground points from non-ground LiDAR points) more difficult. The spatial heterogeneity of urban forests is also a challenge. Unlike trees in natural forests, which tend to have similar canopy characteristics when spatially adjacent, trees in urban settings are often single trees or isolated groups with variable heights and crown widths, and multiple treetops. Urban forests are also rich in mixtures of many different tree species and ages with diverse shapes. Broadleaves and conifers are often blended with each other. These combined factors make automated urban forest inventory very difficult.

A few attempts have been made to detect and model forests in an urban environment using LiDAR data alone or a synergy of LiDAR and optical images (e.g., [27,28,29,30,31,32,33,34]), but application of LiDAR data for urban forest inventory at the individual tree level is still scarce. To this end, the main objective of this study is to develop new methods to isolate individual trees and estimate tree metrics (tree height, base height, crown depth, and crown diameter) from LiDAR point clouds, which can bring the automated urban forest inventory down to the individual tree level.

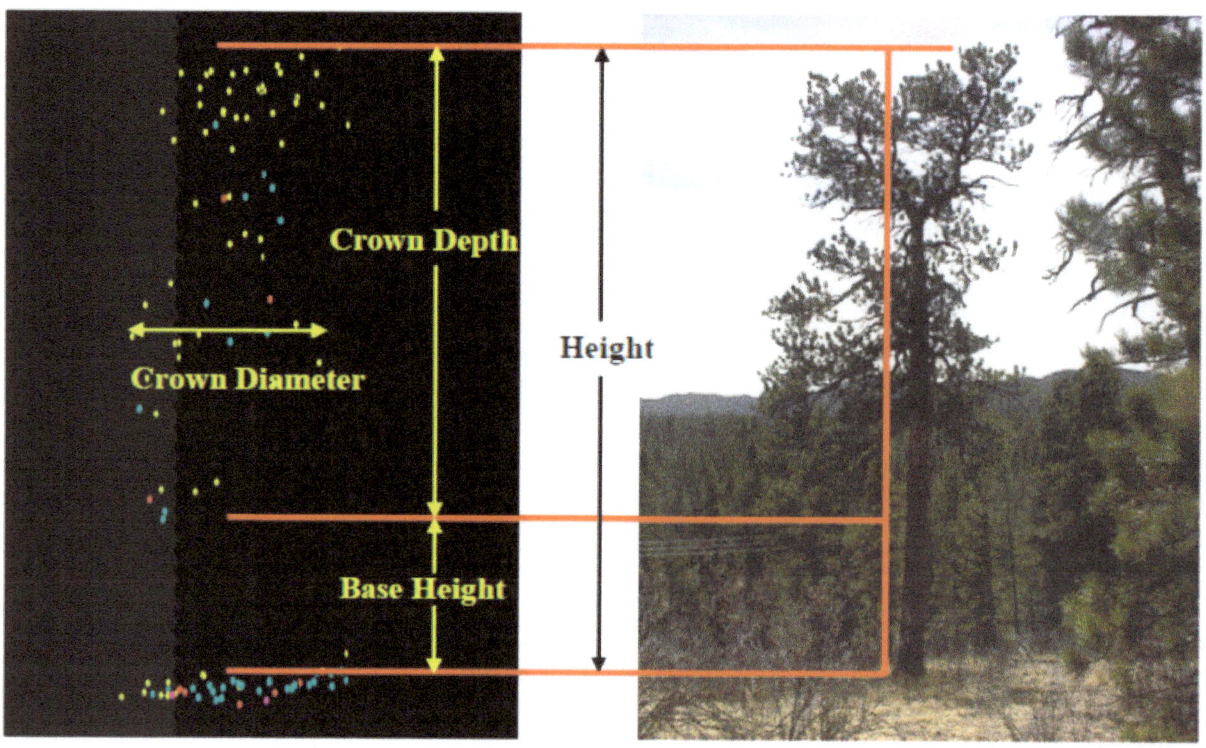

Figure 1: Illustration of tree metrics estimation using raw LiDAR data.

STUDY AREA AND DATA

The Turtle Creek Corridor in north Dallas, Texas was selected as our study area (Figure 2). The topography of the area is dominated by the Turtle Creek, which starts in north central Dallas (at 32°51'N, 96°48'W) and flows southwest five miles through Highland Park and University Park, to its mouth on the Trinity River (at 32°48'N, 96°50'W). Elevation in the study area varies from 112 meters to 156 meters, with the lower elevations found over the creek and higher elevations observed along the bank of the creek. This site is a typical urban area with complex spatial assemblages of vegetation, buildings, roads, creeks, and other man-made features. The Turtle Creek area has a complex ecosystem and has been a center of interest and development in Dallas for over 100 years. To better manage this asset, help understand the current ecosystem, and encourage the development of priorities over this region, the Turtle Creek Association conducted a field survey in August 2008, which was followed by a simultaneous acquisition of small footprint discrete LiDAR data and high spatial resolution hyperspectral images on 24 September 2008.

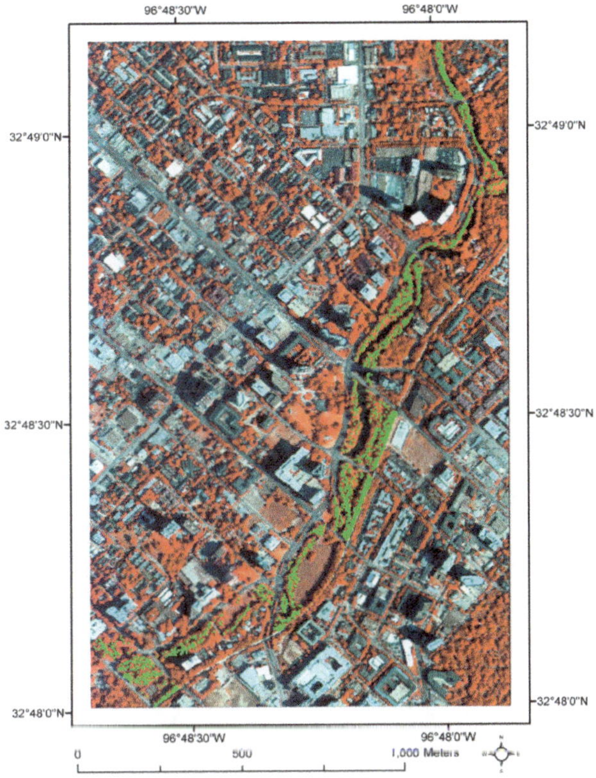

Figure 2: Study site shown in a color composite from the collected hyperspectral data. Measured tree locations are shown in green dots.

The field survey was conducted through a contract with Halff Associates, Inc. Only trees with a diameter at breast height (DBH) greater than four inches were measured and tagged with an identification number, resulting in a total of 2602 trees being surveyed in the Turtle Creek Corridor. Tree location, species attributes, DBH, and health condition were recorded for each measured tree.Figure 2 shows the locations of these surveyed trees in green dots. Unfortunately, tree height, base height, crown depth, and crown radius were not measured during this survey. We thus went to the field and measured 144 trees using a TruPulse 360 Laser Range Finder and a Trimble Geo^{XT} GPS device in August 2010.

LiDAR data were collected by Terra Remote Sensing Inc. (TRSI) using a proprietary Lightwave Model 110 whisk broom scanning LiDAR system. An average point density of 3.5 p/m^2 was obtained due to an intentional 80% overlap between flights. The raw LiDAR point cloud data were processed by TRSI using Microstation Terrasolid and TRSI proprietary software. Final products were provided in the commonly used LAS LiDAR format with a projection of NAD_1983_UTM_Zone_14N. Additionally, fine spatial resolution optical hyperspectral images were simultaneously acquired with the LiDAR data using an AISA Dual hyperspectral sensor. Hyperspectral images with 492 spectral bands were acquired at a spatial resolution of 1.6 m. Radiometric calibration and geometric correction were performed by the vendor. The original collected images were transformed from raw digital numbers to radiance by applying a pixel by pixel correction using the gain and offset. The offset was extracted from data stored in the "dark" files, which are sensor readings with a closed shutter in order to estimate sensor noise. The gain was extracted from the calibration files supplied by the sensor manufacturer. During the transformation from raw to radiance values, further corrections for dropped lines and spectral shift were also applied. A transformation from radiance to surface reflectance of the entire image dataset was then carried out using ATCOR, a Modtran-based code specifically developed to perform atmospheric correction. For the geometric correction, the vendor used the LiDAR positional information to compute the real world coordinates of each pixel in the dataset, followed by re-sampling of the hyperspectral images. The preprocessed images were then mosaicked and clipped for the study area. A false color composite from the hyperspectral data for the study site is shown in Figure 2.

METHODOLOGY

A Framework to Combine LiDAR and Hyperspectral Data for Urban Forest Inventory

In order to detect each individual tree and estimate its associated metrics (tree height, base height, crown depth, crown diameter), multiple data processing steps are required. These are summarized in Figure 3. LiDAR data filtering is the first step to separate ground points from aboveground points. Ground points are then used to build a Digital Terrain Model (DTM), which will assist with the elimination of local terrain impact on tree metrics estimation. Among all the aboveground points, vegetation points then need to be separated from non-vegetation points. This is achieved by using the Normalized Difference Vegetation Index (NDVI) derived from the hyperspectral data. To extract individual trees, LiDAR points hitting on the tree canopy surface need to be extracted by isolating them from the points that penetrate into the canopy. After this, a treetop detection algorithm (referred to as the *tree climbing method*) is applied to identify the treetop of each individual tree, and then a tree boundary extraction algorithm (referred to as the *donut expanding and sliding method*) is used to isolate individual trees. All individual tree level attributes including tree location, height, base height, crown depth, and crown diameter are stored in a tree inventory database. Species for each LiDAR-detected tree can be identified using the hyperspectral data, which has been reported in Zhang and Qiu (2012) [35]. There are three major steps in the framework: LiDAR data filtering, treetop detection, and individual tree extraction. They are described in the following subsections.

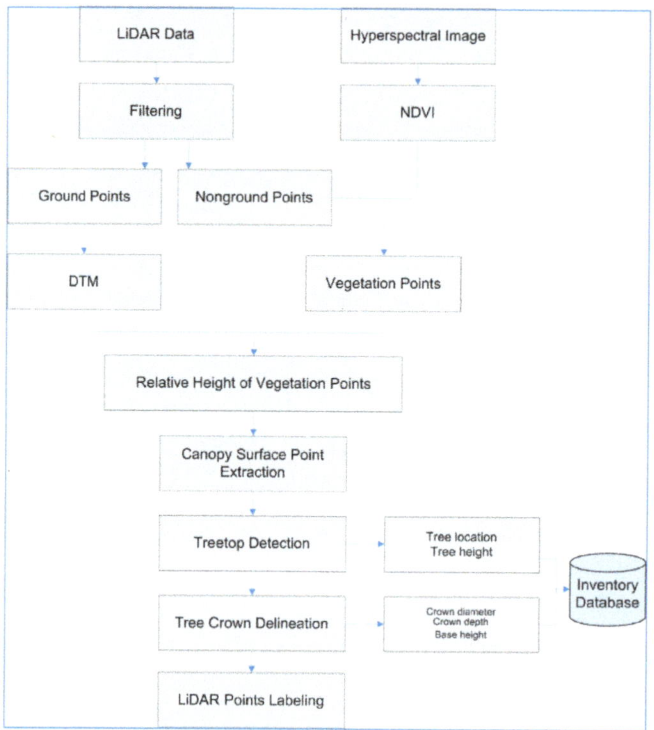

Figure 3: The designed framework to combine LiDAR and hyperspectral data for urban forest inventory at the individual tree level.

LiDAR Data Filtering and Canopy Points Separation

Various methods have been developed to automate LiDAR data filtering procedures. Sithole and Vosselman (2004) [36] compared eight filters and found that these filters performed well in smooth rural landscapes, but they produced larger errors in complex urban areas and rough terrain with vegetation. Since a variety of feature types are present in our study area, none of these eight filters is likely to achieve satisfactory results. In this study, a vector-based filtering technique known as the *k mutual nearest neighborhood clustering algorithm* [37] was used. The idea of this algorithm is that points in a neighborhood sharing similar LiDAR attributes should belong to the same cluster which represents either ground or aboveground. A main advantage of this algorithm is that there is no loss of information by working directly with the raw LiDAR point clouds. More details of this algorithm can be found in Chang (2011) [37]. The accuracy of the filtering result was assessed using the commonly adopted omission and commission error approaches based on randomly selected reference points. After data filtering, the ground derived points were used to generate the DTM that normalized the original LiDAR elevation data to eliminate the terrain effects.

To extract individual trees in the aboveground LiDAR point cloud, points hitting tree crowns need to be identified so that non-tree points will not interfere with individual tree isolation. The NDVI derived from the narrow-band hyperspectral data can be used for this purpose. We derived an NDVI image from the hyperspectral data (bands: 860 nm and 660 nm) and then spatially overlaid the aboveground LiDAR point cloud on it. The NDVI value at the cell in which the LiDAR point falls was then extracted as an additional attribute for the point. A LiDAR point with an NDVI value of more than zero was then considered as the vegetation point.

Treetop Detection

To effectively and efficiently find treetops, the LiDAR points hitting the tree canopy surface need to be extracted by isolating them from the points that penetrate into the canopy. This was accomplished by moving a square window over the study area in a non-overlapping manner and preserving the highest first-return pulse within it. After this, a treetop detection algorithm, *tree climbing algorithm*, was applied to identify the highest point (*i.e.* the treetop) of each individual tree.

Local maximum algorithms have been commonly used for treetop detection by finding the local maxima within a fixed or variable-sized window in a LiDAR-derived CHM (e.g., [5,8,12]). The main problem of this approach is that large commission errors are generated, that is, non-treetop local maxima may be incorrectly classified as treetops [12]. The local maximum method with a variable-sized window for treetop detection using LiDAR point cloud data suffers the same problem as the CHM-based methods [20].

In this study, a treetop detection algorithm using LiDAR point clouds was developed based on an improvement to the local maximum method. The algorithm involves several steps. The first step is to select the first LiDAR point of the study area. The second step is to obtain the local maximum that is the highest point within a circle centered at the selected point. The third step is to designate the local maximum point as the new center and repeat steps two and three if the elevation at the local maximum point is higher than the elevation at the currently selected point (*i.e.*, the current center); otherwise, flag the currently selected point as a potential treetop and stop searching the local maximum point. This process will continue for the remaining point clouds until all treetops are identified. This tree finding process is referred to as a *tree climbing algorithm* because a treetop can be reached from any direction.

The *tree climbing algorithm* described above may work for trees with a single well-defined apex, such as most conifers. However, for trees with multiple tops or a concave crown shape, commission errors may again occur. This is often the case for urban trees because most of them are broadleaves with complex shapes. False treetops may be detected in these situations. A simple and effective solution to this problem is to use a horizontal threshold (*H*) to remove the false treetops. Among all the treetops detected using the *tree climbing algorithm*, if the horizontal Euclidean distance of two treetops is less than the defined threshold, then the lower treetop is dropped and the higher one will be preserved for further analysis. This improved treetop detection method is referred to as the*constrained tree climbing algorithm* because it incorporates a horizontal threshold. Also note that the aboveground LiDAR vegetation points may come from shrubs; thus, if the height of the detected trees is less than 1.5 m, then the detected trees are considered as shrubs to be dropped.

Individual Tree Extraction

The detected treetops will be used as starting points to further segment individual trees. An algorithm based on an inverse operation of the tree climbing method is developed to achieve this. This algorithm is referred to as the *donut expanding and sliding method* and is comprised of the following steps:

(1) Randomly select one of the detected treetops $p_t(x_t, y_t, z_t)$ resulting from the treetop detection algorithm.

(2) Calculate the mean elevation of all points (x_i, y_i, z_i) falling within a horizontal circle centered at the selected treetop (x_t, y_t, z_t):

$$\overline{Z_c} = \frac{\sum_{i=1}^{N} Z_i}{N}, if \sqrt{(x_i - x_t)^2 + (y_i - y_t)^2} < r \quad (1)$$

where $\overline{Z_c}$ is the mean elevation of *N* points falling into the circle with a radius *r*, (x_i, y_i, z_i) is the 3-D coordinate of each point, (x_t, y_t, z_t) is the 3-D coordinate of the circle center, and *N* is the total number of points within the circle. The initial radius (*r*) of the circle is the smallest crown size observed in the ground inventory.

(3) Compare the mean elevation $(\overline{Z_c})$ with the elevation of the selected treetop (Z_t). If $\overline{Z_c} > Z_t$, reduce *r* to a smaller value (r-Δr) and repeat Steps 2 and 3 until $\overline{Z_c} < Z_t$, then export the radius of the circle as the estimated crown radius. Otherwise, for $(\overline{Z_c} < Z_t)$ increase the radius of the circle by Δr and get an expanded circle. Then calculate the mean elevation of the points falling within a horizontal donut $(\overline{Z_{d1}})$ defined by the previous circle and the expanded circle:

$$\overline{Z_{d1}} = \frac{\sum_{i=1}^{N} Z_i}{N}, if \sqrt{(x_i - x_t)^2 + (y_i - y_t)^2} > r \; and \; \sqrt{(x_i - x_t)^2 + (y_i - y_t)^2} < r + \Delta r \; and \; \overline{Z_c} < Z_t \quad (2)$$

where Δr is a positive value specified by a user to define the expanding step.

(4) Compare $\overline{Z_{d1}}$ with $\overline{Z_c}$. If $\overline{Z_{d1}} > \overline{Z_c}$ or $\overline{Z_{d1}} = 0$, indicating the donut expands to the boundary of trees, then export the radius of the circle as the estimated crown radius. Otherwise, for $\overline{Z_{d1}} < \overline{Z_c}$ increase the radius again by Δr and get a newly expanded donut, which means the donut slides downward compared with its proceeding one. Then calculate the mean elevation of the points falling into the expanded new donut $(\overline{Z_{d2}})$ defined by the increased circle and its preceding circle:

$$\overline{Z_{d2}} = \frac{\sum_{i=1}^{N} Z_i}{N}, if \sqrt{(x_i - x_t)^2 + (y_i - y_t)^2} > r + \Delta r \; and \; \sqrt{(x_i - x_t)^2 + (y_i - y_t)^2} < r + 2\Delta r \; and \; \overline{Z_{d1}} < \overline{Z_c} \quad (3)$$

(5) Compare $\overline{Z_{d2}}$ with $\overline{Z_{d1}}$. If $\overline{Z_{d2}} > \overline{Z_{d1}}$ or $\overline{Z_{d2}} = 0$, export the radius of the proceeding donut as the estimated crown radius. Otherwise for $\overline{Z_{d2}} < \overline{Z_{d1}}$, meaning the donut slides again, repeat the process in Step 4 by increasing the radius *r* and getting an expended new donut and again calculating the mean elevation of the points falling into the expanded new donut.

This *donut expanding and sliding algorithm* only uses points hitting on the canopy surface to test the downward direction of the expanding donuts. LiDAR points penetrating the tree canopies must be removed in order to effectively find the boundary of tree crowns. After the derivation of the crown size, LiDAR points both hitting on the crown surface and below the crown surface (penetrating points) that fall within the body of the crown circle are assigned to the detected individual trees. Then an individual tree is isolated from the LiDAR point cloud. To make sure that points from a lower tree are not mistakenly assigned to adjacent taller trees, the algorithm can start with the lowest tree and progress to the highest. This was achieved by sorting the detected treetops based on their corresponding relative heights.

The base height and crown depth of each single tree can also be estimated from this *donut expanding and sliding method*. The base height of a tree is basically derived from the minimum elevation of the LiDAR points falling into the outermost donut. The crown depth can then be estimated by subtracting the base height from the elevation of the treetop.

RESULTS AND DISCUSSION

Filtering Results

A total of 4,310,395 original LiDAR points were first filtered by the *k mutual nearest neighborhood clustering algorithm*, leading to the identification of 969,836 ground points and 3,340,559 aboveground points, as shown in Figure 4. A zoomed view of the filtering results or a selected region is also displayed (see Figure 4, right panel). As can be seen, the LiDAR points falling on roads, bare soil, and lawns were filtered as ground points (Figure 4a), while those hitting buildings and tree canopies were filtered as aboveground points (Figure 4b). A digital surface model (DSM) was generated using all LiDAR points (Figure 5a) and the DTM was generated (Figure 5b). The complexity of this urban view arises from the variety of aboveground features depicted on the DSM. The DTM effectively reveals the bare earth topography and preserves linear features such as roads. The bases of buildings can also be easily detected in the DTM once the aboveground points are removed. Visual comparison of the DSM and DTM illustrates that aboveground features were successfully separated from the bare earth beneath them, suggesting that the filtering algorithm performed well in this complex urban area. Accuracy assessment of the filtering results for the study site showed an omission error of 3.9% and commission error of 1.4% based on randomly selected reference points [37].

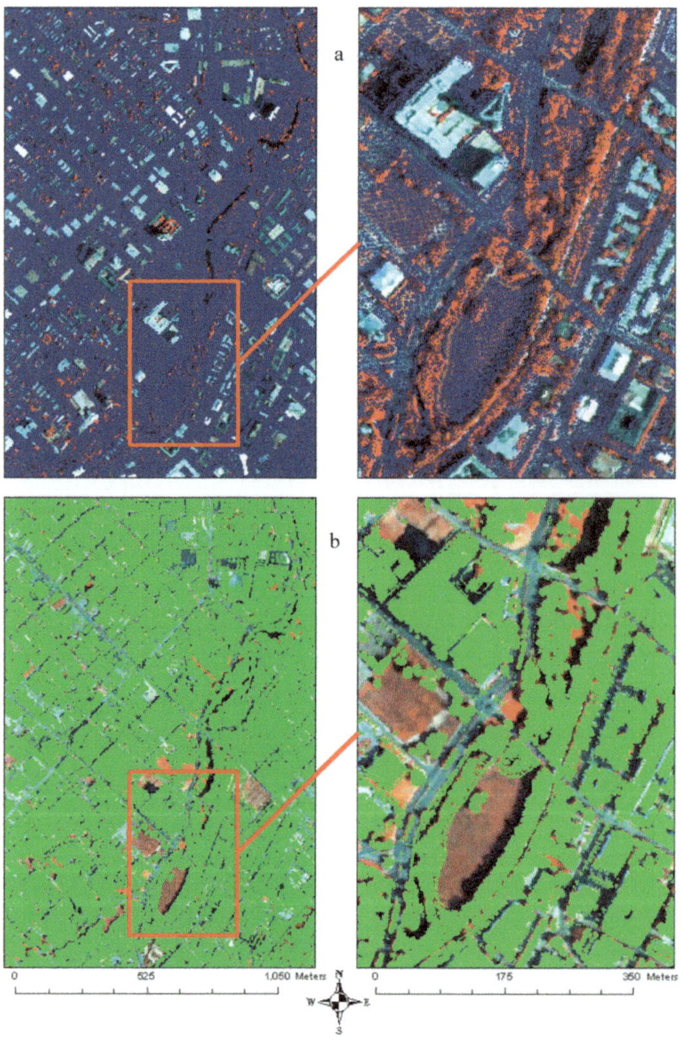

Figure 4: (**a**) Separated ground points in blue dots. (**b**) Separated non-ground points in green dots for the study area.

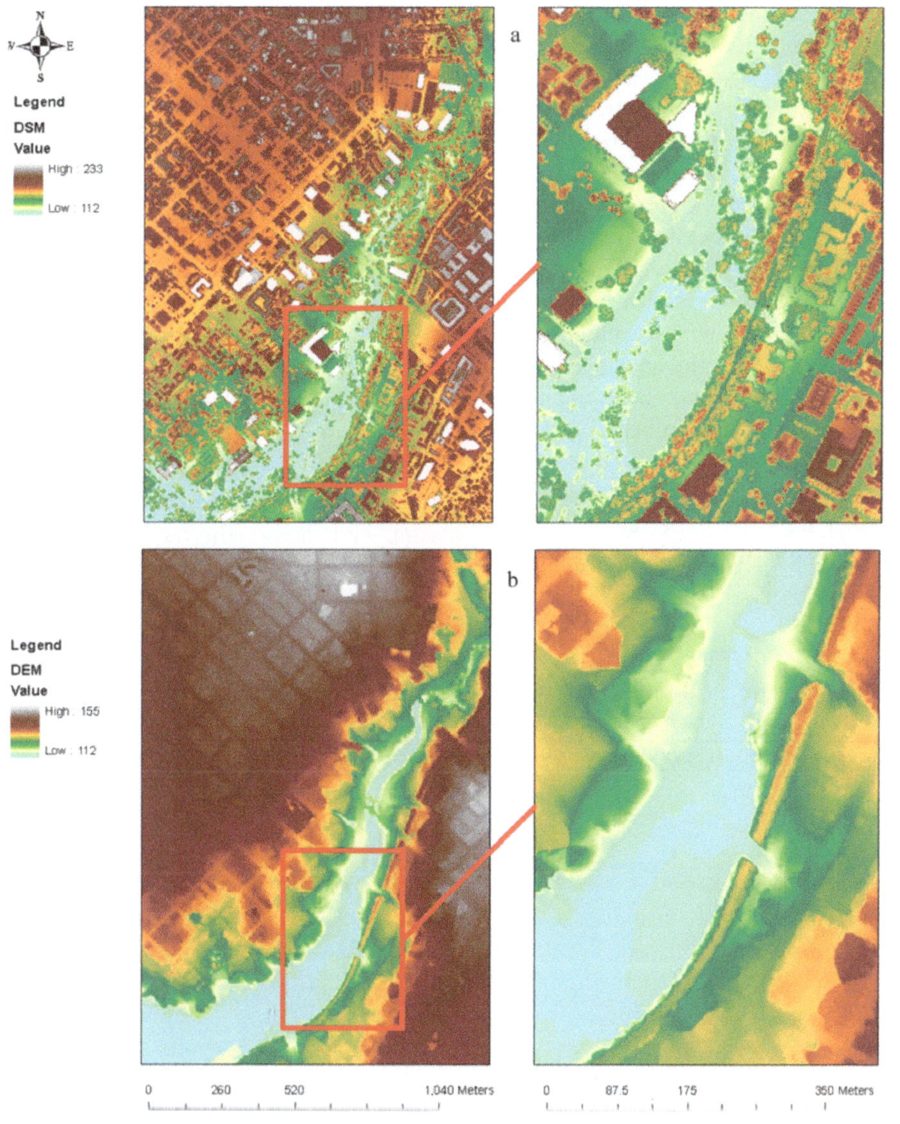

Figure 5: LiDAR-derived (**a**) DSM and (**b**) DTM for the study site (unit: meters; cell size: 1 m).

The vertical accuracy of LiDAR-derived DTM was assessed using 370 field-surveyed elevation points, which are randomly distributed over the study area. In the literature, Root Mean Square Error (RMSE) has often been used to evaluate the accuracy of a LiDAR estimated DTM. An RMSE of 0.4 m was obtained for the DTM, suggestive of a reliable DTM. An accurate DTM is critical for tree parameter extraction because it provides the base from which the relative height of each LiDAR point is calculated, thus the accuracy of the DTM is reflected indirectly in the accuracy of the estimated tree heights.

Detected Tree Numbers

The detection of individual trees from a vegetation point cloud is one of the most difficult estimation tasks using airborne LiDAR. To achieve this, the top of each tree was first identified using the proposed *tree climbing algorithm*, assuming that each tree has only one highest point. The number of trees was then estimated by counting the number of treetops identified by the algorithm. To assess the accuracy of the LiDAR-derived tree numbers, 17 spatially homogenous subareas were selected in the study site. A total of 11,434 trees were estimated for the entire area. The tree count was severely overestimated for each subarea from the original *tree climbing algorithm*. To better detect individual trees and obtain a more accurate tree count, we applied the *constrained tree climbing algorithm*, leading to a total of 2652 detected trees and an average agreement of 93.5%. The number of LiDAR-detected, field-surveyed trees, as well as the calculated agreement for the 17 subareas, are shown in Table 1. The best results were observed in subareas 3 and 17, where a 99.1% agreement was achieved. Subarea 10 had the least agreement, with 32 out of 41 trees being detected. To further assess the commission and omission errors in LiDAR tree detection, a revised tree matching scheme was used based on the strategy in Kaartinen *et al.* (2012) [15]. When a LiDAR-detected tree was within a distance of 5 m to a field tree or a field tree was within a LiDAR-estimated tree crown, this field tree was assumed to be found in LiDAR-detected trees. In this way, a total of 2360 field trees found their corresponding LiDAR trees, leading to an omission error of 9.4%. Any LiDAR-detected trees without corresponding field references were assumed to be extra trees. This led to a total of 633 unmatched trees, resulting in a commission error of 24.3%.

There is one parameter in the *tree climbing algorithm*, the radius of the circle (r) for searching local maximum. Values that are too small or too large will lead to commission errors or omission errors, respectively. For the *constrained tree climbing algorithm*, a small value is preferred because false treetops can be removed by applying the horizontal threshold (H) in the subsequent procedure of the algorithm. The value of H can be empirically set based on field data. For this study, a heuristic method was designed to optimally set this value through a training process using the field survey data. This method involves several steps. Begin with a small value for this threshold such as one equal to the point spacing (d). Next, run the false treetop removal algorithm and obtain the number of newly estimated treetops. If the new number of treetops is more than that of the field survey trees, then the value set for this threshold will be gradually increased and the treetop removal algorithm will be repeated until an expected number for detected trees is obtained for the testing area. Both parameters (r and H) can be locally adjusted based on prior knowledge of the study area, such as whether the area is characterized by well-separated single trees or severely overlapping trees, to obtain better results.

The detection of individual trees is one of the most challenging but at the same time most important tasks for a LiDAR-based urban forest inventory. It is anticipated that errors may occur in attempting to estimate individual trees because of the complex crown shapes of urban trees. However, the estimations demonstrated that the results are in good agreement with the field data. This may be attributed to the higher degree of spatial homogeneity of trees within each subarea than within the entire study areas. As far as the algorithm is concerned, potential sources of error remain. Underestimations might occur if small trees are located next to big ones because lower trees could be considered as the branches of big trees, and thus incorrectly removed during the treetop removal procedure. Overestimations could also occur if tree crowns are so big that some branches extend far from the crown. The algorithm may fail to identify these far-extending branches as false treetops, thus leading to overestimation.

Table 1: The number of LiDAR-detected, field-surveyed trees, and agreement of LiDAR detection for 17 selected testing subareas.

Subarea	No. of LiDAR-Detected Trees	No. of Field-Surveyed Trees	Agreement
1	69	68	98.5%
2	196	186	94.6%
3	117	116	99.1%
4	45	42	92.9%
5	235	242	97.1%
6	502	488	97.1%
7	97	93	95.7%
8	215	227	94.8%
9	131	128	97.7%
10	41	32	71.9%
11	191	163	82.8%
12	161	170	94.8%
13	68	72	94.4%
14	52	48	91.7%
15	41	46	90.1%
16	263	257	97.7%
17	228	226	99.1%
Total	2652	2604	93.5%

Tree Height Estimation

Tree height is one of the most important variables in forest inventory and can be estimated from LiDAR-detected treetops. Tree height for each individual tree is estimated by subtracting the elevation of the terrain from the elevation of treetops. To assess the accuracy of the estimation, field measurements for 144 trees were taken using a Laser Range Finder and then compared with the LiDAR-derived estimations. The summary statistics of the datasets for estimations and field measurements are listed in Table 2. The field-measured heights range from 5.18 m to 26.49 m, with an average of 14.07 m and a standard deviation of 3.95 m, while the LiDAR-estimated tree heights range from 5.43 m to 23.05 m, with an average of 13.9 m and a standard deviation of 3.85 m. These values suggest general agreement between the two datasets. The largest discrepancy between LiDAR-estimated and field-measured heights is 5.97 m, and the difference between their means is 0.17 m. In the literature, RMSE has been used to evaluate the accuracy of LiDAR estimated tree heights. A RMSE of 1.7 m was obtained using 144 field measured tree heights as reference. A linear regression was conducted with the LiDAR-estimated height as the independent variable and the field-measured height as the dependent variable. An R-square value of 0.93 was obtained for the regression (Table 3). A scatter plot (not shown here) of these two variables revealed a few outliers. These outliers were further detected using the statistical studentized residual test, a widely used technique for the diagnosis of outliers. The occurrence of outliers may be attributed to several factors, including the mismatch between the LiDAR-detected treetops and ground-surveyed treetops, as well as errors caused by inappropriate operation of the Laser Range Finder during the field survey procedure. Outliers are most likely to occur where the terrain is rough and in neighborhoods where trees significantly overlap each other. For example, the surveyors may miss a treetop when using the Laser Ranger Finder to measure tree heights and record the readings from nearby trees instead. To minimize the influence of the outliers, a new regression was conducted with the outliers removed from the dataset. Subsequently, the R-square value increases from 0.93 to 0.98 and the RMSE is decreased from 1.11 m to 0.57 m (Table 3).

Table 2: Summary statistics of field-measured *vs.* LiDAR-estimated tree height, crown diameter, base height, and crown depth.

	Field (meter)				LiDAR (meter)			
	Max.	**Min.**	**Mean**	**Std.**	**Max.**	**Min.**	**Mean**	**Std.**
Tree Height	26.49	5.18	14.07	3.95	23.05	5.43	13.90	3.85
Crown Diameter	28.02	4.12	13.37	5.84	25.96	3.86	11.70	4.75
Base Height	8.32	1.49	4.13	1.20	8.50	1.12	4.54	1.53
Crown Depth	21.61	2.38	9.86	3.51	21.36	2.25	8.93	3.25

Table 3: Statistics for regressions of field-measured and LiDAR-estimated tree height, crown diameter, base height, and crown depth with and without outliers.

	With Outliers		Without Outliers	
	R^2	**RMSE (meter)**	R^2	**RMSE (meter)**
Tree Height	0.93	1.11	0.98	0.57
Crown Diameter	0.70	2.58	0.84	1.90
Base Height	0.63	1.79	0.84	0.94
Crown Depth	0.71	2.39	0.84	1.61

Overall, the results demonstrate that tree height can be accurately estimated using LiDAR point cloud data. LiDAR estimations for tree heights can explain 98% of the variance of field measurement after the outliers were detected and removed. The findings are particularly important given that tree heights on the ground are more difficult and costly to collect than species types, especially in densely forested areas. As suggested in the literature, it was anticipated that LiDAR-based estimation tends to underestimate tree heights, which was evidenced by the smaller mean of LiDAR-estimated tree height compared to that of the field-measured height. This is because there is a high probability for the LiDAR pulse to miss a treetop, especially when average spacing of the LiDAR point cloud is large. The LiDAR-estimated tree heights have an average only 0.17 m lower than that of the field measurements. This may be attributed to the fact that most trees in this study area are broadleaved and have a wider top than conifers, increasing the probability that laser pulses hit the vicinity of the treetops. Also note that there is around a two-year time gap between LiDAR and field data collection, which may partially explain the errors in the estimation.

When evaluating the accuracy of LiDAR-estimated tree heights using Laser Range Finder measurements as reference data, one should bear in mind that both LiDAR point cloud data and reference data have measurement uncertainties. When measuring tree heights using the Laser Range Finder, the operators may not be able to locate a treetop exactly, but can often find an approximate location close to the top when firing the device to record its elevation. It was also noted earlier that the error of the Laser Range Finder itself is ± 0.3 m in its distance measurements. The LiDAR point cloud data also has a vertical measurement uncertainty of ± 0.15 m. Leaving out operator-caused uncertainty, the combined uncertainty for LiDAR and Laser Range Finder measurement can be as high as ± 0.45 m. If the 0.45 m measurement uncertainty were able to be excluded, the RMSE of LiDAR-estimated tree heights would be reduced to 0.66 m and 0.12 m with and without the outliers, respectively.

Crown Size Estimation and Individual Tree Extraction

In forestry, crown size is usually represented by the maximum diameter of a crown. In this study, crown diameters were estimated using the *donut expanding and sliding method*, which determines the diameter of the outermost donut and uses it as the estimated crown diameter. A total of 93 trees with field-measured crown diameters from the Laser Range Finder were used to assess the LiDAR estimations. The summary statistics for estimated and field-measured crown diameter are also listed in Table 2. The crown diameters measured by the Laser Range Finder vary from 4.12 m to 28.02 m, with an average of 13.37 m and a standard deviation of 5.84 m. Comparatively, LiDAR-estimated crown diameters vary from 3.86 m to 25.96 m, with an average of 11.70 m and a standard deviation of 4.75 m. The scatter plot with the regression line between the estimated and the measured again reveals the existence of several outliers and seven were diagnosed by the statistical studentized residual test. After deleting these seven outliers, the R-square value increased from 0.7 to 0.84, and the RMSE decreased from 3.58 m to 1.9 m (Table 3). By subtracting the combined uncertainties of the Laser Range Finder and LiDAR, the RMSE could be reduced to 1.45 m.

Crown size is not always measured in urban forest inventories, but it is needed if the canopy cover is to be determined as a parameter for urban ecosystem modeling. When needed, crown size is usually either obtained through field sampling or estimated from other parameters. According to the literature, crown diameter estimated from LiDAR is usually less accurate than tree height. In estimating crown diameters, Gill *et al.* (2000) [38] obtained R-square values in the range of 0.27 to 0.61 for several conifer species. Popescu *et al.* (2003) [39] obtained R-square values of 0.62 and 0.63 for pines and deciduous trees, respectively, and Heurich *et al.* (2004) [40] reported R-square values of 0.48 for conifers and 0.31 for deciduous trees. In this study, an R-square value of 0.84 was achieved. This higher accuracy may be attributed to the use of the original LiDAR point cloud instead of derived raster surface as in the other studies. The challenges in estimating crown diameters using LiDAR include the complexity of crown shapes and the low density of LiDAR sampling. Our method may underestimate the crown size for overlapping trees and produce relatively better results for single trees.

A total of 20,736 trees were detected and segmented. Their treetops and delineated tree crowns using the *isotropic donut expanding and sliding algorithm* are shown in Figure 6. The complexity of this urban forest is well presented with varying crown size, high spatial heterogeneity, and both well-separated and overlapping trees.

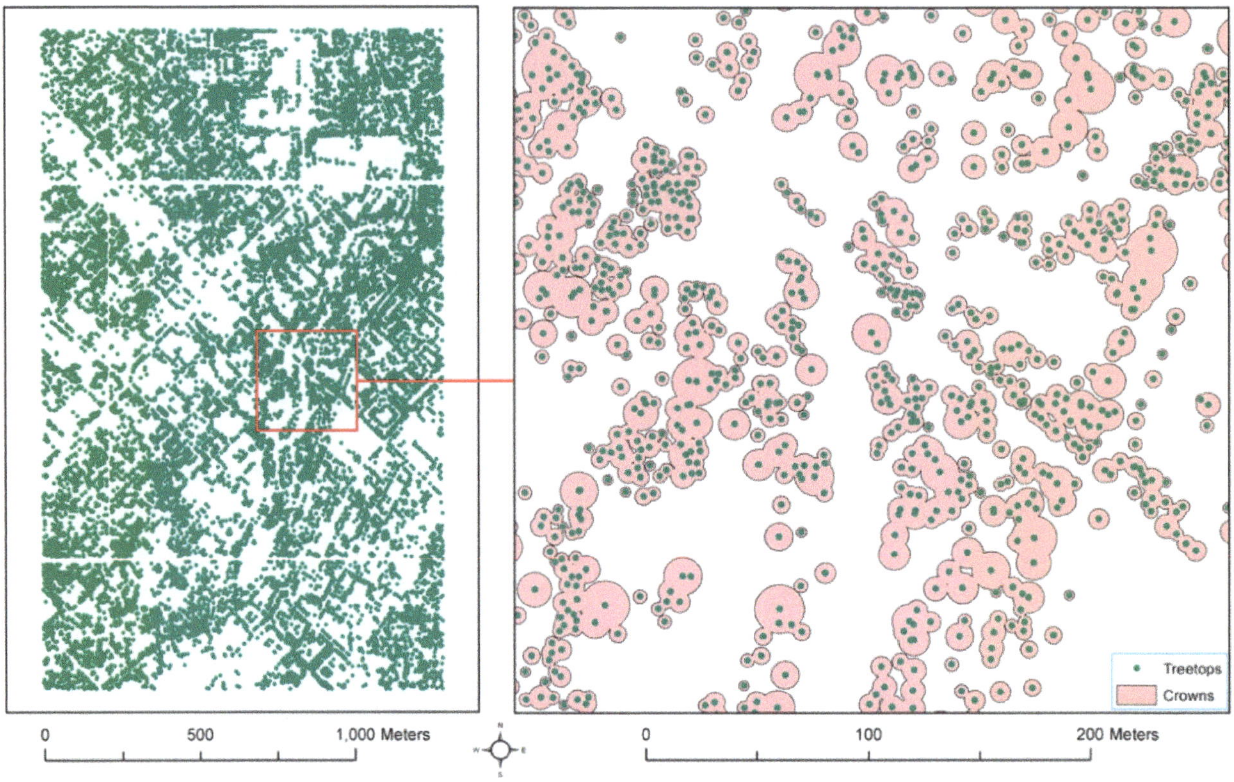

Figure 6: Detected trees shown in green dots for the study area (**left**) and delineated crowns for a zoomed region (**right**).

The *donut expanding and sliding method* assumes that the tree crown is symmetrical in shape and its projection on the ground is a circle, which may not be true in the real world, especially for urban trees that usually have complex geometric shapes. An improvement on this method is to incorporate the concept of anisotropy into the algorithm. Anisotropy is the property of being directionally dependent, as opposed to isotropy, which exhibits homogeneity in all directions. In the improved tree segmentation method, each tree crown boundary is searched from the treetop outward in four or more directions. In the case of searching in four directions, a Cartesian coordinate system from a selected treetop needs to be built first. The tree crown boundary is then searched separately in each quadrant in a similar manner to the *donut expanding and sliding algorithm* but only using LiDAR points falling into that quadrant. The anisotropic *donut expanding and sliding* method delineates crowns in more detail by searching multiple directions in order to model more effectively the complexity of tree crown shapes.

A testing of the isotropic and anisotropic *donut expanding and sliding* algorithms is displayed for a portion of the study area in Figure 7a and Figure 7b, respectively. A total of seven trees were detected, three well separated and four overlapped. The diverse size and complex crown shapes of this testing area are well exhibited by the LiDAR point cloud data. The estimated crown boundary is a circle when the isotropic *donut expanding and sliding* method is applied (Figure 7a), which assumes that the shape of tree crowns is symmetrical. The crown boundary defined for each tree is irregular with quadrants of different radii when four directions are considered in the algorithm (Figure 7b).

Two parameters need to be specified in the *donut expanding and sliding algorithm*: the initial radius of the circle (r) and the expanding radius (Δr). The initial radius was set as the smallest crown size in the ground inventory. The expanding radius cannot be too small or too big. A value that is too small will lead to a small number of points within the donut, which may not be enough to indicate the declining of tree crowns. A value that is too large may include points hitting on neighboring trees. Our experiments, by gradually increasing Δr at an interval of half of the point spacing (d), indicate that $\Delta r = 1.5d$ can produce the best result for crown size estimation.

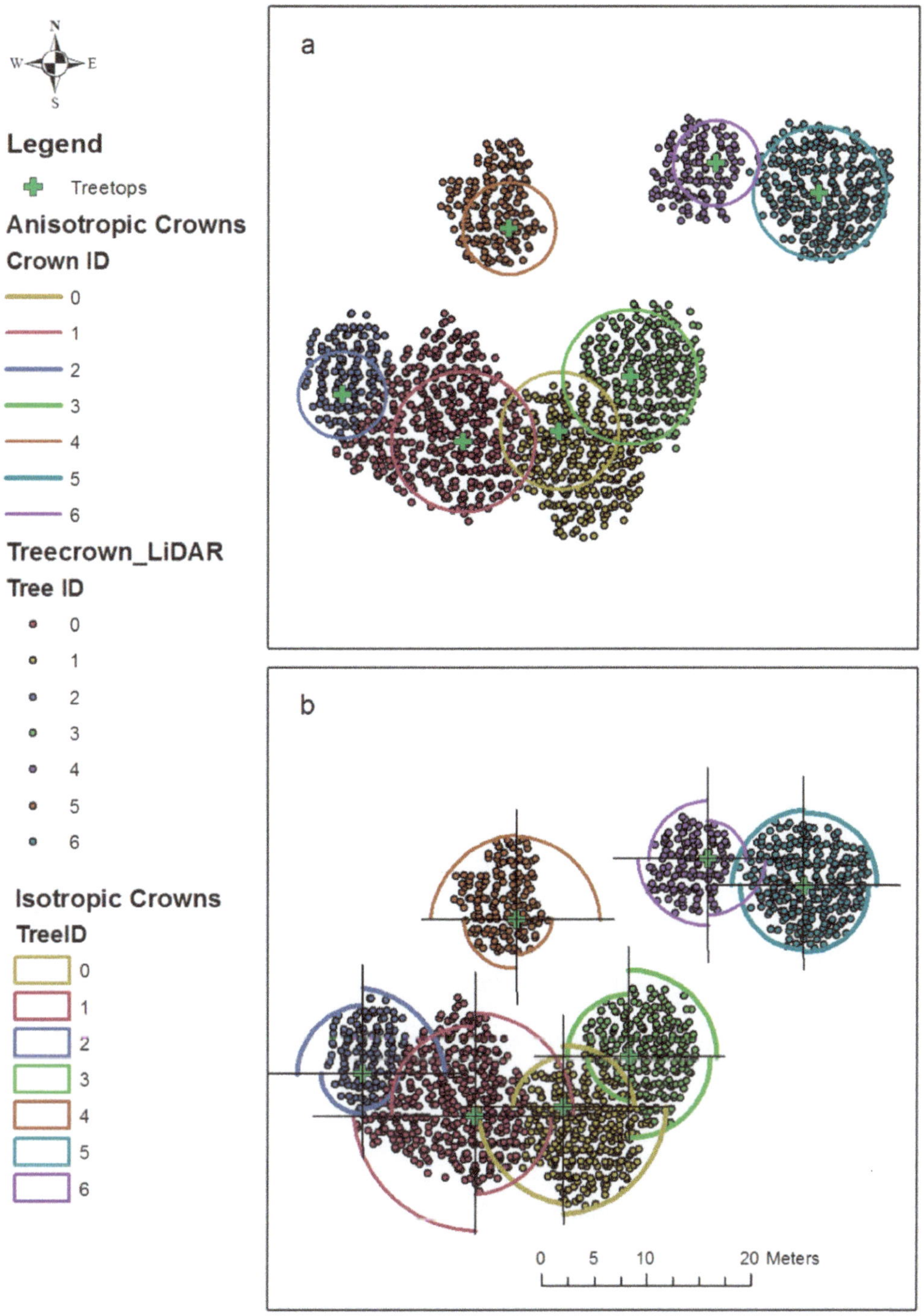

Figure 7: Defined crown boundary and extracted individual trees for a portion of the study area from (**a**) the isotropic and (**b**) anisotropic *donut expanding and sliding method.*

Crown Base Height and Depth Estimation

The base height and crown depth were measured for 140 trees in the field using the Laser Range Finder, which were then used to assess the accuracy of the LiDAR estimations for these two variables. The summary statistics of the two datasets show that the field-measured base heights vary from 1.49 m to 8.32 m, with an average of 4.13 m and a standard deviation of 1.2 m, while LiDAR-estimated base heights range from 1.12 m to 8.5 m, with an average of 4.54 m and a standard deviation of 1.53 m (Table 2). These values are again similar in magnitude. A regression was conducted with the field-measured base height as the dependent variable and LiDAR-estimated base height as the independent variable. The results show that the LiDAR estimations can explain 63% of the field-measured base height variance (Table 3). Six outliers were observed in the regression scatter plot and they were again diagnosed using the statistical studentized residual test. After deleting these diagnosed outliers, the R-square value increases to 0.84 from 0.63, and the RMSE decreases to 0.94 m from 1.79 m (Table 3). Similarly, if the combined uncertainty of 0.45 m from LiDAR and Laser Range Finder measurement was excluded, the RMSE of LiDAR-estimated base heights could be reduced to 0.49 m without the outliers.

The accuracy of the LiDAR-estimated crown depths was analyzed using the Laser Range Finder measured depths for 140 trees. The LiDAR-estimated *vs.* field-measured crown depths and the difference between them are given in Table 2.

Large variation of crown depth is observed in this area, with a range from 2.38 m to 21.61 m for the field measurements, and a range from 2.25 m to 21.36 m for the LiDAR estimations. The average crown depth for field measurements is 9.86 m, and the average crown depth for LiDAR estimations is 8.93 m. The standard deviation of crown depth for field measurements is 3.51 m and for the LiDAR estimations the standard deviation is 3.25 m (Table 2). On average, LiDAR underestimated the crown depth by 0.94 m but overall the estimates exhibit a close correspondence to the field data. Regression results demonstrate that LiDAR can explain 71% of the variance of the field measurements (Table 3). The scatter plot reveals the existence of a few outliers falling far away from the regression line. A statistical studentized residual test detected a total of 10 outliers. After deleting these outliers, the RMSE decreases from 2.39 m to 1.61 m, and the R-square value increases from 0.71 to 0.84. The RMSE could be further reduced to 1.16 m if the combined measurement uncertainty of 0.45 m from the LiDAR and Laser Range Finder devices were able to be removed. Again, the time gap between LiDAR data and field data collection may explain part of the disagreement in the estimation.

The results for the base height and crown depth estimation are encouraging. Similar to the tree height, these two parameters are also difficult to measure in the field, especially for densely forested areas. Part of the unexplained variance may be attributed to the ground irregularities of the bare earth surface, which may lead to uncertainties in measuring the elevation of tree base using the Laser Range Finder. Furthermore, the lowest point of a canopy used to determine the canopy base height may be blocked by thick foliage cover above it and missed by any LiDAR pulse, especially for broadleaved trees with complex crown shape. Higher LiDAR sampling density may lead to better estimation of base height accuracy because more LiDAR echoes from lower branches could be received and the probability of hitting the lowest branches would be increased. Also note that the errors in tree height estimations are the propagated estimations of base height and crown depth since the latter is derived partly based on the former.

By jointly analyzing the LiDAR tree detection and tree metrics estimation results and the tree species map produced from hyperspectral data [28], we found that better results were produced for areas with relatively homogeneous characteristics (*i.e.*, similar species and tree height), or regions where trees are well-separated such as street trees. This is expected given that heterogeneous areas are more complex in tree size, structure, and shape or dominated by overlapping trees. Increasing of LiDAR point density may produce better results. In addition, availability of more prior information for each specific region in the study area could improve the performance of the algorithms.

CONCLUSIONS

In this study, we developed new algorithms for detecting individual trees and estimating related tree metrics from LiDAR point clouds for urban forest inventory purpose. These algorithms were tested in a typical urban forest. The results illustrated that individual trees can be detected from LiDAR point cloud data using the proposed *tree climbing algorithm*, and tree crowns can be delineated using the *donut expanding and sliding method*. Tree metrics (tree height, crown diameter, crown depth, and base height) can be estimated from original LiDAR data and the accuracy is reasonable. Seeing individual trees in a forest is important and should be the first step in the urban forest inventory. LiDAR point cloud data provide the possibility of effectively estimating the number of trees and their locations for a forest. The ability of the proposed algorithms to automatically and effectively estimate tree metrics is important because they are key variables for the prediction of other forest parameters. The research proved that LiDAR is a suitable technology in determining these parameters.

Note that only the NDVI imagery produced from the hyperspectral data was used in the LiDAR tree processing. A further integration of LiDAR and hyperspectral data using data fusion techniques may improve the accuracy of individual tree delineation, metrics estimation, and species classification. For example, the spectral information may be used to provide a better tree extraction by excluding LiDAR points with very different spectra at the edge of the tree. This could bring striking gains in characterizing urban forests, which would consequently accelerate the application of LiDAR and hyperspectral in their respective fields from scientific interests to commercial operational implementations. A complete fusion of the two data sources for forest applications is one of the major directions for future research.

Similar to other LiDAR algorithms using point cloud data, the proposed methods for treetop detection and individual tree extraction are time-consuming because a large volume of point cloud data needs to be processed. To address this problem, two solutions could be adopted in the future. The first one is software-based optimization of the LiDAR algorithm. The other solution is hardware-based. Using a graphics processing unit (GPU)-based computation systems to parallelize the proposed algorithms should greatly reduce the processing time. Indeed, combining these software and hardware solutions may, in fact, completely solve the computational bottle-neck problem for raw LiDAR data-based tree inventory.

The designed algorithms have so far only been tested over the Dallas urban area. Considerable additional research is needed in other urban areas with different species, forest compositions, and sensors in order to examine the robustness and extensibility of these techniques. It is also anticipated that the algorithm may work in non-urban forests with modifications to adapt to the natural environment, which would be a major task for future studies.

ACKNOWLEDGMENTS

We appreciate the Dallas Urban Forest Advisory Committee providing funding and data for this research. We also acknowledge Steve Houser at Arborilogical Services, Inc. for his assistant in the field investigation of the trees in the study area.

AUTHOR CONTRIBUTIONS

This work was completed as part of the Ph.D. dissertation research of Caiyun Zhang, supervised by the coauthor Fang Qiu. Yuhong Zhou assisted with data processing and analysis. All of the authors contributed to editing and reviewing the manuscript.

CONFLICTS OF INTEREST

The authors declare no conflict of interest.

REFERENCES

1. Lim, K.; Treitz, P.; Wulder, M.; St-Onge, B.; Flood, M. LiDAR remote sensing of forest structure. *Prog. Phys. Geogr.* **2003**, *27*, 88–106.
2. Hyyppä, J.; Hyyppä, H.; Leckie, D.; Gougeon, F.; Yu, X.; Maltamo, M. Review of methods of small-footprint airborne laser scanning for extracting forest inventory data in boreal forests.*Int. J. Remote Sens.* **2008**, *29*, 1339–1366.
3. Hyyppä, J.; Inkinen, M. Detecting and estimating attributes for single trees using laser scanner. *Photogramm. J. Finl.* **1999**, *16*, 27–42.
4. Huang, Y.; Yu, B.; Zhou, J.; Hu, C.; Tan, W.; Hu, Z.; Wu, J. Toward automatic estimation of urban green volume using airborne LiDAR data and high resolution remote sensing images.*Front. Earth Sci.* **2013**, *7*, 43–54.
5. Persson, Å.; Holmgren, J.; Söderman, U. Detecting and measuring individual trees using an airborne laser scanner. *Photogramm. Eng. Remote Sens.* **2002**, *68*, 925–932.
6. Brandtberg, T.; Warner, T.; Landenberger, R.; Mcgraw, J. Detection and analysis of individual leaf-off tree crowns in small footprint, high sampling density LiDAR data from the eastern deciduous forest in North America. *Remote Sens. Environ.* **2003**, *85*, 290–303.
7. Leckie, D.; Gougeon, F.; Hill, D.; Quinn, R.; Armstrong, L.; Shreenan, R. Combined high-density lidar and multispectral imagery for individual tree crown analysis. *Can. J. Remote Sens.* **2003**, *29*, 1–17.
8. Popescu, S.C.; Wynne, R.H. Seeing the trees in the forest: Using lidar and multispectral data fusion with local filtering and variable window size for estimating tree height. *Photogramm. Eng. Remote Sens.* **2004**, *70*, 589–604.
9. Falkowski, M.J.; Smith, A.M.S.; Hudak, A.T.; Gessler, P.E.; Vierling, L.A.; Crookston, N.L. Automated estimation of individual conifer tree height and crown diameter via two-dimensional spatial wavelet analysis of LiDAR data. *Can. J. Remote Sens.* **2006**, *32*, 153–161.
10. Solberg, S.; Naesset, E.; Bollandsas, O.M. Single tree segmentation using airborne laser scanner data in a structurally heterogeneous spruce forest. *Photogramm. Eng. Remote Sens.***2006**, *72*, 1369–1378.
11. Koch, B.; Heyder, U.; Weinacher, H. Detection of individual tree crowns in airborne Lidar data. *Photogramm. Eng. Remote Sens.* **2006**, *72*, 357–363.
12. Chen, Q.; Baldocchi, D.; Gong, P.; Kelly, M. Isolating individual trees in a Savanna woodland using small footprint LiDAR data. *Photogramm. Eng. Remote Sens.* **2006**, *72*, 923–932.
13. Kwak, D.; Lee, W.; Lee, J.; Biging, G.S.; Gong, P. Detection of individual trees and estimation of tree height using LiDAR data. *J. For. Res.* **2007**, *12*, 425–434.
14. Hyyppä, J.; Yu, X.; Hyyppä, H.; Vastaranta, M.; Holopainen, M.; Kukko, A.; Kaartinen, H.; Jaakkola, A.; Vaaja, M.; Koskinen, J.; Alho, P. Advances in forest inventory using airborne laser scanning. *Remote Sens.* **2012**, *4*, 1190–1207.
15. Kaartinen, H.; Hyyppä, J.; Yu, X.; Vastaranta, M.; Hyyppä, H.; Kukko, A.; Markus, H.; Heipke, C.; Hirschmugl, M.; Morsdorf, F.; *et al.* An international comparison of individual tree detection and extraction using airborne laser scanning. *Remote Sens.* **2012**, *4*, 950–974.
16. Vauhkonen, J.; Ene, L.; Gupta, S.; Heinzel, J.; Holmgren, J.; Pitkänen, J.; Solberg, S.; Wang, Y.; Weinacker, H.; Hauglin, K.M.; *et al.* Comparative testing of single-tree detection algorithms under different types of forest. *Forestry* **2012**, *85*, 27–40.
17. Jakubowski, M.K.; Li, W.; Guo, Q.; Kelly, M. Delineating individual trees from LiDAR data: A comparison of vector- and raster-based segmentation approaches. *Remote Sens.* **2013**, *5*, 4163–4186.
18. Smith, S.L.; Holland, D.A.; Longley, P.A. The importance of understanding error in LiDAR digital elevation models. *Int. Arch. Photogramm. Remote Sens. Spat. Inf. Sci.* **2004**, *35*, 996–1001.
19. Suárez, J.C.; Ontiveros, C.; Smith, S.; Snape, S. Use of airborne LiDAR and aerial photography in the estimation of individual tree heights in forestry. *Comput. Geosci.* **2005**, *31*, 253–262.
20. Tiede, D.; Hochleitner, G.; Blaschke, T. A full GIS-based workflow for tree identification and tree crown delineation using laser scanning. In Proceedings of the ISPRS Workshop CMRT 2005: Object Extraction for 3D City Models, Road Databases and Traffic Monitoring—Concepts, Algorithms, and Evaluation, Vienna, Austria, 29–30 August 2005.
21. Morsdorf, F.; Meier, E.; Kötz, B.; Itten, K.I.; Dobbertin, M.; Allgöwer, B. LiDAR-based geometric reconstruction of boreal type forest stands at single tree level for forest and wildland fire management. *Remote Sens. Environ.* **2004**, *92*, 353–362.
22. Reitberger, J.; Schnörr, Cl.; Krzystek, P.; Stilla, U. 3D segmentation of single trees exploiting full waveform LiDAR data. *ISPRS J. Photogramm. Remote Sens.* **2009**, *64*, 561–574.
23. Lee, H.; Slatton, K.C.; Roth, B.E.; Cropper, J.R.W.P. Adaptive clustering of airborne LiDAR data to segment individual tree crowns in managed pine forests. *Int. J. Remote Sens.* **2010**, *31*, 117–139.
24. Tittmann, P.; Shafii, S.; Hartsough, B.; Hamann, B. Tree detection and delineation from LiDAR point clouds using RANSAC. In Proceedings of SilviLaser 2011, Hobart, TAS, Australia, 16–19 October 2011.
25. Li, W.; Guo, Q.; Jakubowski, M.K.; Kelly, M. A new method for segmenting individual trees from the LiDAR point cloud. *Photogramm. Eng. Remote Sens.* **2012**, *78*, 75–84.
26. Sima, M.C.; Nüchter, A. An extension of the Felzenszwalb-Huttenlocher segmentation to 3D point clouds. In Proceedings of 2012 5th International

Conference on Machine Vision (ICMV 12), Wuhan, China, 20–21 October 2012.

27. Haala, N.; Brenner, C. Extraction of buildings and trees in urban environments. *ISPRS J. Photogramm. Remote Sens.* **1999**, *54*, 130–137.
28. Wu, B.; Yu, B.; Yue, W.; Shu, S.; Tan, W.; Hu, C.; Huang, Y.; Wu, J.; Liu, H. A voxel-based method for automated identification and morphological parameters estimation of individual street trees from mobile laser scanning data. *Remote Sens.* **2013**, *5*, 584–611.
29. Liu, J.; Shen, J.; Zhao, R.; Xu, S. Extraction of individual tree crowns from airborne LiDAR data in human settlements. *Math. Comput. Model.* **2013**, *58*, 524–535.
30. Holopainena, M.; Kankarea, V.; Vastarantaa, M.; Liang, X.; Lin, Y.; Vaajac, M.; Yu, X.; Hyyppä, J.; Hyyppä, H.; Kaartinen, H.; *et al.* Tree mapping using airborne, terrestrial and mobile laser scanning—A case study in heterogeneous urban forest. *Urban For. Urban Green.***2013**, *12*, 546–553.
31. Saarinen, N.; Vastaranta, M.; Kankare, V.; Tanhuanpää, T.; Holopainen, M.; Hyyppä, J.; Hyyppä, H. Urban-tree-attribute update using multisource single-tree inventory. *Forests***2014**, *5*, 1032–1052.
32. Secord, J.; Zakhor, A. Tree detection in urban regions using aerial LiDAR and image data.*IEEE Geosci. Remote Sens. Lett.* **2007**, *4*, 196–200.
33. Kim, J.; Muller, J.-P. Tree and building detection in dense urban environments using automated processing of IKONOS image and LiDAR data. *Int. J. Remote Sens.* **2011**, *32*, 2245–2273.
34. Lafarge, E.; Mallet, C. Creating large-scale city models from 3D-point clouds: A robust approach with hybrid representation. *Int. J. Comput. Vis.* **2012**, *99*, 69–85.
35. Zhang, C.; Qiu, F. Mapping individual tree species for an urban forest using airborne LiDAR and hyperspectral imagery. *Photogramm. Eng. Remote Sens.* **2012**, *78*, 1079–1087.
36. Sithole, G.; Vosselman, G. Experimental comparison of filter algorithms for bare-earth extraction from airborne laser scanning point clouds. *ISPRS J. Photogramm. Remote Sens.* **2004**,*59*, 85–101.
37. Chang, J. Segmentation-Based Filtering and Object-Based Feature Extraction from Airbornc LiDAR data. Ph.D. Thesis, University of Texas at Dallas, Richardson, TX, USA, 2011.
38. Gill, S.J.; Biging, G.S.; Murphy, E.C. Modeling conifer tree crown radius and estimating canopy cover. *For. Ecol. Manag.* **2000**, *126*, 405–416.
39. Popescu, S.C.; Wynne, R.H.; Nelson, R.F. Measuring individual tree crown diameter with LiDAR and assessing its influence on estimating forest volume and biomass. *Can. J. Remote Sens.* **2003**, *29*, 564–577.
40. Heurich, M.; Persson, Å.; Holmgren, J.; Kennel, E. Detecting and measuring individual trees with laser scanning in mixed mountain forest of central Europe using an algorithm developed for Swedish boreal forest conditions. *Int. Arch. Photogramm. Remote Sens. Spat. Inf. Sci.* **2004**, *XXXVI-8/W2*, 307–312.

CHAPTER 6 MONITORING OF RECENTLY-PLANTED URBAN TREES

Jessica M. Vogt and Burnell C. Fischer

Keywords: Urban tree monitoring, tree growth, tree survival, citizen science, tree inventory methods, Planted Tree ReInventory Protocol

INTRODUCTION

In the last decade, efforts are beginning to converge to monitor the survival, growth, and longevity of planted urban trees. In a comprehensive review of published single-tree inventory methodologies used in urban forestry (including aerial and satellite methods as well as traditional ground survey inventory methods), Nielsen et al. (2014) found that traditional "field survey," or on-the-ground, inventory methods constituted the vast majority of single-tree inventory studies (46 of 57 articles reviewed). Several recent large-scale, single-city tree-monitoring efforts have used field survey methods to measure the survival rates of urban trees. In the summer of 2006, the Parks and Recreation Department of New York City conducted a large-scale young street tree mortality study to examine the many factors in the city influencing the survival of over 14,000 newly planted street trees (NYC Parks 2014). The site assessment tools used in this study included factors measuring the surrounding social and physical environment of each tree (NYC Parks et al. 2010). Other recent regional monitoring efforts include Sacramento, California, where Roman monitored the survival rates over 5 years of over 400 trees that were handed out as part of a utility company tree distribution program (Roman 2013); Milwaukee, Wisconsin, where, most recently, Koeser et al. (2013) use 25 years of monitoring data for a cohort of nearly 800 trees to determine the impacts of a variety of factors on tree survival rates; and New Haven, Connecticut, where Jack-Scott et al. (2013) evaluate the impact of community and other characteristics on survival rates for almost 1,400 trees planted between 1995 and 2007. To our knowledge, large-scale, multi-city planted tree monitoring studies do not seem to exist.

Standards for monitoring tree survival and growth over time are important for comparing the data obtained through different monitoring efforts across multiple locations and years (Leibowitz 2012; Roman et al. 2013; Nielsen et al. 2014). In 2011, the International Society of Arboriculture and The Morton Arboretum convened an international meeting on the subject of urban tree growth and longevity (Leibowitz 2012). This meeting organized four topic areas around descriptive studies of tree growth and longevity, plus three categories of factors influencing urban tree outcomes: tree production and sales, site design and tree selection, and tree and site management (Liebowitz 2012). The Urban Tree Growth and Longevity (UTGL) Working Group that emerged out of this meeting has undertaken to develop of a set of standards for monitoring the survival and growth of planted urban trees, as well as the factors that may influence survival and growth (UTGL Working Group 2014a). The Urban Tree Monitoring

Citation: Vogt, Jessica M. and Fischer, Burnell C. , Monitoring of Recently-Planted Urban Trees, Cities and the Environment (CATE): Vol. 7: Iss. 2, Article 4.

Protocol, as these standards are called, considers the factors of the tree, site, community, and management that may relate to tree survival and growth (UTGL Working Group 2014b).

We present in this paper the Planted Tree Re-Inventory Protocol for citizen science-based monitoring of recently-planted urban trees. Although we are members of the UTGL Working Group and the Urban Tree Monitoring Protocol committees, the protocol presented here was originally developed prior to the creation of the UTGL Working Group. Although our protocol and the in-progress UTGL monitoring protocol are informed by one another, our protocol is distinct in that it explicitly presents a data collection methodology for use by non-experts (i.e., citizen scientists) to measure trees in the urban landscape that have been planted relatively recently (trees in the establishment1 and semi-mature phase)

This paper proceeds as follows: First, we reflect on tree planting organizations and their desire and capacity for monitoring. Then we define citizen science and review its use in urban forestry to date. Next, we discuss the measurement of urban tree outcomes (survival and growth) and summarize the literature on factors influencing tree success and urban forest outcomes. Finally, we present an overview of the main categories of variables included the protocol. The entire protocol is available on the Bloomington Urban Forestry Research Group website (http://www.indiana.edu/~cipec/research/bufrg_protocol) and as an appendix to this paper.

THE TREE-PLANTING ORGANIZATION CONTEXT

In 2010, our research group (the Bloomington Urban Forestry Research Group [BUFRG] at the Center for the Study of Institutions, Population and Environmental Change at Indiana University) was approached by the nonprofit urban greening organization, Keep Indianapolis Beautiful, Inc. (KIB), who was curious about the survival and growth of their planted trees. KIB works with neighborhoods and other groups to plant 1-2" (2-5 cm) caliper trees in the greater Indianapolis and Marion County, Indiana, area. They collect information about the location of each planted trees using global positioning system (GPS) units, and combine this with information obtained from the nursery about the species, planting packaging, and size (caliper, container size, etc.) of the trees they plant using a custom, self-designed Microsoft Access-based data management system. KIB lacked the resources to follow-up and monitor the survival, growth, and condition of these planted trees over the trees' early years (i.e., during the establishment and semi-mature phases before the trees reached their mature size). Their interest was twofold: First, KIB wanted to learn more about the survival and growth of trees they plant, and about the factors influencing the success of these trees. Second, and more importantly, KIB was looking for a way to expand capacity to monitor their planted trees into the future.

With KIB and other urban tree-planting organizations (including citizen groups, municipalities, etc.)3 in mind, BUFRG embarked on the task of designing a method for reinventorying recently-planted urban trees that could be used by minimally-trained data collectors, ranging from high school students to casual adult volunteers. That our methods for inventorying planted trees be usable by non-expert individuals with minimal to no training in urban forestry or arboriculture—i.e., citizen scientists—was of key importance to our research group and to our main stakeholder, KIB. The resulting Planted Tree Re-Inventory Protocol enables citizen scientists to collect information about planted tree success (survival, growth and condition) as well as the factors that may influence tree success. Usability by citizen scientists makes our Protocol unique from existing urban forestry inventory protocols or standards.

CITIZEN SCIENCE

Citizen science, broadly defined, is the involvement of nonprofessional and amateur scientists— the average citizen—in scientific research efforts (Dickinson et al. 2012; Miller-Rushing et al. 2012; Shirk et al. 2012; Bonney et al. 2014). Citizen scientists can be paid interns, temporary workers or unpaid volunteers, and their efforts can augment data collection efforts undertaken by trained researchers, and thus expand the production of knowledge. Citizen science can involve a wide range of activities and various relationships between scientists and the general public. Miller-Rushing et al. (2012) describe three types of citizen science efforts, based on the level of public participation in the research process:4 contributory (public contributes to data collection efforts only), collaborative (involving the public in data collection and also some parts of data analysis and results reporting), and co-created (public involved in all or most parts of the research process, from generating research questions to analyzing and reporting results).

True citizen science—like all science—involves a research question. Most projects in urban forestry are versions of Miller-Rushing et al.'s (2012) contributory citizen science that may or may not involve the processing and analysis of data to answer a true research question. These projects typically involve the public as members of urban forest inventory teams or in other monitoring efforts that might otherwise have been undertaken by urban forestry practitioners and certified arborists. Practitioners undertake inventories for a number of management purposes, including monitoring the success (survival and growth) of a group of trees over time, generating information about survival rates for use planning future tree planting efforts, providing information about the maintenance needs of a tree population, and more. All of these uses of inventory data center on the idea of adaptive management. Adaptive management occurs when the strategies used by resource managers are almost viewed as experiments or means of testing predictions about the relationships between management and a desired outcome (Holling 1996). Nonprofits or municipal forester managers that change the management strategies they use to plant or maintain trees based on the observed conditions of the urban forest as seen in urban tree inventory data are using adaptive management.

The use of volunteers to collect inventory data is not new in urban forestry. Tretheway et al. (1999) summarize the results of workshop on "Volunteer-Based Urban Forest Inventory and Monitoring Programs" convened by the U.S. Forest Service Pacific Southwest Research Station in 1999. Workshop participants identified three purposes for involving volunteers (i.e., citizen scientists) in urban forestry: to "provide a direct connection" between the community and the urban forest, to increase public awareness of the benefits and value of the urban forest, and to enhance support for urban forest "planning, management and stewardship" (Tretheway et al. 1999: p. 2). Cowett and Bassuk (2012) make the case for using university students at land grant colleges to conduct inventories; their "Student Weekend Arborist Teams" conducted more than 40 street tree inventories in small communities across New York State. Bancks (2014) discussed a University of Minnesota extension program that trains volunteers in communities of all sizes in urban forest rapid inventory methods, with the intent of assessing preparedness for emerald ash borer (see also http://mytreesource.com; University of Minnesota et al. 2014). Clarke (2009) describes the use of citizen science to track phenological trends in the urban forest as part of a larger citizen science program, Project BudBurst, managed by the U.S. Forest Service.

When research relies on citizen science for data collection, there can be concerns with the quality of the data collected. Several authors raise concerns about the accuracy of data collected by non-professionals (e.g., Dickenson et al. 2012; Roman et al. 2013). Bloniarz and Ryan (1996) evaluated the accuracy of inventory data collected by volunteers and found it to have similar levels of accuracy and consistency as data collected by certified arborists. In a more recent similar study, Bancks (2014) also found acceptable levels of accuracy for urban forest inventory data collected by volunteers. Future citizen science data collection efforts should continue to monitor the accuracy of data collected to ensure that it meets the quality required for good research. Citizen science has the potential to substantially expand our ability to not only measure and monitor planted urban trees through time, but to also learn more about the factors influencing tree outcomes. Forty-two percent of practitioner-driven tree monitoring organizations surveyed by Roman et al. (2013) already make use of volunteers. And many tree-planting organizations already keep records with at least some information about the trees they plant (Roman et al. 2013). Rigorous citizen science tools that allow the public to record additional information about planted urban trees could help enhance both the quantity and quality of data on the urban forest available to tree planting organizations, tree managers, researchers, and decision makers. For instance, PhillyTreeMap (http://www.phillytreemap.org) is an urban tree mapping and monitoring project involving collaboration between multiple stakeholders in the Philadelphia area, including Azavea (a geographic information systems software and analysis company), Pennsylvania Horticultural Society (a tree-planting nonprofit organization), and the City of Philadelphia Parks and Recreation department, among other partners (Urban Forest Map et al. 2014). The PhillyTreeMap website and mobile applications allow individuals to enter information about a tree, including species, diameter, and height, and to view the amount of ecosystem services that tree and other trees in the database provide.

The implementation of similar tree-monitoring projects across multiple cities and regions and the integration of data collection methods for more information about each tree would enhance the appeal of volunteer-generated datasets to researchers interested in answering explicit research questions. More direct connections and collaborations between practitioner-driven inventory efforts and researchers would truly launch urban forestry into the land of citizen science. New technologies for monitoring may even allow urban tree monitoring to eventually rival "big data" citizen science projects like Galaxy Zoo (http://www.galaxyzoo.org; Zooniverse 2014) and the Christmas Bird Count (http://birds.audubon.org/christmas-bird-count; National Audubon Society 2014)

MEASURING URBAN TREE OUTCOMES

Whether as trained experts or citizen scientists, when we measure urban forest outcomes at the level of the individual tree, there are two different general approaches: place-based inventories and cohort studies. Place-based inventories aim to capture information about a particular type of trees in a given area (e.g., street trees on a major street, public trees in a single neighborhood, all trees on a particular piece of property). Inventories are the more common approach to measuring the urban forest, and street tree inventories in particular have been the norm for capturing the information necessary to calculate the benefits of the urban forest. Cohort studies take a different approach: instead of measuring a particular type of trees in a single area, these studies monitor a cohort—or group of trees planted at the same time—through multiple years or at multiple future points in time. Cohort studies may follow all the trees planted as part of a neighborhood treeplanting project, annual tree-planting program by a municipality or nonprofit, tree distribution program, or other event where multiple trees were planted at the same time, and there is interest in tracking the outcomes of the planted trees over time. For cohort studies, we usually know the actual date, season, or year of planting for each tree, whereas for inventories the date of planting is likely unknown. Whether tracking a single cohort of trees planted at the same time or inventorying all the street trees in an entire city, we are measuring features of each individual tree in the inventory. At the level of the individual tree, urban forest outcomes can be operationalized several ways: we could measure tree health, vigor, or condition; the amount or value of benefits produced by a tree; tree size or growth rate; or, most simply, whether or not a tree lives or dies. Here, we discuss tree survival (or conversely, mortality) and tree growth rates, as two of the more common tree-level outcomes.

Urban tree survival (and mortality)

A common urban forest axiom is that the expected life of a street tree is only 7 or 10 years, but Roman and Scatena (2011) acknowledge that it's unclear where this life expectancy estimate comes from, and provide a more empirical estimate of 19 to 28 years. There are a number of types of mortality for trees in urban areas. Clark and Matheny (1991) identify three primary reasons that trees die in urban areas: structural failure, environmental degradation, and parasitic attack. Different types of mortality may be more closely linked to certain stages in a tree's lifecycle, and so another typology of mortality might be establishment-related mortality, damage-related mortality, and age-related mortality. Establishment-related mortality is connected to the tree's failure to establish in the landscape after transplanting, either due to inadequate care (i.e., not watered after planting), poor tree stock, or improper site selection (not the "right tree" in the "right place"). Damage-related mortality is the death of a tree directly due to damage by humans, either during construction of roads, buildings, or other urban infrastructure that results in removal of the tree during or after the construction, or other damage (due to an automobile, lawnmower, etc.) that necessitates the tree's removal. Age-related mortality is the typical cause of death for non-urban trees; age-related death results from the natural senescing process undergone by trees, through which first small branches and then large branches and then the whole tree stop producing new growth or green leaves every season. Age-related mortality is closely connected to mortality caused by pests or diseases, which are more likely to affect declining or already dying trees.

When calculating a mortality rate for a group of planted trees, unless the cause of tree mortality or failure was recorded for each tree (i.e., as in the case of trees in the International Tree Failure Database; ITFD 2014), most of the time we cannot distinguish the types of mortality from one another. Especially in cases where the tree has been removed, the only thing monitors can know is that where there was once a tree, there is no longer a tree. For this reason, defining "mortality rates" for a cohort of planted trees or for an inventory becomes rather muddled. We cannot know, for instance, what portion of the calculated mortality rate is due to the planting of poor nursery stock relative to what is due to activities undertaken (or not) post-planting in the name of tree care. Long-term data on the same trees at multiple points in time generated through citizen science-based monitoring efforts can help fill this gap in our knowledge

Growth

Urban tree growth is another measurable urban forest outcome. Large, mature trees provide many more benefits than small or immature trees; thus, the faster a tree grows, the sooner it will yield a return on investment (Nowak et al. 1990). Growth rates are measured a number of different ways in the urban forestry literature, including change in tree height (e.g., Stoffberg et al. 2008; Jutras et al. 2009), amount of new shoot growth at the ends of branches (e.g., Solfjeld and Hansen 2004), change in diameter at breast height (e.g., Nowak, McBride & Beatty 1990; Jack-Scott et al. 2013), change in caliper (diameter at 6 in [15 cm] above the first lateral root; e.g., Struve et al. 2000), and the width of annual growth rings as obtained from tree cores (e.g., Iakovoglou et al. 2001). Peper and McPherson (2003) evaluated several methods for measuring leaf area of urban trees that could be used to measure or model canopy growth and change. There are relatively few studies of urban tree growth—particularly longitudinal studies (Liebowitz 2012). And although tree growth has been examined in nursery and experimental settings, few researchers have examined urban tree growth in situ in actual cities.

FACTORS THAT INFLUENCE URBAN TREE OUTCOMES

Tree survival (mortality) and growth is influenced by a large number of factors. Several existing organizing frameworks can be helpful in identifying categories of variables that might influence urban tree outcomes. The social-ecological system (SES) framework developed in rural natural resource management settings states that the characteristics of the resource itself (for example, a forest), the resource system (the trees), the resource users or actors (timber harvesters), and their governance system (rules about when and how to cut trees) influence outcomes observed in coupled human-natural systems (e.g., Ostrom 2009; Ostrom and Cox 2010). In urban forestry, the Clark et al. (1997) "Model of Urban Forest Sustainability" states that sustainable urban forest outcomes are predicated on "a healthy tree and forest resource, community-wide support and a comprehensive management approach" (Clark et al. 1997, 17). Tree biologists and plant physiologists also delineate categories of variables that influence plant growth. In Growth Control in Woody Plants, Kozlowski and Pallardy (1997) review the numerous factors influencing tree and shrub growth. These authors outline categories of physiological factors, environmental factors, and "cultural practices," and describe how each category influences the reproductive (production of flowers and pollen, fertilization and eventually fruiting) and vegetative (root and shoot) growth of woody plants (Kozlowski and Pallardy 1997).

We combine these ideas into an interdisciplinary5 social-ecological systems perspective of urban forest outcomes (Table 1, Figure 1). Adapted from SES theory (Ostrom 2009) and the Clark et al. (1997) model, and informed by tree physiology research (Kozlowski and Pallardy 1997), Table 1 presents urban forest outcomes as the product of interactions between the components of the urban forest social-ecological system. Thus, urban forest outcomes— including tree survival, growth, condition, etc.—are influenced by the interactions between the characteristics of the tree itself, the biophysical environment, the community, and the institutions and management strategies (Figure 1).6 The following section describes the current state of knowledge for each of the four main categories of variables that influence tree outcomes. This abbreviated literature review uses the three key sources from Table 1 (Clark et al. 1997; Kozlowski & Pallardy 1997; Ostrom 2009) as well as other relevant literature from the fields of urban forestry/arboriculture, urban ecology, natural resource management, coupled human-natural systems, and more.

Table 1: The urban forests as social-ecological systems perspective draws on several organizing frameworks, including the Model of Urban Forest Sustainability (Clark et al. 1997), the Social Ecological Systems (SES) Framework (first developed by Ostrom [2009], but see also Ostrom & Cox [2010]), and Kozlowski and Pallardy's (1997) Growth Control in Woody Plants. *"Institutions" refers to the rules and shared strategies (per Ostrom 2005) used by people to manage and maintain trees as well as the surrounding biophysical environment in the urban forest. [Modified from http://www.indiana.edu/~cipec/research/bufrg_about.php).

Social-Ecological Systems Framework	Model of Urban Forest Sustainability	Growth Control of Woody Plants	Urban Forests as Social-Ecological Systems
Resource Units	Vegetative Resource	Physiology	Trees
Resource System		Environment	Biophysical Environment
Governance System	Resource Management	Cultural Practices	Institutions & Management
Resource Users or Actors	Community Framework	--	Community

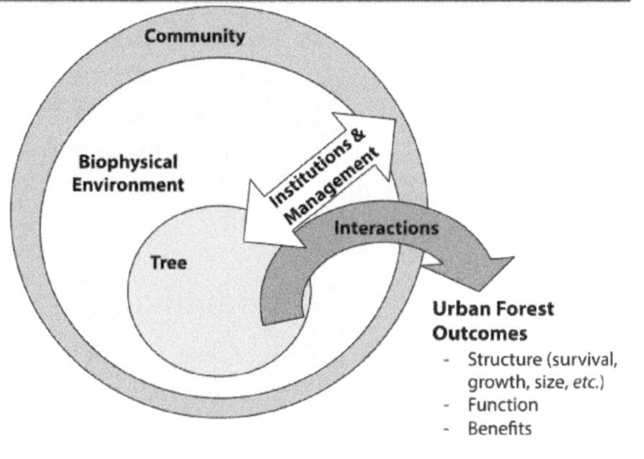

Figure 1: The urban forests social-ecological systems perspective emphasizes that the community interacts with trees and the biophysical environment through institutions and management to produce outcomes in the urban forest.

Characteristics of the tree

The characteristics of the tree itself obviously impact its survival and growth. Clark et al. (1997) use vegetation resource to refer to the trees in the urban forest, listing canopy cover, age distribution, species mix, and native vegetation as the key features of the urban forest that influence its sustainability. Here, we focus on the characteristics of an individual tree—including physiology—that influences its success. Kozlowski and Pallardy (1997) discuss the following key physiological processes as they relate to tree growth: production of carbohydrates via photosynthesis, mineral uptake and use, internal water relations and evapotranspiration, and hormone regulation. Clark and Matheny (1991) note that a tree's growth rate depends significantly on the availability of resources (carbohydrates, minerals, water, etc.) and that when resources become limiting growth is reduced. Because these physiological processes that manage resources are clearly connected to tree genetics, it should come as no surprise that different species exhibit different survival and growth rates (e.g., Iakovoglou et al. 2001; Grabosky and Gilman, 2004). For transplanted trees, the physiological processes that impact tree establishment, survival, and growth in the landscape are affected by characteristics of the tree at the time of transplanting. The size of the tree at planting has been linked to subsequent survival and growth (Lambert et al. 2010). Nursery production method and the type of plant packaging can also impact transplanted tree success (Gilman and Beeson 1996; Buckstrup and Bassuk 2000). Trees planted too deeply or with excessive mulch covering the rootball exhibit higher mortality rates than trees planted at the proper depth (Gilman and Grabosky 2004). Tree condition and health are also linked to tree success. Lower tree condition ratings are associated with decreased odds of tree survival (Koeser et al. 2013) and lower growth rates (Berrang et al. 1985).

Biophysical environment

Factors in the surrounding biophysical environment also influence tree outcomes. Environmental factors include variables that might be studied by a plant ecologist, such as light availability and intensity, water relations (including drought and flood conditions), temperature, soil nutrient content and physical structure (e.g., compaction), pollution, and other abiotic (e.g., wind, fire) and biotic (pests and diseases) factors (Kozlowski and Pallardy 1997). The biophysical environment may have a particularly strong effect on urban tree success, and street trees in particular experience stressful growing conditions. The most influential environmental factors are significantly different for trees in urban areas compared to rural, more natural growing environments. Urbanization increases impervious surfaces, buildings, and other built or grey infrastructure, resulting in radical changes in the water, temperature, and other abiotic conditions across the urbanized landscape (Arnold and Gibbons 1996; US EPA 2008). Water stress is commonly cited as a limiting factor for urban tree growth (Kramer 1987; Krizek and Dubik 1987; Graves et al. 1991), particularly in arid regions (Costello 2013; Symes and Connellan 2013). High air temperatures can disrupt tree phenology and reproductive growth, and higher soil temperatures can change seasonal root growth patterns (Kozlowski and Pallardy 1997). Because water availability, temperature, and other characteristics of the biophysical environment vary throughout the year for most locales, the season of planting may also impact tree outcomes (Anella et al. 2008; Solfjeld and Hansen 2004). Additionally, several authors have found that smaller available rooting volume leads to constrained root, trunk, and shoot growth (Krizek and Dubik 1987; Grabosky and Gilman 2004; Day et al. 2010). Competition with other trees for space, nutrients, light, water and more can also limit tree growth and survival (Nowak et al. 1990; Rhoades and Stipes 1999; Iakovoglou et al. 2001). Compounding space constraints are the generally poor soil conditions in urban areas (Scharenbroch et al. 2005; Smith et al. 2001)

The urban forest axiom right tree, right time, right place is often on the minds of tree planters, and even sometimes a piece of urban forest policies, plans, or ordinances. Several efforts are currently underway to develop a more empirical foundation to the linkages between tree outcomes and site and soil characteristics, including work led by Bryant Scharenbroch at the The Morton Arboretum (MASS Laboratory 2014). Our protocol includes measurement of variables that are proxies or indicators for available growing space above and below ground and the quality of the site.

Institutions and management

Tree success is also impacted by the institutions—i.e., management strategies and maintenance practices—that arborists, urban foresters and other members of the community use to care for urban trees. Kozlowski and Pallardy (1997) refer to these activities as cultural practices that influence tree growth, and their list includes typical tree maintenance activities such as pruning and watering, use of fertilizers, growth regulators, or other chemicals, spacing of trees (both initial arrangement of planted trees and thinning of existing forest stands), and, even protection from freezing. The Clark et al. (1997) resource management component includes mostly variables representing administrative or organizational features of city government as these might relate to adequacy of resources for urban tree management: city-wide management plan, funding, staffing, assessment tools, protection of existing trees, species and site selection, standards for tree care, citizen safety, and recycling. The SES framework (e.g., Ostrom 2009) uses the term institutions to refer to the formal and informal rules and shared strategies that structure the interactions among individuals and groups of people and between people and their environment (Ostrom 2005).

Much of the research on institutions emerges from studies on common pool resource (CPR) management conducted in the disciplines of political science, economics, and anthropology (e.g., Ostrom 1990, 2005). Theory on CPR management states that several principles are likely to be linked to persistent or sustainable systems, including effective monitoring, appropriate sanctioning of rule-breakers, rules allowing individuals impacted by the resource and rules to change those rules, and strategies for effective conflict resolution (Cox et al. 2010). Institutions as rules have only been cursorily examined in urban ecosystems, and not at all for urban forest outcomes (Mincey et al. 2012). Larson et al. (2008) describe rules of homeowners' associations that limited the appearance and management strategies used for residential vegetation, including pest and water management methods and species composition. Mincey and Vogt (2014) find that watering strategy used by the neighborhood impacts tree survival rates.

Tree maintenance strategies can be characterized by the type of maintenance (e.g., pruning, watering), intensity (how much maintenance is performed, i.e., training pruning, 15 gallons of water), frequency (how often the activity is performed, e.g., annually, once per every week it does not rain), duration (how long the activity is performed, e.g., for the first 5 years after transplanting), and extent (which trees or what part of each tree is maintained, e.g., pruning up lower branches, watering all trees in the State St. right-of-way) (Vogt, Hauer, and Fischer in review). Maintenance type, intensity, frequency, duration, and extent all influence tree and urban forest outcomes; the impact of watering (Gilman 2001, 2004), pruning (Whitcomb 1979; Miller and Sylvester 1981; Evans and Klett 1985), and mulching (Gilman and Grabosky 2004) varies depending on the particulars of the maintenance strategy. Maintenance strategies or institutions or rules about tree care may not always be visible on the tree itself or in the area nearby. Our Protocol includes a few key maintenance practices— pruning, mulching, staking—of which evidence can be seen on the tree itself.

Community

Because urban trees are surrounded by people, the characteristics of the community of people living in and around the urban forest influence tree outcomes. For instance, Boyce (2010) observed that the designation of volunteer tree stewards in the community dramatically reduced urban tree mortality rates. The components of community framework included in the Clark et al (1997) model are public agency cooperation, involvement of large private and institutional landholders, green industry cooperation, neighborhood action, citizen-government-business interaction, general awareness of trees as a community resource, and regional cooperation.

Most of the empirical evidence for the influence of community characteristics on environmental outcomes emerges from the research that informed development of the SES framework. Because of its emphasis on rural natural resource management, the SES framework uses the terms "resource users" or "actors" to describe the community of people that manage and use a resource (Ostrom 2009; Ostrom and Cox 2010; Epstein et al. 2013). Features of the community that impact resource management outcomes according to the SES framework include community size (population or number of people involved), history using or managing the system (i.e., experience), demographic or socioeconomic characteristics, individual knowledge (of the resource system), norms (individual perceptions of socially-acceptable practices), and the location of the community (Ostrom 2009; Ostrom and Cox 2010; Epstein et al. 2013). Some of the resource user or actor characteristics listed above have been examined for urban forest social-ecological systems. Iakovoglou et al. (2002) find no significant difference in growth rates between different-sized communities. Jack-Scott et al. (2013) found that a greater number of participants in tree planting events during a year is associated with higher survival and growth rates. Land use type is a factor partially indicative of the features of the biophysical environment but perhaps more closely captures community characteristics. Several authors have found an effect from adjacent or surrounding land use type on tree success (Nowak et al. 1990; Lu et al. 2011). A few studies have found that demographic characteristics (i.e., variables from the U.S. Census) are related to tree outcomes (Nowak et al. 1990; Grove et al. 2006). Lastly, studies from the field of urban ecology have observed that norms or individual motivations impact landscape outcomes (Austin 2002; Grove et al. 2006; Nassauer et al. 2009). Like institutions, characteristics of the community of people are difficult to observe during on-the-ground inventory. Our Protocol adapts several of the stewardship factors collected by the New York Young Street Tree Mortality Study (NYC Parks et al. 2010) as indicators of a care ethic in the community surrounding the tree.

Interactions and endogeneity

Complex coupled human-natural systems are inherently filled with endogeneity, or simultaneous interactions between variables that complicate and sometimes obscures our understandings of the causal impact of variables on observed outcomes (Liu et al. 2007; Schlüter et al. 2014). The urban forest social-ecological system is no exception: interactions within and between tree, biophysical environment, community, and institutional factors can influence urban forest outcomes as much as the influence of a single factor. For instance, proper, proactive maintenance strategies may actually mitigate the impact of sub-optimal growing conditions. Additionally, alignment between rules, the characteristics of the community and local conditions has been demonstrated to impact common-pool resource outcomes (Cox et al. 2010). And characteristics of the community such as individual preferences and knowledge may impact choice of management strategies. A study of residential yards in Minnesota found that homeowners' application of water, fertilizers, and weed killers, as well as other yard management techniques was strongly influenced by resident knowledge and perception of the yard as a relatively closed system (Dahmus and Nelson 2013). Additionally, Vogt et al. (in review) observed an interaction between watering strategy and planting season.

THE PLANTED TREE RE-INVENTORY PROTOCOL

In light of these four main categories of variables that influence urban forest outcomes, we present here the Planted Tree Re-Inventory Protocol (see the Appendix of this paper for Version 1.1; refined from an earlier version of the protocol: Vogt et al. 2013). The protocol describes standardized methods that can be used by non-professional inventory personnel to gather data necessary to evaluate the survival and growth of recently-planted8 urban trees, as well as the many factors influencing survival and growth. Selection of variables to include in the protocol was informed by the literature review summarized above as well as existing urban tree inventory methods, including the i-Tree Eco field methods (i-Tree version 4.0 of the user's manual was consulted for this work), the Standards for Urban Forestry Data Collection (IUFRO et al. 2010), and the methods of New York City's Young Street Tree Mortality Study (NYC Parks et al. 2010; results summarized by Lu et al. 2010). Individual variables and values of each variable were debated by members of the Bloomington Urban Forestry Research Group (BUFRG) over the course of a 6-month period following the review of literature and inventory methods. Table 2 lists each of the variables in the final protocol and, if applicable, the original source for their methods. We adapted and modified variables from other inventory methods to make sure that each variable could be successfully assessed by minimally-trained data collectors. To this end, many variables in the protocol require only simple, qualitative, visual assessments of the tree and its environment, and not precise measurements. For instance, a simple presence or absence assessment method, where the data collector only has to determine whether or not a particular feature is present or absent on the tree or nearby surrounding environment, is used for many variables. Variables that do ask for more precise quantification (e.g., measurements of diameter, height, or distance) require use of only two or three simple tools: a diameter tape and a digital range finder (hypsometer) or clinometer and measuring tape. The protocol was tested by several different parties (Table 3). A preliminary list of variables was tested by members of BUFRG in the summer of 2011. Since the final users of the protocol were to be minimally-trained, non-professional data collectors, high school members of KIB's Youth Tree Team (YTT) tested the protocol during the summer of 2012; YTT used a version of the protocol adapted for use on ESRI's ArcGIS iPhone mobile application to collect data for more than 700 recently-planted street trees. YTT data collection team members were trained in data collection methods during two 6-hour training days, and overseen by a collegeaged YTT Leader who had participated in approximately 15 additional hours of data collection activities with members of BUFRG during Protocol development. The YTT training procedures described above are similar to those used in studies that have found high accuracy for volunteercollected data (Bloniarz and Ryan 1996; Bancks 2014). The protocol was also tested on slightly more mature trees planted between 2000 and 2011 on City of Bloomington right-of-ways; IU master's students collected data on over 1,000 street trees using paper-and-pencil in the summer of 2012.

In addition to collection and evaluation of tree data using the protocol, testing also consisted of written daily field notes taken by YTT members (Vogt et al. 2012) as well as extensive informal discussion between members of the YTT team engaging in data collection and the researchers. For instance, the original protocol called for collecting presence or absence information on several different leaf conditions (evidence of insects, rust, chlorosis, and other leaf condition notes); however, based on written field notes from YTT members, we reduced leaf condition variables to just one: chlorosis. We also clarified that to be considered "present," chlorosis must be evident on at least 25% of the leaf surface area of the tree, and provided pictures and sketches of chlorosis to help with identification and estimation. Written field notes feedback also encouraged us to clarify instructions provided for locating each tree. Additionally, at the end of the data collection season YTT members narrated their thinking while collecting data into an audio recorder. This recording was used to verify that data collection methods had not changed between the beginning and end of the summer, and slight modifications were made to variable descriptions and instructions in the protocol based on decisions and strategies that data collectors were using in the field. (For example, narration revealed that data collectors were marking "incorrect mulching" for trees with very old, degraded mulch, where only few bark chips were still visible. The definitions of correct, incorrect, and no mulching in the protocol were updated to clarify that this case would actually better be classified as "no mulch," given the biophysical implications of capturing information about correct versus incorrect mulching.) Version 1.1 is presented here. In the remainder of this paper, we briefly describe the variables included in the protocol. The entire protocol (in PDF form) is available as a supplementary online appendix to this article, as well as on the BUFRG website (http://www.indiana.edu/~cipec/research/bufrg_protocol.php) in both greyscale and color versions, along with a quick reference guide for the field and customizable and printable data collection sheets.

Tree characteristics

Biophysical variables (tree characteristics and local environmental variables) compose the majority of the variables in most tree inventory protocols, including this one, for a couple reasons: first, factors about the tree and immediate surroundings are most easily observed by data collectors. Second, most tree inventory methods used by urban foresters and arborists are informed by forest mensuration methods used in traditional forestry. Third, as noted above, most research on urban tree survival and growth has emerged from the fields of horticulture and arboriculture, and these fields are strongest in their assessment of the impact of tree and environmental factors on growth.

Table 2: Original sources for variables included in the Planted Tree Re-Inventory Protocol. Complete citations in Literature Cited.

VARIABLE NAME		ADAPTED/MODIFIED FROM *(if applicable)*
Tree characteristics		
Identifying information		
V1	Tree ID#	
V2	Location	
V3	Species	IUFRO et al. 2010: p. 1
Size		
V4	DBH	IUFRO et al. 2010: p. 2-3
V5	Caliper	
V6	Total height	IUFRO et al. 2010: p. 3
V7	Height to crown	IUFRO et al. 2010: p. 3-4
Canopy		
V8	Crown dieback	IUFRO et al. 2010: p. 8
V9	Crown exposure	IUFRO et al. 2010: p. 4-5
V10	Chlorosis	
Trunk		
V11	Root flare	IUFRO et al. 2010: p. 23
V12	Lower trunk damage	
Overall condition		
V13	Other damage	
V14	Overall tree condition	Fischer et al. 2007: appendix
Local environment		
Near tree		
V15	Utility interference	IUFRO et al. 2010: p. 9
V16	Building interference	IUFRO et al. 2010: p. 9
V17	Fences interference	IUFRO et al. 2010: p. 9
V18	Sign interference	IUFRO et al. 2010: p. 9
V19	Lighting interference	IUFRO et al. 2010: p. 9
V20	Pedestrian traffic interference	IUFRO et al. 2010: p. 9
V21	Road traffic interference	IUFRO et al. 2010: p. 9
V22	Ground cover at base	IUFRO et al. 2010: p. 14
V23	Ground cover under canopy	IUFRO et al. 2010: p. 14
Planting area		
V24	Planting area type	
V25	Planting area relative to road	
V26	Planting area width	IUFRO et al. 2010: p. 15-16
V27	Planting area length	
V28	Curb presence	NYC Parks and Recreation et al. 2010: p. 20
Proximity to other things		
V29	Number of trees in 10-m radius	Iakovoglou et al. 2001: p. 75
V30	Number of trees in 20-m radius	Iakovoglou et al. 2001: p. 75
V31	Number of trees in same planting area	
V32	Distance to road	IUFRO et al. 2010: p. 16
V33	Distance to building	IUFRO et al. 2010: p. 9

VARIABLE NAME		ADAPTED/MODIFIED FROM (if applicable)
Management		
Maintenance		
V34	Pruning	NYC Parks and Recreation et al. 2010: p. 22
V35	Mulching	
V36	Staking	
Community		
Evidence of care		
V37	Water bag	NYC Parks and Recreation et al. 2010: p. 22
V38	Bench	NYC Parks and Recreation et al. 2010: p. 22
V39	Bird feeder	NYC Parks and Recreation et al. 2010: p. 22
V40	Yard art	NYC Parks and Recreation et al. 2010: p. 22
V41	Trash/debris	NYC Parks and Recreation et al. 2010: p. 22

Table 3: Protocol testing sites, trees, and data collectors. *Living trees indicates that only trees remaining at the time of re-inventory were assessed using the Protocol. Planted trees indicates that all trees planted were assessed (i.e., for trees removed since planting, the Overall tree condition was assessed as "Missing" and only select biophysical environment variables were collected).

Site	Number of trees*	Tree planting years	Trees planted by	Data collectors	Data collection dates
Indianapolis	120 living trees	2006-2007	Volunteers of Keep Indianapolis Beautiful	IU BUFRG researchers	June-Sept 2011
Bloomington	1,097 planted trees	2000-2011	City of Bloomington Parks and Recreation Division of Urban Forestry	IU Master's of Science in Environmental Science students	May 2012
Indianapolis	714 planted trees	2006-2009	Volunteers of Keep Indianapolis Beautiful	High-school aged Keep Indianapolis Beautiful's Youth Tree Team (YTT) members led by a college-age YTT leader	June-July 2012

Identifying information

The most critical information collected in any inventory protocol is basic identifying information about the tree. This includes a tree identification number, some sort of location information, and species. An identification number is a unique value for each tree in the inventory, useful for tracking the same tree over time through multiple inventory years. Location information should include enough information so that the physical location of the tree in space can be found. Location may be an address number and street name of the property adjacent to the tree, geographic coordinates (I.e., GPS latitude and longitude), distance and direction of the tree from the nearest street intersection, or any other way to precisely locate the tree. Species is the biological name for the type of tree that was planted. Species can be detailed, and include the cultivar or variety (e.g., autumn blaze maple, Acer x freemanii 'Jeffersred'), or could be limited to just the genus (e.g., Acer spp.) of the tree planted, depending on the level of detail desired for the inventory and the tree identification skills of data collectors

Size

In order to measure growth of trees over time, we need information about trees' size. Size information included in the protocol is diameter at breast height, caliper, total height, and height to crown. Diameter at breast height (DBH, or diameter measured at 4.5 ft or 1.3 m off the ground) is one of the most commonly used metrics of size for trees in rural or urban areas. The change in DBH over time is one way to calculate tree growth, and DBH can also be used to calculate the total benefits provided by the tree (e.g., carbon storage). Caliper, or tree diameter 6 inches (15 cm) from the first lateral root, can also be used to calculate tree growth. This is a particularly convenient measure for recently-planted trees, because trees are often sold from the nursery by caliper size; comparing current caliper with that from the tree at the time of planting is another means of calculating tree growth. Total tree height and height to crown provide a metric of above ground size, and can be combined to provide a simple proxy for crown or canopy volume and potential for photosynthesis and growth.

Canopy

Tree health and condition includes information about the canopy, trunk, and entire tree. Information about the condition of the canopy (or leafy top of the tree, also called the crown) is important for assessing the health of the tree. Canopy information included in the protocol is crown dieback rating, crown exposure rating, and presence of chlorosis. Crown dieback and exposure are qualitative visual assessments, recorded on simple point rating scales, using methods modified from the Urban Forestry Data Standards (IUFRO et al. 2010). Crown dieback is a qualitative assessment of the percent of dead branches in the canopy relative to the total living crown, assessed on a 0-6 scale. Crown exposure is a rating of how much of the tree's canopy is exposed to sunlight, based on how many sides of the canopy are shaded by buildings or other trees, assessed on a 0-5 scale. Chlorosis is a presence or absence metric, where "presence" implies that leaf chlorosis is evident on at least 25% of the leaf surface area of the entire tree.

Trunk

Trunk condition metrics are equally as important as canopy condition in assessing overall health of the tree. Trunk condition is related to the health of its vascular tissue and the ability of a tree to successfully transfer nutrients and water between the root system and canopy. Trunk information included in the protocol is presence of a root flare and presence of lower trunk damage. A root flare, or gradual taper of the trunk of a tree as it enters the ground, may be indicative of how deeply the tree was planted.9 The roots of trees planted too deeply may lack sufficient access to oxygen, may be more at risk of water stress (e.g., Gilman 2004) or may be prone to root girdling of the tree. Trees exhibiting lower trunk damage—such as that caused by a lawn mower, weed-whacker, or even animals—may be at greater risk of infection by fungus or disease. Repeated damage over time and on all sides of the lower trunk, such as from a lawn mower, may even girdle the tree, severing the vascular tissue and preventing water and nutrient transfer.

Tree condition

Presence of any other damage and determining an overall tree condition rating are the final assessments of tree-level variables, made after both canopy and trunk condition as well as all other aspects of the individual tree have been examined. Other damage to the tree that may impact its health, condition, survival or growth include: broken branches, branches stripped of leaves or bark, damage to the upper trunk of the tree, a wire or other item choking or girdling the tree, etc. Overall tree condition takes into account the condition of the trunk and canopy. A deciduous tree in good health and condition exhibits a full canopy of dark green leaves that are not undersized for the current season, and a growth form appropriate for its species, without dead branches or excessive water sprouts growing out of the base or main trunk of the tree. Conifers in good health have full boughs with dark green needles. Tree condition ratings should consider a tree from all angles and from top to bottom. The protocol condition ratings range from good to dead and include categories for stumps, sprouts, or absent trees.

Local environment

Near tree environment

In the local environment immediately around the tree, we can assess the quality and quantity of growing space by assessing interference with infrastructure (utility, building, fences, sign, lighting, pedestrian traffic, and road traffic) and type of ground cover (at the base of the tree, and under the canopy). Interference with infrastructure is assessed according to whether or not the tree is in conflict with aboveground utility wires or poles, buildings, fences, signs, or lighting at the time of re-inventory. Interference with traffic refers to the presence of branches more than ½ inch (1 cm) in diameter at or below 8 ft (2.4 m) above a pedestrian walkway or sidewalk for pedestrian traffic, or, for road traffic interference, at or below 14 ft (4.3 m) above an active lane of traffic (i.e., not a parking lane). Trees that are located in close enough proximity to infrastructure so as to conflict with it may compete with this infrastructure for aboveground growing space, or may require more frequent pruning to limit conflicts between branches and the built environment. The type of ground cover around the tree is a qualitative assessment of the type of cover (e.g., bare soil, mulch, grass, etc.) at the base of as well as under the canopy of the tree. Ground cover reflects the surface conditions of the belowground growing environment, including potential competition with other plants for water and nutrients, the permeability of the area to infiltration of water, or even the likelihood of surface soil disturbing activities (such as digging in an annual flowerbed).

Planting area characteristics

The quality and quantity of growing space is also related to the planting area type, its position relative to the road, its length and width, and the presence of a curb at the edge of the planting area. Planting area type refers to the type of physical space in which the tree as planted; types of planting areas include a tree lawn, median, shoulder, tree grate, tree pit, bumpout, front yard, side yard, or other open area. Sketches of each type of planting area are provided in the protocol. The size of the planting area as measured by its surface area (length and width) is a proxy for available rooting space below ground. In addition to the type and size, the position of the planting area relative to the road (i.e., above, even, or below the surface of the road) as well as whether or not the planting area has a curb may impact the quantity and quality of any runoff into the tree planting area.

Proximity to other things

Other living and nonliving things in the larger growing area of the tree can also impact tree success. The protocol considers the number of trees in a 10-meter (33-ft) radius, a 20-meter (67-ft) radius, and the same planting area, as well as the distance to the nearest road and building. The number of other trees near the sample tree influence the amount of competition a tree experiences, both above and below ground, for light, nutrients, water, and growing space. The distance to the nearest road can tell us about potential exposure to factors that may influence a tree's health, condition or growing potential, including the potential for automobile injury or road spray contaminated by fuels, salts and other particles. The distance to the nearest building can tell us about the potential exposure to radiant building or for shading by the building.

Management variables

Most management and maintenance cannot be captured using on-the-ground tree inventory methods, but might be better captured through surveys or interviews of the individuals or groups responsible for the trees. However, some maintenance is visible when looking at the tree during an on-the-ground inventory. The protocol includes variables that consider evidence of pruning, mulching, and staking on the tree, as well as whether the maintenance activity appears to have been performed correctly or incorrectly. For instance, correct pruning cuts should be a smooth, flat cut, made just outside the branch collar for a branch off the main trunk of the tree, or just after the branching for secondary branches in the crown. The protocol includes sketches with examples of correct and incorrect pruning and mulching, and complete text descriptions for correct and incorrect pruning, mulching, and staking.

Community variables

The last suite of variables included in the protocol considers the surrounding community as it is manifested in evidence of care around the tree. The protocol includes four indicators of positive norms of care—presence or absence of a water bag, bench, bird feeder, or yard art (adapted from the list considered by the New York City Young Street Tree Mortality study [NYC Parks et al. 2010])—and one indicator of a lack of care—presence of trash or debris.

CONCLUSION

Data collected via the protocol has many uses, depending on the end user. Tree planting organizations might use the data to help plan the locations and management of future tree planting efforts. Municipal urban foresters might use data on cohort survival rates to help determine an annual budget for planting new trees. Researchers might use data to better understand the myriad factors that influence urban tree outcomes and to create better models of tree growth and survival over time and to improve estimates of the benefits of the urban forest. As urban areas continue to develop and redevelop, to expand and infill, the number of non-planted (i.e., remnant) trees in cities will continue to decrease, as relatively natural areas are replaced by designed landscapes of buildings, roads, planted trees, and other infrastructure (both green and grey). While cities and developers often maintain complete and detailed plans of buildings and roads, detailed records of planted trees rarely exist. However, trees are an integral part of urban infrastructure. In order to ensure they continue providing benefits to urban residents, we should keep track of the location, survival and growth of the trees we plant so that they can be efficiently managed and maintained throughout their lifetimes, and then removed and replaced after they die. With better data about planted urban trees, we can more efficiently allocate limited resources for managing and maintaining the urban forest. The protocol methods presented in this paper can serve as a beginning of a conversation between researchers, urban forestry practitioners, and the public about the measurement of the factors that influence the success of recently-planted urban trees. The protocol will continue to be used and tested by various groups, and accuracy assessments of data collected by citizen scientists should be conducted. We expect to continue to publish new and updated versions of the protocol on the BUFRG website.

APPENDIX

Planted Tree Re-Inventory Protocol, Version 1.1 booklet (PDF available for download here: http://www.indiana.edu/~cipec/research/bufrg_protocol.php)

ACKNOWLEDGEMENTS

The authors thank Sarah K. Mincey and Matt Patterson, co-creators with the authors of the Protocol who participated in the deliberation over and design of variables. We are exceptionally grateful to current and former staff at Keep Indianapolis Beautiful who were supportive of Protocol development and testing (David Forsell, Jerome Delbridge, Nate Faris, Andrew Hart, Bob Neary, and Molly Wilson), and to the members of the KIB Youth Tree Team who tested the Protocol data collection methods and kept excellent field notes. We are also grateful to current and former members of the Bloomington Urban Forestry Research Group and Center for the Study of Institutions, Population, and Environmental Change, who contributed to refinement of Version 1.0: Rachael Bergmann, Kaitlyn McClain, Nick Myers, Shannon Lea Watkins, and Sarah Widney. Development of the Protocol and pilot research was provided by the Efroymson Family Fund, Keep Indianapolis Beautiful, Inc., the City of Bloomington Parks and Recreation Department, the State of Indiana Division of Forestry Community and Urban Forestry Program, and the Garden Club of America Zone IV Fellowship in Urban Forestry. Funding for printing of Version 1.0 was provided by the Alliance for Community Trees.

REFERENCES

1. Anella, L., Hennessey, T. C., and Lorenzi, E. M. (2008). Growth of balled-and-burlapped versus bare-root Trees in Oklahoma , U.S. Arboriculture & Urban Forestry, 34(3), 200–203.
2. Arnold, C. L., & Gibbons, C. J. (1996). Impervious surface coverage: The emergence of a key environmental indicator. Journal of the American Planning Association, 62(2), 243–258. doi:10.1080/01944369608975688
3. Austin, M. E. (2002). Partnership opportunities in neighborhood tree planting initiatives: Building from local knowledge. Journal of Arboriculture, 28(4), 178–186.
4. Bancks, N. (2014). Count us in! Accuracy of volunteers in inventory (survey) data collection. 2014 Minnesota Shade Tree Short Course, 18-19 March 2014, St. Paul, MN.
5. Berrang, P., Karnosky, D. F., and Stanton, B. J. (1985). Environmental factors affecting tree health in New York City. Journal of Arboriculture, 11(6), 185–189.
6. Bloniarz, D. V, & Ryan, H. D. P. (1996). The use of volunteer initiatives in conducting urban forest resource inventories. Journal of Arboriculture, 22(2), 75–82.
7. Bonney, R., Shirk, J. L., Phillips, T. B., Wiggins, A., Ballard, H. L., Miller-Rushing, A. J., and Parrish, J. K. (2014). Next steps for citizen science. Science, 343, 1436–1437.
8. Boyce, S. (2010). It Takes a Stewardship Village: Effect of Volunteer Tree Stewardship on Urban Street Tree Mortality Rates, Cities and the Environment, 3(1), 1–8.
9. Buckstrup, M. J., and Bassuk, N. L. (2000). Transplanting success of balled-and-burlapped versus bare-root trees in the urban landscape. Journal of Arboriculture, 26(6), 298–308.
10. Clark, J. R., and Matheny, N. (1991). Management of mature trees. Journal of Arboriculture, 17(7), 173–184. doi:10.1016/0006-3207(92)90962-M
11. Clark, J. R., Matheny, N. P., Cross, G., and Wake, V. (1997). A model of urban forest sustainability. Journal of Arboriculture, 23(1), 17–30.
12. Clarke, K. C. (2009). A citizen science campaign encouraging urban forest professionals to engage the public in the collection of tree phenological data.

American Geophysical Union Fall Meeting, 5-9 December 2014, San Francisco, CA.

13. Costello, L. (2013). Urban trees and water: An overview of studies on irrigation needs in the Western United States and a discussion regarding future research. Arboriculture & Urban Forestry, 39(3), 132–135.

14. Cowett, F. D., Associate, P., and Bassuk, N. L. (2012). SWAT (Student Weekend Arborist Team): A model for land grant institutions and cooperative extension systems to conduct street tree inventories. Journal of Extension, 50(3), art. 3FEA9.

15. Cox, M., Arnold, G., and Villamayor Tomas, S. (2010). A review of design principles for community-based natural resource. Ecology and Society, 15(4), art. 38. http://www.ecologyandsociety.org/vol15/iss4/art38/

16. Dahmus, M. E., and Nelson, K. C. (2013). Yard stories: Examining residents' conceptions of their yards as part of the urban ecosystem in Minnesota. Urban Ecosystems, 17(1), 173– 194. doi:10.1007/s11252-013-0306-3

17. Day, S., Wiseman, P., Dickinson, S., and Harris, J. R. (2010). Tree root ecology in the urban environment and implications for a sustainable rhizosphere. Arboriculture & Urban Forestry, 36(4), 193–204.

18. Dickinson, J. L., Shirk, J., Bonter, D., Bonney, R., Crain, R. L., Martin, J., Phillips, T., and Purcell, K. (2012). The current state of citizen science as a tool for ecological research and public engagement. Frontiers in Ecology and the Environment, 10(6), 291–297. doi:10.1890/110236

19. Epstein, G., Vogt, J. M., Mincey, S. K., Cox, M., and Fischer, B. (2013). Missing ecology: integrating ecological perspectives with the social-ecological system framework. International Journal of the Commons, 7(2), 432–453. http://www.thecommonsjournal.org/index.php/ijc/article/view/371/331.

20. Evans, P. S., and Klett, J. E. (1985). The effects of dormant branch thinning on total leaf, shoot, and root production from bare-root Prunus cerasifera "Newportii." Journal of Arboriculture, 11(5), 149–151.

21. Gilman, E. F. (2001). Effect of nursery production method, irrigation, and inoculation with mycorrhizae-forming fungi on establishment of Quercus virginiana. Journal of Arboriculture, 27(1), 30–39.

22. Gilman, E. F. (2004). Effects of Amendments, soil additives, and irrigation on tree survival and growth. Journal of Arboriculture, 30(5), 301–310.

23. Gilman, E. F., and Beeson Jr., R. C. (1996). Production Method Affects Tree Establishment in the Landscape. Journal of Environmental Horticulture, 14(2), 81–87.

24. Grabosky, J., and Gilman, E. (2004). Measurement and prediction of tree growth reduction from tree planting space design in established parking lots. Journal of Arboriculture, 30(3), 154– 164.

25. Graves, W. R., Dana, M. N., and Joly, R. J. (1989). Influence of root-zone temperature on growth of Ailanthus altissima (Mill.) Swingle. Journal of Environmental Horticulture, 7(2), 79–82.

26. Grove, J. M., Cadenasso, M., Burch, W., Pickett, S., Schwarz, K., O'Neil-Dunne, J., Wilson, M., Troy, A., and Boone, C. (2006). Data and methods comparing social structure and vegetation structure of urban neighborhoods in Baltimore, Maryland. Society & Natural Resources, 19(2), 117–136. doi:10.1080/08941920500394501

27. Holling, C.S. (1996). Surprise for science, resilience for ecosystems, and incentives for people. Ecological Applications, 6(3), 733-735.

28. Iakovoglou, V., Thompson, J., and Burras, L. (2002). Characteristics of trees according to community population level and by land use in the U.S. Midwest. Journal of Arboriculture, 28(2), 59–69.

29. Iakovoglou, V., Thompson, J., Burras, L., and Kipper, R. (2001). Factors related to tree growth across urban-rural gradients in the Midwest, USA. Urban Ecosystems, 5, 71–85.

30. International Tree Failure Database (ITFD). 2014. http://svinetfc8.fs.fed.us/natfdb/. Accessed 18 April 2014.

31. International Union of Forest Research Organizations, International Society of Arboriculture, United States Forest Service, & Urban Natural Resources Institute. (2010). Standards for Urban Forestry Data Collection: A Field Guide, Draft 2.0. Forestry (2.0 ed., p. 29). http://www.unri.org/standards/wp-content/uploads/2010/08/Version-2.0-082010.pdf. Accessed 18 April 2014.

32. i-Tree. 2014. http://www.itreetools.org. Accessed 18 April 2014.

33. Jack-Scott, E., Piana, M., Troxel, B., Murphy-Dunning, C., and Ashton, M. S. (2013). Stewardship success: How community group dynamics affect urban street tree survival and growth. Arboriculture & Urban Forestry, 39(4), 189–196.

34. Jutras, P., Prasher, S. O., and Mehuys, G. R. (2009). Prediction of street tree morphological parameters using artificial neural networks. Computers and Electronics in Agriculture, 67(1-2), 9–17. doi:10.1016/j.compag.2009.02.008

35. Kenney, W. A., van Wassenaer, P. J. E., and Satel, A. L. (2011). Criteria and indicators for strategic urban forest planning and management. Arboriculture & Urban Forestry, 37(3), 108–117.

36. Koeser, A., Hauer, R., Norris, K., and Krouse, R. (2013). Factors influencing long-term street tree survival in Milwaukee, WI, USA. Urban Forestry & Urban Greening, 12(4), 562-568. doi:10.1016/j.ufug.2013.05.006

37. Kozlowski, T. T., and Pallardy, S. G. (1997). Growth Control in Woody Plants. San Diego, CA: Academic Press, Inc.

38. Kramer, P. (1987). The role of water stress in tree growth. Journal of Arboriculture, 13(2), 33– 38.

39. Krizek, D. T., and Dubik, S. P. (1987). Influence of water stress and restricted root volume on growth and development of urban trees. Journal of Arboriculture, 13(2), 47–55.

40. Lambert, B. B., Harper, S. J., and Robinson, S. D. (2010). Effect of container size at time of planting on tree growth rates for baldcypress (Taxodium distichum (L.) Rich), red maple (Acer rubrum L.), and longleaf pine (Pinus palustris Mill.). Arboriculture & Urban Forestry, 36(2), 93–99.

41. Larson, K. L., Hall, S. J., Cook, E. M., Funke, B., Strawhacker, C. A., and Turner, V. K. (2008). Social-ecological dynamics of residential landscapes: Human drivers of management practices and ecological structure in an urban ecosystem context. Tempe, AZ: Central Arizona Phoenix Long-Term Ecological Research. http://caplter.asu.edu/docs/papers/2008/CAPLTER/Larson_etal_2008.pdf. Accessed 18 April 2014.

42. Leibowitz, R. (2012). Urban tree growth and longevity: An international meeting and research symposium white paper, Arboriculture & Urban Forestry, 38(5), 237–241.

43. Liu, J., Dietz, T., Carpenter, S. R., Alberti, M., Folke, C., Moran, E., Pell, A. N., Deadman, P., Kratz, T., Lubchenco, J., Ostrom, E., Ouyang, Z., Provencher, W., Redman, C. L., Schneider, S. H., and Taylor, W. W. (2007). Complexity of coupled human and natural systems. Science, 317(5844), 1513–6. doi:10.1126/science.1144004

44. Lu, J. W. T., Svendsen, E. S., Campbell, L. K., Greenfeld, J., Braden, J., King, K. L., and FalxaRaymond, N. (2011). Biological, social, and urban design factors affecting young street tree mortality in New York City. Cities and the Environment, 3(1), 1–15. http://digitalcommons.lmu.edu/cate/vol3/iss1/5/

45. Morton Arboretum Soil Science (MASS) Laboratory. 2014. Morton Arboretum Soil Science. http://www.masslaboratory.org/urban-site-index.html. Accessed 18 April 2014.

46. Miller, R. W., and Sylvester, W. A. (1981). An economic evaluation of the pruning cycle. Journal of Arboriculture, 7(4), 109–112.

47. Miller-Rushing, A., Primack, R., and Bonney, R. (2012). The history of public participation in ecological research. Frontiers in Ecology and the Environment, 10(6), 285–290. doi:10.1890/110278

48. Mincey, S. K., Hutten, M., Fischer, B. C., Evans, T. P., Stewart, S. I., and Vogt, J. M. (2013). Structuring institutional analysis for urban ecosystems: A key to sustainable urban forest management. Urban Ecosystems, 16(3), 553–571. doi:10.1007/s11252-013-0286-3

49. Mincey, S. K., and Vogt, J. M. (2014). Watering strategy, collective action, and neighborhoodplanted trees: A case study of Indianapolis, Indiana, U.S. Arboriculture & Urban Forestry, 40(2), 84–95.

50. Nassauer, J. I., Wang, Z., and Dayrell, E. (2009). What will the neighbors think? Cultural norms and ecological design. Landscape and Urban Planning, 92(3-4), 282–292. doi:10.1016/j.landurbplan.2009.05.010

51. National Audubon Society. 2014. Christmas Bird Count. http://birds.audubon.org/christmasbird-count. Accessed 18 April 2014.

52. Nielsen, A. B., Östberg, J., and Delshammar, T. (2014). Review of urban tree inventory methods used to collect data at single-tree level. Arboriculture & Urban Forestry, 40(2), 96–111.

53. Nowak, D. J., McBride, J. R., and Beatty, R. A. (1990). Newly planted street tree growth and mortality. Journal of Arboriculture, 16(5), 124–129.

54. NYC Parks. 2014. Young Street Tree Mortality. http://www.nycgovparks.org/trees/ystm. Accessed 18 April 2014.

55. NYC Parks, USDA Forest Service Northern Research Station, Rutgers University, & Parsons, T. N. S. F. D. (2010). New York City's Young Street Tree Mortality Study Site Assessment Tools Description. October. New York City, NY. Retrieved from http://www.nycgovparks.org/sub_your_park/trees_greenstreets/images/how_to_assess_your _citys_st_survival_10-6-2010.pdf

56. Ostrom, E. (1990). Governing the Commons. Cambridge, UK: Cambridge University Press.

57. Ostrom, E. (2005). Understanding Institutional Diversity. Princeton, NJ: Princeton University Press.

58. Ostrom, E. (2009). A general framework for analyzing sustainability of social-ecological systems. Science, 325(5939), 419–422. doi:10.1126/science.1172133

59. Ostrom, E., and Cox, M. (2010). Moving beyond panaceas: A multi-tiered diagnostic approach for social-ecological analysis. Environmental Conservation, 37(04), 451–463. doi:10.1017/S0376892910000834

60. Peper, P. P., and McPherson, E. G. (2003). Evaluation of four methods for estimating leaf area of isolated trees. Urban Forestry & Urban Greening, 2, 19–29.

61. Rhoades, R. W., and Stipes, R. J. (1999). Growth of trees on the Virginia Tech campus in response to various factors. Journal of Arboriculture, 25(4), 211–217.

62. Roman, L. A. (2013). Urban tree mortality. Dissertation. University of California, Berkeley.

63. Roman, L. A., McPherson, E. G., Scharenbroch, B. C., and Bartens, J. (2013). Identifying common practices and challenges for local urban tree monitoring programs across the United States, Arboriculture & Urban Forestry, 39(6), 292–299.

64. Roman, L. A., and Scatena, F. N. (2011). Street tree survival rates: Meta-analysis of previous studies and application to a field survey in Philadelphia, PA, USA. Urban Forestry & Urban Greening, 10(4), 269–274. doi:10.1016/j.ufug.2011.05.008

65. Scharenbroch, B. C., Lloyd, J. E., and Johnson-Maynard, J. L. (2005). Distinguishing urban soils with physical, chemical, and biological properties. Pedobiologia, 49, 283–296. doi:10.1016/j.pedobi.2004.12.002

66. Schlüter, M., Hinkel, J., Bots, P., and Arlinghaus, R. (2014). Application of the SES framework for model-based analysis of the dynamics of social-ecological systems. Ecology & Society, 19(1), art. 36. http://www.ecologyandsociety.org/vol19/iss1/art36/.

67. Shirk, J. L., H. L. Ballard, C. C. Wilderman, T. Phillips, A. Wiggins, R. Jordan, E. McCallie, M. Minarchek, B. V. Lewenstein, M. E. Krasny, and R. Bonney. (2012). Public participation in scientific research: a framework for deliberate design. Ecology and Society, 17(2), 29. http://dx.doi.org/10.5751/ES-04705-170229

68. Smith, K. D., May, P. B., and Moore, G. M. (2001). The influence of compaction and soil strength on the establishment of four Australian landscape trees. Journal of Arboriculture, 27(1), 1–7.

69. Solfjeld, I., and Hansen, O. B. (2004). Post-transplant growth of five deciduous Nordic tree species as affected by transplanting date and root pruning. Urban Forestry & Urban Greening, 2, 129–137.

70. Stoffberg, G. H., van Rooyen, M. W., van der Linde, M. J., and Groeneveld, H. T. (2008).

71. Predicting the growth in tree height and crown size of three street tree species in the City of Tshwane, South Africa. Urban Forestry & Urban Greening, 7(4), 259–264. doi:10.1016/j.ufug.2008.05.002

72. Struve, D. K., Burchfield, L., and Maupin, C. (2000). Survival and growth of transplanted largeand small-caliper red oaks. Journal of Arboriculture, 26(3), 162–169.

73. Symes, P., and Connellan, G. (2013). Water management strategies for urban trees in dry environments: Lessons for the future. Arboriculture & Urban Forestry, 39(3), 116–124.

74. Tattar, T. A. (1980). Non-infectious diseases of trees. Journal of Arboriculture, 6(1), 1-4.

75. Tretheway, R., Simon, M., McPherson, E. G., and Mathis, S. (1999). Volunteer-Based Urban Forest Inventory and Monitoring Programs: Results from a Two-Day Workshop, 17-18 February 1999, Sacramento, CA. http://www.fs.fed.us/psw/programs/uesd/uep/products/5/cufr_90.pdf. Accessed 18 April 2014.

76. United States Environmental Protection Agency (US EPA). (2008). Reducing Urban Heat Islands: Compendium of Strategies: Urban Heat Island Basics. http://www.epa.gov/heatisland/resources/compendium.htm. Accessed 18 April 2014.

77. Vogt J. M., Hauer, R. J., and Fischer, B. C. (In review). The costs of maintaining and not maintaining trees: A review of the urban forestry and arboriculture literature.

78. Vogt, J. M., Mincey, S. K., Fischer, B. C., and Patterson, M. (2013). Planted Tree Re-inventory Protocol, Version 1.0. Bloomington, IN: Bloomington Urban Forestry Research Group at the Center for the study of Institutions, Population and Environmental Change, Indiana University. http://www.indiana.edu/~cipec/research/bufrg_protocol Accessed 18 April 2014.

79. Vogt, J. M., Watkins, S. L., Mincey, S. K., Patterson, M., and Fischer, B. C. (In review). Explaining planted-tree survival and growth in urban neighborhoods: A social-ecological approach to studying recently-planted trees in Indianapolis.

80. Vogt, J. M., Wilson, M., and Swilik, J. (2012). Engaging high school students in urban tree growth research. Conference on Public Participation in Scientific Research at the Ecological Society of American Annual Meeting, 3-4 August 2012, Portland, OR.

81. University of Minnesota (UM) Department of Forest Resources, UM Extension Service, and Minnesota Department of Agriculture. (2014). Urban & Community Forestry: Community Engagement and Preparedness Program. http://www.mntreesource.com. Accessed 18 April 2014.

82. Urban Forest Map, Pennsylvania Horticultural Society, Philadelphia Parks and Recreation, Township of Lower Merion, Azavea, and Delaware Valley Regional Planning Commission. (2014). PhillyTreeMap Beta. http://www.phillytreemap.org. Accessed 18 April 2014.

83. Urban Tree Growth and Longevity (UTGL) Working Group. (2014a). Urban Tree Growth and Longevity. http://www.urbantreegrowth.org. Accessed 18 April 2014.

84. Urban Tree Growth and Longevity (UTGL) Working Group. (2014b). Urban Tree Monitoring Protocol. http://www.urbantreegrowth.org/urban-tree-monitoring-protocol.html. Accessed 18 April 2014.

85. Whitcomb, C. E. (1979). Factors affecting the establishment of urban trees. Journal of Arboriculture, 5(10), 217–219.

86. Zooniverse. 2014. GalaxyZoo. http://www.galaxyzoo.org. Accessed 18 April 2014.

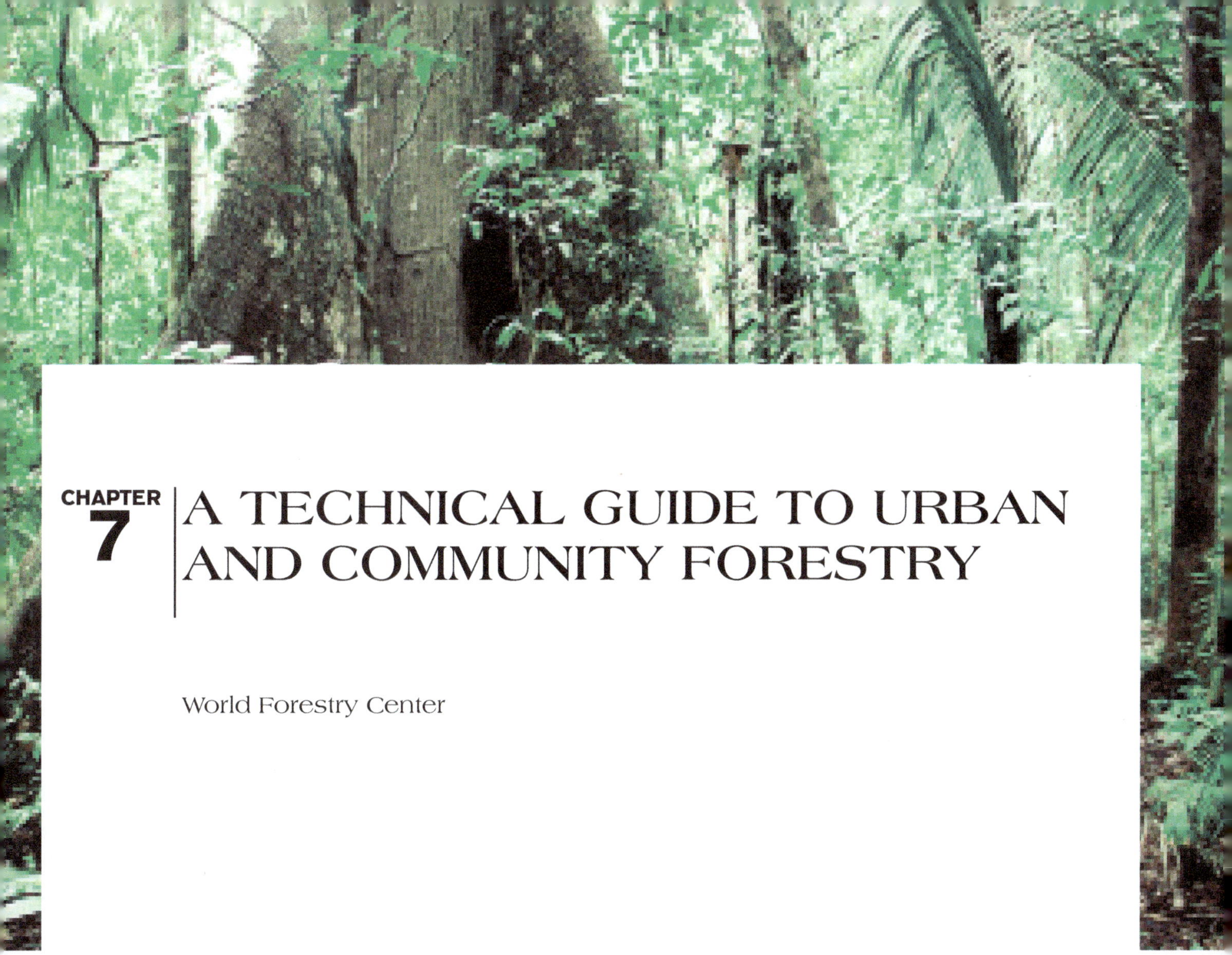

CHAPTER 7 A TECHNICAL GUIDE TO URBAN AND COMMUNITY FORESTRY

World Forestry Center

INTRODUCTION

Trees growing within cities and towns form a forest-an urban forest. But urban trees require special attention, because they are expected to exist within the urban environment. With its infrastructure of streets, sidewalks, curbs, buried utilities, overhead power lines and buildings, the urban environment places tremendous stresses on trees. With proper care, trees become assets which grow in value over time. Without care, tree value declines, eventually becoming a liability to the community. The Technical Guide to Urban Forestry was produced to help communities develop, manage and protect their urban forest resources. The purpose of this publication is to provide technical information about growing trees in the relatively adverse urban environment. This guide is a sequel to An Introductory Guide to Community and Urban Forestry in Washington, Oregon and California. The two publications have been designed to complement each other. Together they provide introductory and technical information regarding urban forestry programs and projects. Both guides were produced in response to requests from local officials and professional tree managers. The need for tree management is never clearer than when urban development eliminates a significant stand of trees, or even a single well placed specimen. Sometimes trees are lost inadvertently, resulting from a lack of information. For example, roots are a tree's most vulnerable part, yet heavy equipment routinely runs over roots. By the time the tree starts to show signs of physical deterioration, five to ten years may have passed. Other times trees are killed intentionally as developers "landscape" before construction. This approach portrays an attitude that trees have little or no value in developed areas, those who guide local development need to understand urban stress factors and the role they play in the premature death of trees. The topics covered in this guide include: tree values, principles of tree growth, planning, tree selection, site preparation, planting and early care of trees, maintenance of established trees, handling trees affected by development, utility forestry, and the politics of municipal tree care. This guide is intended for use by citizens, government decision makers, land developers, and workers out on the streets. Like the original Introductory Guide, this Technical Guide is written for anyone in the private or public sectors who cares enough about their environment to see that trees in their community are planted, valued and protected. The Technical Guide was written by a team of professionals identified on the last page. This group of researchers, educators and practitioners from throughout Washington, Oregon and California are dedicated to providing information on the urban forest resource. Their volunteer efforts reflect the need for people everywhere to

Citation: World Forestry Center, A Technical Guide to Urban and Community Forestry, http://digitalcommons.unl.edu/nebforestpubs/70

contribute their energy and knowledge to make cities more livable for trees

Tree pruning involves not only knowledge of tree growth, but technical expertise with equipment and supplies as well as strength and athletic agility.

VALUES OF URBAN TREES

The basis for the value of an urban tree could be emotional, aesthetic, or it could be strictly utilitarian. However, people seldom perceive value as strictly aesthetic or monetary. There is often substantial overlap that makes "value" difficulty to classify. In many communities, public spending on tree care and management reflects an approximate value of trees. Spending patterns that go unchallenged, especially among an informed public, indicate the value people associate with trees.

The following categories describe different values that people place on trees. They are arranged primarily by their measurability. The least measurable values are discussed first.

Psychological and Aesthetic Values

Although difficult to gauge, uplifted spirits is one important benefit of trees. Some of the difficulty in measuring these benefits may grow out of society's decision to exclude tree values from the marketplace. Other emotion-based commodities, such as flowers, perfume, view property, prestige automobiles, and entertainment, are readily assigned monetary values. But with proper treatment, researchers can tie monetary values to the emotional benefits of trees.

The pleasure and good feelings we associate with trees may be far more practical than generally believed. Data on the connection between vegetation and human health are beginning to accumulate. For example, surgery patients who could see a grove of deciduous trees recuperated faster and required less pain-killing medicine than matched patients who viewed only brick walls. And, prisoners with cells overlooking green landscapes used prison health facilities significantly less than prisoners whose cells provided views of other prison facilities. The vaguely expressed "enjoyment" people associate with trees may be partly a subconscious sign of substantial health benefits.

Social Values

In Oakland, California, a neighborhood tree planting program generated community identity, cooperation, and benefits similar to those reported for urban gardening. After coming together to plant trees, Oakland residents continued working together with "paint-up-fix-up parties", neighborhood protective societies, and community gardens.

Historic Values

Trees provide important symbolic links with the past. If a living tree is associated with important events, the tree takes on historical values unrelated to aesthetics or usefulness. For example, a community would normally value a tree that shaded the deliberations of the community's founders. A tree would also be valuable if planted by George Washington or some other important figure in history. Aside from specific events, old trees may be regarded as important simply because they have lived through eras with which we have few other connections.

As for emotional and aesthetic values, historic values of trees depend primarily on community attitudes. If historic trees are threatened by changes, such as new buildings and street widening, the issue will usually be settled by public pressure not by market forces.

Photo Credit - Lane County Museum, Eugene, Oregon, First Christian Church, 1922.

Environmental Values

People value both the aesthetic and physical quality of our environment. Trees contribute to this quality by modifying local climates, reducing noise and air pollution, and by protecting soil and water.

Climate control is one important service trees provide naturally in the landscape, but the urban landscape is far from natural. Streets, parking lots and buildings have changed the climate of urban areas by absorbing solar radiation. Water that once percolated into the soil and later evapotranspired from soil and plants now drains away or dries on the hard surfaces. These changes have increased the temperatures of cities. Compared to the surrounding rural areas, the urban "heat islands" are five to nine degrees Fahrenheit warmer (three to five degrees Celsius).

Trees help moderate the "heat island" effect. They also greatly increase human comfort: indoors or outdoors. On hot days, trees pump hundreds of gallons of water through their foliage. This water evaporates, keeping the tree and its immediate surroundings cool.

While groves of trees reduce local air temperatures, individual trees increase human comfort primarily by controlling solar radiation, not air temperature. (Radiation is the movement of heat from a warmer body, the Sun, to a cooler body, the Earth.) Trees and other vegetation shield people from direct sunlight. Trees also shade soil, pavement, buildings, and other surfaces that would absorb solar energy and then radiate that heat back to the surroundings. Without the protection of trees, city dwellers are literally surrounded by radiant heat.

At night, radiation moves heat in the opposite direction: from the relatively warm earth to the relatively cool sky. Again, tree cover steps in by blocking radiant heat loss from homes and people. Icy mornings provide evidence of this process, lawns otherwise white with frost often have green circles under the trees.

Indoor air temperatures are also affected by trees growing around buildings. During hot weather, trees reduce cooling costs by buffering high air temperatures and blocking unwanted solar energy. But during winter months, solar gain is desirable, because it cuts heating costs. To get the best balance, on the south and west sides of buildings plant deciduous trees that have thin, open branches to allow winter sun penetrate into the building. In addition, the schedule of leaf growth and leaf drop should coincide with the need for heating and cooling. Few, if any, species will meet these requirements perfectly, but it's wise to select species that give the best possible match.

Air pollution control is another way that trees improve the urban environment. The reductions in air pollution are modest, and air pollution poses some risk to the trees themselves.

Trees are fairly effective at removing both solid and gaseous particulates from the air. In one study, stands of trees reduced particulates by 9 to 13 percent, and the amount of dust reaching the ground was 27 to 42 percent less under a stand of trees than in an open area. Among gaseous pollutants, ozone, chlorine, fluorine, sulphur dioxide and PAN (peroxyacetylnitrate, a photochemical component of smog) are all absorbed by trees. In most cases, these gases also damage the trees. Unfortunately, trees remove little, if any, carbon monoxide which amounts to roughly one-half the total weight of air pollutants in this country.

Increasingly, carbon dioxide is being recognized as a "greenhouse gas" pollutant with potentially devastating consequences, such as global warming, dramatic changes in rainfall patterns, and rising sea levels that threaten flooding in coastal cities Since photosynthesis in green plants consumes carbon dioxide, plants could help to counteract the increase of this gas in the atmosphere. Rosenfeld, Martin, and Rainer report that planting urban trees could reduce heating and cooling demands enough to significantly cut fossil fuel consumption. They suggest that urban trees could be about 10 times as effective as forest trees for lowering carbon dioxide in cities.

Noise pollution from highways and other sources can be reduced with trees. Used alone, trees must be planted in belts 35 to 100 feet wide to create noticeable reductions. However, earth berms can cut traffic noise by up to half, if they are tall enough to hide the source of noise and are planted with trees, shrubs, and grasses. Where this kind of adjustment to the topography is not possible, a row of trees and a solid wall reaching up to the base of the crowns will provide a similar reduction.

First Christian Church, Eugene, Oregon, 1986. Although these trees are protected under the Historic Tree Ordinance, it is a tree that is no longer recommended for planting by the current administration because of thorns.

Soil and water quality are protected by trees. In urban settings, large areas are covered by buildings, pavement, and other impervious surfaces. Instead of percolating into the soil, rainwater and snowmelt are concentrated and accelerated, increasing soil erosion and silt accumulation in streams. Trees and other vegetation protect the soil from erosion. Along watercourses, roots and fallen leaves help hold the soil together and shield it against the cutting forces of surface water. Vegetation also absorbs some of the force of failing rain, so soil particles are not dislodged. And, the leaf litter that accumulates under trees creates an environment for earthworms and other organisms that help maintain soil porosity.

In studies at Pennsylvania State University, tracts of trees in municipal watersheds were used to purify partly treated sewage and protect surface waters. By adjusting sewage water application rates researchers prevented the ground water from becoming contaminated with nitrates. Ninety percent of the water applied went into recharging the underlying aquifer. Heavy metals, a worrisome component of municipal sewage, did not become a problem.

Monetary Values

Urban trees often have substantial monetary values. A number of studies have shown that real estate agents and home buyers assign between 10 and 23 percent of the value of a residence to the trees on the property. Local governments capture some of this monetary value because enhanced property values increase assessed values and the tax base.

Appraisal methods have been developed for landscape plants, including trees. The standard for estimating the monetary value of landscape vegetation, usually accepted by insurance companies, courts and public agencies, is Valuation of Landscape Trees, Shrubs, and Other Plants: A Guide to the Methods and Procedures for Appraising Amenity Plants. This guide was prepared by the Council of Tree and Landscape Appraisers and published by the International Society of Arboriculture, P.O. Box 71, Urbana, Illinois 61801. It is now in its seventh edition. The guide describes two methods to estimate tree value.

The first method simply identifies the amount of money needed to replace the tree. Replacement value can be used for trees less than or equal to eight inches in diameter and commercially available in appropriate sizes. Adjust the cost of the replacement tree for condition and location. Then add the cost of planting and establishment, including labor, equipment, materials and maintenance. Costs of guarantees and a reasonable profit margin could also be added. If the purpose of the estimate is to settle a claim for the death or damage of a tree, removal and cleanup costs for the dead or damaged tree are also included.

Technical competence and experience are required for accurate tree valuations

The second method uses a formula that multiplies the cross-sectional area of the trunk by a value (currently $27) per square inch. It then adjusts this value for species, condition, and location. Trees of different sizes are measured at various distances from the ground. Tree diameter determines where the cross-sectional area is measured.

Diameter of Tree	Height of Measurement
4 inches	6 inches
8 inches	12 inches
over 8 inches	4 ½

Three factors affect tree value: species, condition, and location. Each of these factors are rated from 0 to 100 percent. This sample calculation was taken from the valuation guide:
Circumference of trunk at 4½ feet = 47 (inches)
Circumference multiplied by 0.08 = 177 (square inches)
Multiplied by current value per square = $27/inch x 177 square inches = $4,779
Multiplied by species factor, e.g., 85 percent = $4,779 x .85 = $4,062
Multiplied by condition factor, e.g., 75 percent = $4,062 x .75 = $3,046
Multiplied by location factor, e.g., .80 = $3,046 x .80 = $2,437
Value of tree = $2,437

Regardless of the method, determining the adjustment factors for species, condition, and location requires technical competence and experience. A tree valuation is only as credible as the arborist who makes the determination.

The value of a particular species or variety can be looked up in an established tree listings or judged by a local arborist. Value is often affected by local usage.

Physical condition is also a factor' regardless of the kind or size of the plant. Parts deformed by crowding, storm damage, fire, insect damage, disease or other cause are primary concerns when judging condition. Careful inspection should reveal these and other factors.

Plant location also affects value, especially when it is near buildings, utility lines or other plants. Plants used in planned landscapes often have higher appraised values than those growing wild.

Appraising a landscape plant requires considerable time on the site for looking at the tree, its environment, measuring and examining its parts, collecting samples, taking photo graphs, and asking questions of people living nearby. In addition, the process requires decisions based on specialized knowledge of plant pathology, aesthetics, ornamental horticulture, and tree physiology. Appraisers often call on other experts for consultation in some of these areas.

Monetary value can also be assigned to some of the benefits trees bring to the urban environment.

As mentioned earlier, sewage treatment costs could be cut by applying treated effluent to trees and other landscape plantings. Irrigating with recycled waste water reduces the need for costly evaporation ponds and sewage lagoons, as well as the dumping of municipal wastes into rivers, lakes and oceans. And, water-loving species could serve as natural water pumps for areas with high water tables.

Trees can cut energy costs, too. Trees shade buildings and pavement, which reduces the temperatures in and around buildings. By cutting air-conditioning costs, trees indirectly reduce carbon dioxide emissions equivalent to fifteen times the amount the tree alone could absorb. According to Akbari, et al., planting trees to reduce peak-load energy demand by one kilowatt-hour costs about one cent.

Three well placed trees can cut air-conditioning costs by ten to fifteen percent. For every ton of new wood that grows, about 1.8 tons of carbon dioxide is removed from the air and 1 .3 tons of oxygen is produced. Reducing carbon dioxide through tree planting costs about 0.3 to 1.3 cents per pound.

Akbari, et al., estimate that there are 100 million available tree planting spaces around American homes and businesses. Planting trees in those spaces could reduce atmospheric carbon dioxide by an estimated 18 million tons per year, and save consumers $4 billion each year.

Economics and Decision Making

In simpler times, municipal tree budgets were often increased routinely based on inflation and additional trees without close scrutiny for efficiency. increasingly, however, city budgets are strained, and tree managers must justify their funding requests.

This native vegetation was preserved to serve as a valuable buffer between buildings and parking areas.

Sound records on the costs of tree management operations help urban foresters justify budgets. However, many cities do not keep adequate records or the cost of retrieving the information manually has been too high. Fortunately, the cost of microcomputers has plummeted while their capabilities have soared. The hardware and software capable of maintaining records for thirty thousand or more trees can now be purchased for between $3,000 and $5,000.

A number of cities have computerized data systems for their trees, and increasingly good tree management software is coming on the market at reasonable prices. These systems allow tree managers to improve their organization in four ways.

First, computerized records allow managers to increase efficiency. They can identify the costs of planting and maintaining various species. Some species are much more costly to maintain than others, but differences may not be obvious without good records. Operational costs can also be tracked, and adjustments made to increase efficiency. For example, scheduled maintenance of all trees on a block costs much less per tree than requested maintenance of single trees. These cost differences may persuade decision-makers to provide the funding needed to permit more scheduled maintenance.

Second, workload projections permit jobs to be scheduled efficiently, coordinating the appropriate personnel and equipment.

Third, the value of a community's trees can be calculated. When maintenance is shown as a percentage of value, the cost of tree management should compare favorably with the costs of maintaining streets, sewers, and other assets. Over time, records should show that trees grow in value, while most other urban assets decline in value.

Fourth, budget needs can be projected and supported with local data. Most urban tree managers excel in the care and treatment of individual trees. Many have mastered the administrative and political aspects of local government. it is now time for them to use modern tools to track costs and increase efficiency.

Urban forest managers must become adept at communicating the value of urban trees, if their requests are to compete for an appropriate share of local budgets. Powerful computers that retrieve and summarize information can help them generate and communicate this information (Fig. 1).

As per your request, the following summary of the 17th Street Planting Project has been prepared:

Date of Start:	June 23, 1989
Date of Completion:	July 12, 1989
Location:	17th Street Medians from Irvington Blvd. To Frisco Street
Total Project Cost:	$19,696
Cost Breakdown:	Plant Material — $13,141 Supplies — $ 1,209 Contractual Services — $ 5,346
Total Project Man-hours:	312.75

Hours Per Activity:

Activity	Hours
Pit Preparation	54.75
Planting	52.00
Staking	48.00
Berm Building	26.50
Planning and Organization	23.25
Watering	21.25
Stake Repair	20.00
Mobilization	16.75
Mulching	13.00
Cleanup	11.25
Site Preparation	9.25
Herbicide Application	4.50
Pickup/Delivery	4.00
Fertilizing	3.50
Berm Repair	2.50
Equipment Preparation	1.00
Equipment Maintenance	.75
Tree Removal	.50
TOTAL	**312.75**

Landscape Inventory as of August 7, 1989:

Planting	Number
Sweet Gum	68
Red Maple	15
Holly	11*
Palm Island	3

*Two hollies are broken and need to be re-planted. Exact cause of breakage assumed to be vandalism.

Figure 1:

BASIC HORTICULTURAL PRINCIPLES

Basic Needs

Plant growth is a very complicated process, but it is not necessary to understand all the intricate details in order to successfully manage the urban forest. However, understanding the basic requirements of tree growth will help. The primary environmental factors are:

- Soil is a reservoir for nutrients and moisture, and provides mechanical support.
- Sunshine furnishes heat and light.
- Air supplies carbon dioxide and oxygen.

Plants in the wild obtain these basic needs from a complex, but finely balanced biological system. Cultivation often disturbs the natural balance. Any factor in the plant's environment that becomes less than optimum will limit its growth.

Naturally, quantity is important, but quality also affects the vigor of trees. Air pollution, construction activity, and soil compaction place stresses on trees. Healthy trees can better with stand these stresses year after year. This section describes principles of tree growth and their relationship with the environment. Knowing these basics will help the people responsible for tree management make wise decisions.

Relationship of Air and Water in the Soil

Soil provides a foundation for tree growth-structurally and biologically. Soil supports a tree's physical weight and resists the forces of wind. Soil al supplies water, air, and nutrients. The value of the soil as structural and biological support is related to soil structure.

Soil structure influences plant growth, because it affects moisture, aeration, heat transfer, and the mechanical resistance to root growth. Most people think of soil as being a mixture of solid elements, such as minerals and organic matter. But, open spaces in soil, called pores, are equally important. The size and distribution of soil pores affect the movement and availability of moisture and air through the soil.

The air-water relationship is a critical one for optimum plant growth. The number of pore spaces must be adequate and the variety of sizes appropriate. Pore size determines the relative amounts of air and water that soil will hold. Gravity quickly drains water from larger pores, making them good sources of air. Smaller pores tend to hold water against the pull of gravity, making them good for water storage. Because plants need both air and water in the soil, a good balance of large and small pore spaces is required for optimum growth.

Sand, with its open structure of many large pores, is called a "coarse" soil. Sandy soils generally drain well. They contain ample amounts of air, but little water. "Fine" clay soils drain poorly and tend to hold water in their many small pore spaces. But, pores in clay soils can be so small and hold water so tightly that tiny feeder roots cannot extract the moisture.

Organic matter in the soil affects both the physical and chemical properties of the soil. A fertile soil rich in organic matter is literally alive. Although insects and earthworms are the most obvious inhabitants, microorganisms, such as bacteria and fungi, constitute the largest population by weight in the soil community. Organic matter in the soil comes from decomposed plant and animal tissue as well as the micro-organisms themselves. Organic matter enhances the aeration and water-holding capacity of the soil. It also affects the soil's chemical properties by supplying necessary plant nutrients.

The phrase "effective rooting depth" describes the portion of the soil where conditions are favorable for root growth-most often in the top three feet of the soil. Effective rooting depth may be limited by circumstances that restrict soil porosity or hinder plant growth. Four relatively common problems are surface crusting, high water tables, poor sub-surface drainage and claypans or hardpans.

In extremely compacted soil, or soil without adequate aeration, roots are found growing on the surface of the soil where conditions are more favorable for growth

Soil Compaction

Soil compaction is the major cause of death or decline of mature trees where efforts have been made to save them. It poses a very serious threat to good soil structure. Delicate soil pores are easily crushed, decreasing their capacity for water and air movement and hindering root growth. Wet soil is particularly vulnerable, because water lubricates soil particles and loosens binding agents. Small particles slip between the larger particles, filling the pore spaces. Loose soils will compact more than tight soils, and soils that have a broad range of particle sizes can be more severely compacted than more uniform soils.

Few soils can withstand traffic without becoming severely compacted. Compaction depends not only on the amount of pressure exerted, but also on the duration and frequency of exertion. For example, the heel of a shoe exerts force per unit of area equal to that of heavy equipment (although the turning, starting and stopping of heavy equipment increases the force). Pressure spreads with depth, so the compacting effects of pedestrians and vehicles usually occur to the top four inches of the soil.

Compaction is easier to prevent than to remedy. Avoid cultivating wet soils. Keep foot and vehicle traffic over the roots to a minimum. Where traffic is necessary, confine it to a few paths that stay well away from the driplines of trees. Lightweight vehicles with large, smooth, low pressure tires compact the soil less than heavy vehicles with smaller, heavily treaded, high pressure tires. A thick, coarse mulch spread on the soil surface where travel must occur helps to disperse the load.

The Root System

A healthy root system serves multiple functions for the tree. Roots collect water and minerals from the soil, store nutrients, and anchor the tree. Fibrous roots are primarily responsible for the absorption of water and minerals from the soil. They most often grow in the top six inches of the soil. Woody roots provide a framework for anchoring the tree in the ground. They support the weight of the trunk and branches, including leaves, snow and ice. And, wind pressure adds to the load.

Woody roots radiate out horizontally and vertically, decreasing in size as they get farther from the trunk. Woody roots seldom penetrate the soil deeper than three feet. Because they are relatively shallow, roots are vulnerable to damage from activities on the surface. It's important to be aware of roots, because they are not easily seen, and structural deficiencies often remain unnoticed until the top of the tree dies back or the entire tree falls over.

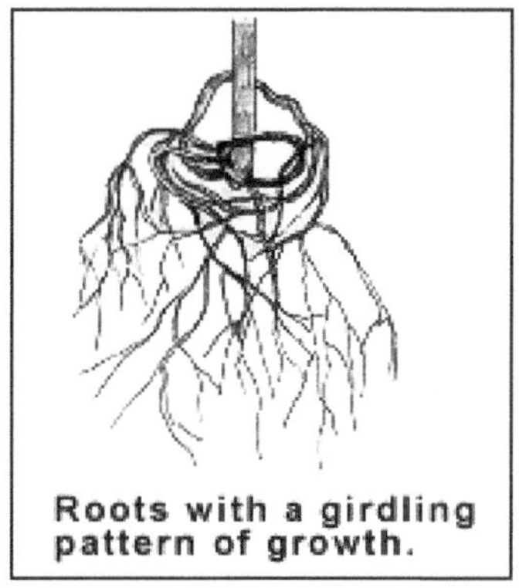

(a)

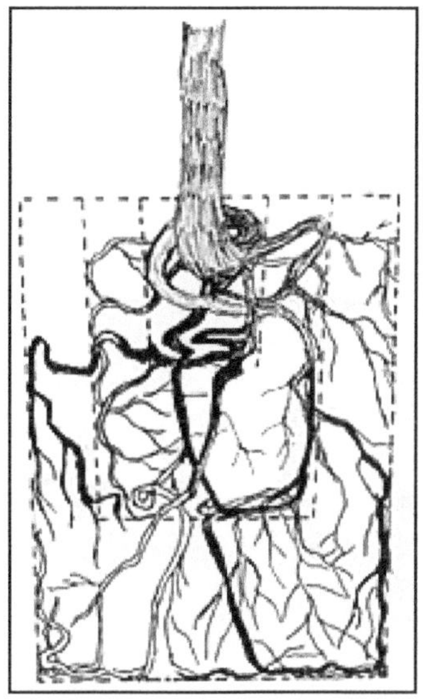

(b)

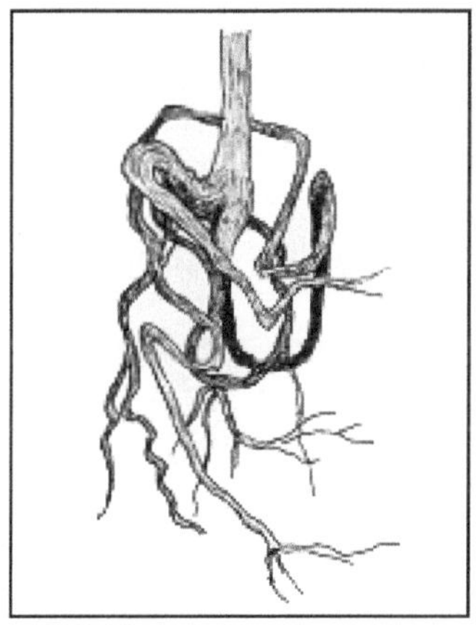

(c)

Figure 2: Types of root defects and the zones within the root mass where they may be found.

Structural Strength of Tree Crowns

A tree's crown must be strong enough to withstand wind forces, and carry the weight of snow, ice, and branches covered with thousands of leaves. To assess the structural strength of a crown, look at these characteristics: the spacing of lateral branches along the main trunk, the presence of a clearly dominant leader or main trunk, trunk taper near the ground, and angle of branch attachment.

A tree with branches spread out along the trunk generally has better structure than one with all its branches growing close together on the trunk. A clearly dominant leader or main trunk provides a strong framework-called a scaffold-that supports lateral branches. Several leaders of similar size offer a weaker scaffold. Even round, oval, and vase shaped trees that commonly have more than one leader should show a clear hierarchy of leader sizes. This distinction of leader size is a sign of a good scaffold. Trunk taper at the base of the trunk helps support for the entire crown. Young trees with no trunk taper usually have to be staked, so they don't topple over in high winds or under heavy loads of ice and snow. Finally, branches that have a wide angle of attachment or have connective tissue in the crotch are usually more structurally sound than those with sharp angles of attachment or imbedded bark in the crotch.

Plant Stress

Trees, regardless of species or location, have the same basic needs that must be met for the tree to remain healthy and grow. Basic requirements for plant growth include water, carbon dioxide, oxygen, nutrients, sunlight, appropriate temperature, and sufficient space. A balance must be maintained between the amount of water lost through the leaves and that taken up through the roots. Soil conditions must allow enough root growth to supply the leaves and branches with nutrients, gaseous exchange, and water. Photosynthesis (the manufacture of sugar by the green parts in the presence of sunlight) must be able to supply the energy needs of the tree. And this energy must be conveyed from the leaves to the stem and roots.

The trees planted in the median strip outside this row of fraternal houses near the University of Washington campus are suffering primarily from soil compaction associated with the recreational use of this space.

Last but not least, the strength with which the tree grows, known as vigor, must remain high enough to prevent attack by disease-causing agents, such as bacteria, fungi, and insects.

The urban environment often places tremendous stress on trees that natural defenses may not adequately protect against. The urban environment changes rapidly, at least in comparison to the life span of most trees. The amount of stress experienced by a tree is directly related to the rate of unfavorable change in its environment, the quality of the planting location, and the level of maintenance it receives. Noninfectious diseases, which are not transmittable from a diseased plant to a healthy plant, will be especially threatening to trees whose defenses are weakened by urban pressures. Most urban stresses can be divided into the broad categories of environmental stress, animal injury, and people pressure.

Most urban stresses can be divided into the broad categories of environmental stress, animal injury, and people pressure.

Environmental stress has the most severe effects on newly planted trees that have not yet adjusted to their new conditions and on established trees whose surroundings have changed, usually from the activities of people. Temperature stress results from extreme heat and cold as well as rapid changes. High light levels are usually associated with high temperatures. Thinning or clearing operations can expose remaining trees to excessive sunlight threatening them with light and temperature stress. Minimize the negative impacts of high temperature and light by planting shrubs and ground covers to create an "understory" beneath and around the trees. These new plants will help keep the area cooler. In areas surrounded by concrete or asphalt group the trees together or plant an understory to buffer high temperatures

Soil compaction, altered surface water drainage patterns, and mechanical damage to this tree's roots are three stress factors represented at this freeway construction site

Low temperature can also injure trees. "Hardiness" refers to the adaptations of woody plants to cold conditions through patterns of growth and dormancy. While plants can often tolerate extreme cold during the dormant season, they could be injured by the same temperature during the growing season. Select plants with the extreme rather than the average temperature in mind. Plants in containers are more susceptible to extreme and rapidly changing temperatures, because their roots are not protected by the "thermal mass" of the earth.

Moisture stress is one of the most common problems. Too much or too little water both inhibit root function, and therefore, the growth of the entire plant. Physical damage from snow and ice, poor water quality, and erosion could also be classified as moisture stress.

One of the most important factors contributing to moisture deficiency is the amount of space available for the roots. If roots are restricted, they cannot spread through enough soil to gather the moisture needed to support the crown.

Another type of moisture stress occurs mostly in broadleaved and needled evergreens. During the winter animals --both pets and wildlife and spring, the air may be windy and warm, but the soil frozen or dry. Known as "winter drying" conditions, more water is lost through leaves than is taken up by the roots.

Excess water from flooding, saturated soils or over watering can also place stress on the tree. Like other parts of the tree, roots need to exchange gases, such as oxygen and carbon dioxide, in order to survive. This exchange of gases occurs between the fibrous roots and the air in the soil's large pores. If the pores are filled with water, the roots suffocate. To avoid problems with excess moisture, determine the soil moisture and drainage patterns of the planting site, select a tree that thrives under, or at least tolerates, those conditions, or modify the soil situation.

Road construction is one of the most damaging forms of tree stress in urban environments.

Soil stress can be classified as either chemical, biological, or physical. Chemical stress involves unfavorable pH, an imbalance in nutrients, or even the presence of toxic materials discharged into the soil by other plants or human recklessness. In cities, soil stress is usually related to physical changes affecting soil aeration and moisture. Physical stress involves something that restricts root growth, including streets, sidewalks, buildings, construction activities, construction debris in the soil, large boulders, trenching, grade changes, soil compaction, improperly used herbicides, and the roots of other trees.

Cultivation of wet soil leads to soil compaction and should be avoided.

Animals injure trees in several ways. Animal waste, especially urine, adds toxic compounds to the soil. Animal traffic compacts the soil. A host of animals --both pets and wildlife inflict direct injuries to the bark: birds peck it, cats shred it, and rodents chew it. Dogs are particularly destructive when chained to a tree.

People apply a variety of pressures that adversely affect the health of trees. Trees are easily disturbed by construction of buildings and roads, soil compaction, chemicals, air pollution, and improper tree maintenance also fall into this category.

In addition to the stresses that come from the direct or indirect actions of people, urban trees also suffer from human inaction. Well-meaning people sometimes impede appropriate urban forest management. Education is one way to increase the knowledge of urban forestry and to get people to work together to support it

Restricted root space is capable of causing roots to grow in a girdling pattern, creating a structural deficiency in the anchorage of the tree.

URBAN FORESTRY PLANNING

Planning is a method for achieving an end, a detailed formulation of a program of action, or an orderly arrangement of parts of an overall design or objective. Planning is the process of establishing goals, policies, and procedures for a social or economic unit. Some people believe the planning process culminates with the publication of a document that presents detailed instructions for reaching a goal, but it does not end there. The final result of planning is the attainment of the goals, not publication of the plan.

Planning for the infinite future could be a daunting task, so the process is generally divided into smaller components: long-range, short-range, and site-specific. Long-range planning tends to be extensive, broadly worded, and enforced through a legally adopted comprehensive plan. Often, these comprehensive plans are refined to specifically address development with regional impacts. Short-range planning, also known as current planning, tends to be explicit, and is enforced through zoning ordinances. Site specific land development regulations may overlap zoning ordinances.

Long-range Planning

Long-range planning mainly concerns growth management, and encompasses the preparing, maintaining, and updating of a comprehensive plan. State government establishes guidelines that must be met by comprehensive plans at the state, region, and county levels. Urban forestry concerns must be reflected in comprehensive planning. By adopting the plan, government officials signify their support for trees in the urban environment.

Comprehensive plans generally contain, among other things, elements that address the environment and transportation. Both of these elements should contain references to trees as they relate to the urban infrastructure. The environmental element should identify the need for preservation of the natural environment within developed areas. This element should also contain language calling for planting, maintenance, and preservation of native and introduced species along streets and in other open spaces. The transportation element should incorporate aesthetic considerations in the development of traffic circulation systems, and in providing for adequate right-of-way for tree planting.

Short-range Planning

These plans help achieve desirable land-use allocation and distribution based on an assessment of the cumulative impact to a given area. Short-range plans usually serve to refine comprehensive plans and development proposals. They are enforced through zoning ordinances. Supporters of urban forestry can use the short-range plans to document areas in need of special attention. This may include preservation of critical habitat, development of tree-lined corridors or beautification of major entrances to the community. Including these concerns in the short-range planning process helps solidify support from planning staff and local decision makers.

Site Specific Planning

Planning for a specific site also has several levels: conceptual, schematic, and master. Conceptual planning illustrates possible physical forms and relationships between various elements of the project. This phase may even be done before a site is selected. Schematic planning may develop several design alternatives for a specific site, but not in enough detail to implement them. Once a schematic design is selected, the master plan refines the design and adds details. Based on an approved master plan, designers produce development plans, which may be reviewed by building, zoning or transportation departments.

Site specific planning includes efforts concentrated on site plan review. Enforced through land development regulations, this level of planning affects preservation and planting of trees on particular sites. Significant individual trees or entire stands may be planted, saved or removed based on a plan at this level. So regardless of the contents of higher level plans, site specific plans should be scrutinized before approval is given. Before development, sites should be inspected to verify that plans have been drawn correctly and that tree protection measures comply with development regulations.

Management Planning

Then there are management plans, which define the overall scope and methodology of certain operations, such as urban forestry or transportation service.

Urban forestry management plans generally follow the planning process mentioned previously. They may also include: maintenance standards, tree inventories, work record processing, planting, removals, reforestation, rotation planting, phased removal, tree selection processes, design criteria, personnel training and development, budgets, coordination within an agency, as well as with other agencies and citizens.

Two specific types of management plans-tree inventory system planning and long-range tree rotation planning-deserve further description. These types of management plans are important because of their effect on the efficiency of urban forestry operations, their impact on budget justifications, and the potential for conflict within the community.

Tree inventory system planning is a method for obtaining and organizing information about the number, condition, and distribution of urban trees. Information that is accurate, accessible, and simple is one of the best tools for making planning and management decisions. With tree inventory information, program resources can be allocated appropriately among the various tree management functions, work can be scheduled for maximum efficiency, and financial decision-makers can evaluate various work plan proposals by comparing expected results with projected budgets.

Several inventory systems have been developed by cities, universities, extension services, and consulting firms. They range from quick, inexpensive survey methods that provide basic information to sophisticated, computerized systems that are integrated with daily tree care activities. A Guide to Urban Tree Inventory Systems was developed by the School of Forest Resources at Pennsylvania State University in 1979. It contains a general review of the characteristics of urban tree inventory systems, as well as profiles of 25 systems and references to 24 other inventories.

All inventories share the same general goal-to provide information about the nature of the urban forest. Most inventories have several objectives. The simplest systems might provide information to support the establishment of a tree care budget, start a community tree program, or at least, initiate a tree advisory board. At this level, the desired information may be as simple as three estimates: the total number of trees, their average condition, and their monetary value.

Computerized systems may be used to justify and prepare annual budgets, organize daily work assignments, keep records on individual trees, aid long-range planning, and support management analyses. These types of inventory systems link day-to-day operations with long-range planning.

Long-range tree replacement or rotation plans should be based on inventory information. Because tree removal or replacement generates community concern, citizens and policy makers should both be involved in the process.

The basic elements of a rotation plan

- Criteria for tree removal;
- Complete versus phased removal;
- Diversity versus monoculture in replacement species;
- Management plan for replacement activities; and
- Fiscal budget.

Criteria for tree removal form the basis for objectivity in the midst of the emotional furor that often develops over tree removal in urban areas. Objective criteria would include: current and future maintenance costs, years of estimated useful life, structural integrity, and public welfare. Because they are more subjective, the following items should be considered only as secondary criteria: diminishing aesthetics, amenities, and engineering values, such as noise abatement and wind reduction.

Removal recommendations must clearly identify priorities for tree removal. These priorities could become a sequence for removal. Dead or dying trees might be first. Second would be trees representing a potential hazard to adjacent property, buildings, parked cars or people. Next might be stumps from trees cut previously. The final category could be trees growing in undesirable locations.

Complete versus phased removal is one important decision that must be made before a management plan for tree removal is developed. Complete removal, as the name suggests, is the removal of all trees within a given area in one operation. In phased removal, a predetermined portion of a stand is removed on a schedule covering a period of years. Where space allows, new trees can be planted among existing ones that have been scheduled for removal. This approach, known as interplanting, encourages age diversity in the stand and minimizes aesthetic and environmental impacts of large-scale tree removal. Interplanting copies nature by providing new trees to take over before older trees come down.

Diversity or monoculture in tree species replacement will usually be an issue in most communities. Species diversity stabilizes the urban forest and tends to reduce losses due to harmful insects or disease. While having similar maintenance requirements is a primary consideration in species selection, large-scale monocultures should be avoided. Diversity also offers the adjacent property owners some individuality. Again, citizen involvement is valuable in developing the rotation plan.

Management planning for replacement trees must ensure the long-range maintenance for these replacement trees. Young trees require considerable care, such as watering, fertilizing, insect and disease control, and regularly scheduled pruning. Any plan must ensure long-range maintenance to protect the investment of public funds in replanting. For example, scheduled trimming of young trees reduces the need for expensive corrective pruning of mature trees.

Budget and long-range funding for tree replacement is an obvious element of the basic rotation plan. Many of the criteria mentioned previously will provide meaningful budget information. Cost/benefit approaches are helpful in justifying tree rotation plans, because policy makers often reduce programs to dollars and cents.

Work Plans

Work plans fill in the details by guiding operations through specific periods of time (from several months to several years) and through specific locations (from a park to a watershed). Because on-going projects need funds, allocated budgets are an essential element of work plans. Because work plans generally address active projects, they change periodically as the projects change. Work plans might address specific planting projects, new tree care, care of established trees or implementation of a tree inventory.

TREE SELECTION AND SITE DESIGN

During this century, the form and character of cities and towns have changed dramatically. Fifty to one hundred years ago urban areas were different; cities were smaller and skylines rarely rose above eight to ten stories. Automobiles were not the major method of personal transportation, and the ratio of open space to buildings was dramatically different. Even the lifestyles of urban residents have changed.

Urban Change

Cities are dynamic, with ever-changing land uses. This is particularly true today. In planning the urban forest, it must be assumed that cities and towns will continue to change. The likelihood of changes and the character of potential environments should be carefully considered. Tree selection must assume that the urban forest will be dynamic, if it is to respond to the changing conditions of the community.

Space for Trees

The stereotypical image of the "city" as a place of continuous tall buildings and intensive, pedestrian -oriented streets is accurate for only a small percentage of urban areas. Most urban environments consist of automobile oriented, low density land uses and recently constructed buildings.

Cities are actually more spread out than the stereotype suggests, with a higher proportion of open space to buildings. However, the presence of open space does not necessarily imply more room for the urban forest. The majority of open space is devoted to transportation corridors and car storage. Space is almost always a limiting factor in some way. It could be a conflict between vehicles, pedestrians and trees for space at ground level. Above ground, the canopies of trees interfere with overhead utilities and views. Beneath the ground, tree roots compete with utilities and paving.

In urban environments, natural areas with continuous vegetation are replaced by discontinuous and fragmented vegetation. The space available for trees occurs in small pieces rather than large ones. In some areas, such as downtown cores, fragmentation of planting sites is extreme. Greater continuity can be found in residential neighborhoods, stream corridors, steep slopes or around other extensive land forms.

There appears to be a trend toward increasing density in many new residential and commercial developments. This threatens a corresponding reduction in the space available for trees.

Climate And Microclimate

Environmental conditions tend to be more varied in urban areas than in less developed areas. For example, asphalt surfaces surrounding one site make it hot and dry. But around the corner, buildings cast shadows that cut temperatures and minimize evaporation of soil moisture. Light, wind, temperature, and soil conditions can change abruptly from one spot to another.

Urban conditions are frequently more severe for the growth of plants than in natural environments. On the other hand, conditions could also be better due to irrigation or lack of competition.

Seasonal variation may also be extreme compared to more natural areas. A site may be hotter in the summer and colder in the winter. Or, it may be drier in the summer and wetter in the winter due to soil compaction and the presence of paved surfaces. However, the city as a whole generates heat that usually keeps temperatures warmer in all seasons than more natural areas. This is called the "urban heat island" effect.

While it's not possible to generalize about the severity of urban conditions, one thing is certain. Conditions vary considerably, and it's wise to assess each site individually. The best way to identify site conditions is to tour on foot. Be sure to consider changes that are planned in the vicinity, even if plans are not yet final. The construction of streets and buildings, rechanneling surface water, and routing utility corridors can all affect the ability of a site to provide for the needs of trees. While it's best to anticipate urban evolution, it's safe to assume that plants and people will always have to adapt to unforeseen changes.

Land Use

One goal of urban forest management is to create and maintain the maximum amount of visual and biological diversity. Preserving the uniqueness of different areas involves selecting trees that compliment the activities occurring there. This may require drawing on a broad range of plant species to create a distinct character. The urban landscape can be divided into four broad land use categories: natural areas, parklands and campuses, residential property, and fragmented spaces. These categories are based primarily on activities that take place in them and the mood created by those activities.

Natural areas are relatively undisturbed. The typical urban resident also considers them to be "natural," because of their size and location. These areas tend to occupy land considered unbuildable due to poor accessibility, rugged topography, poor soil, or inappropriate hydrology. They are frequently linear, such as ravines, stream corridors and steep slopes. Other types of natural area in this category may include: "greenbelts," flood plains, wetlands, ponds, waste areas, abandoned land and "buffer zones" around undesirable land uses, such as land fills.

Activities occurring in natural areas range from casual recreation to no activity at all. Increasingly, these areas are seen as valuable components of the urban environment, because of their aesthetic qualities, their symbolic value as reminders of the natural environment and their value as critical natural habitat.

Coherent management policies for these types of areas have rarely been developed. They typically receive little or no maintenance, and are left to their own devices. They have almost always been disturbed by human activities at some point in their history. In most cases, these areas are in some stage of plant succession as wild plants and animals re-establish themselves. For these areas, select native trees that enhance or maintain the feeling of a "natural" environment by preserving an "uncontrolled" appearance.

Parklands and campuses

include traditional parks that host recreational activities, community open space in planned residential developments, freeway interchanges, as well as business and industrial parks with a campus-like atmosphere. These developments are normally large, with a high percentage of open space to paved or built-up areas. They typically contain large lawns and clearly defined beds of flowers or shrubs. Trees are planted as individuals, groves or small woodlands. Although they usually receive relatively high levels of maintenance, some people consider their character to be "natural." This feeling may grow from traditional landscape maintenance techniques and "naturalistic" plant arrangements.

These areas are intensively used. Even as a campus of industrial or commercial buildings, the landscape may simply be an area viewed by workers and visitors-but rarely entered. If it is only looking out an office window, these people are using the outdoor areas and benefitting from the presence of well-maintained trees and lawns. So, observation represents a high level of "use."

Attitudes toward these areas may be changing, prompting a change of management. Measures are sometimes taken to reduce maintenance costs, encourage wildlife habitat or allow for a wider variety of uses to occur within these campus areas. The long term implications of these trends must be considered when trees are selected.

Residential land

comprises a very large portion of most urban environments. In these areas small parcels are typically under the private ownership of different people. Several issues grow out of this pattern of ownership.

Buildings divide the open space into small, tight fragments. The visual character and the level of maintenance varies dramatically from properties with a relatively "natural" appearance to those that are manicured and open. The intensity of uses include passive visual appreciation, active recreation and even storage of personal possessions. There may not be a consistent pattern of open space. In some cases, buildings and paved surfaces occupy most of a lot, while on others, they cover only a small portion of the total area.

Because of the diversity found in residential areas, planting guidelines should be general rather than specific.

Streets, plazas, and fragmented planting areas have the greatest variety of physical conditions. They range from the narrow, canyon-like streets of the urban core to wide open areas around suburban streets. Between these extremes are streets and boulevards in residential, industrial or commercial areas. Fragmented open space includes parking lot islands and narrow planting strips around parking lots or next to buildings. Fragmented areas and streets fall into the same category, because plant selection criteria and environmental conditions are similar.

In these areas, available space is restricted and the land is divided into small pieces. Urban rooftop gardens where plants must be grown in containers offers an extreme example.

When selecting plants for this component of the urban forest, remember that it is essentially a symbolic forest it is also the most visible part of the urban forest, and the largest part in public ownership. Rather than designing a natural forest, the urban tree manager chooses plants that will bear the burden of this symbolism under the most stressful of the urban environments.

Tree Selection Considerations

One of the Most important aspects of urban forestry is selecting and acquiring trees. As stewards of the urban forest, today's tree planters have an obligation to contribute the best possible trees to future generations.

Tree selection and acquisition are also among the most challenging activities of urban foresters. Thousands of species are available, many of them genetic hybrids that have not been tested by time. Few people know all of the species. Objective information on long-term maintenance requirements and other potential problems is not readily available. In the absence of hard data, some tree managers assume that maintenance needs correlate directly to the rate of growth. In other words, a fast growing species will require more long-term maintenance than species with moderate or slow rates of growth. It may also be reasonable to assume that larger trees require greater expense for maintenance and replacement. With proper care, a tree could live for centuries. It makes little sense to ignore known shortcomings, simply because the problems will become apparent only when the tree matures.

Although it may be tempting to choose species that appear frequently in the landscape, popularity should not determine suitability for urban uses. Many desirable trees that require minimal maintenance seldom find their way into the urban landscape and may not be available from nurseries. Finding the best species requires research. Nursery catalogs are one source of information. The book, Plants that Merit Attention, Volume 1, Trees, may be useful for identifying nursery sources for a selected group of species. It was prepared by the Horticulture Committee of the Garden Clubs of America, edited by Janet M. Poor and published in 1984 by Timber Press.

Picking the best tree species for a particular site is similar to separating gold ore from the surrounding rock. Beginning with an extensive list of trees, the forester must sift out unwanted material until only the right choice remains. To be selected, trees must pass through three levels of selection similar to a series of sieves. These considerations can be grouped into three categories: site, design, and maintenance.

Site Considerations - The environment that will support trees encompasses space, moisture, soil, and other physical conditions. As mentioned earlier, the environment is affected greatly by the land use of a particular area.

Space could limit growth now or in the future. Consider the way that a particular tree will interact with the surroundings. Remember that trees occupy space above the ground, at ground level, and below ground.

The tree canopy offers the most immediate and visible concern. Most people would like to see a tree fill the space available. However, canopies that grow too large will require heavy pruning and possibly removal. Crowns in streetscapes could interfere with adjacent buildings, overhead utility lines, and views. Trees could also shade devices designed to capture solar energy or gardens that need ample sunlight.

Ground level refers to the space that begins at the soil surface and extends up to fourteen feet above the ground. Some areas require no clearance at ground level; sidewalks need ten feet for pedestrians; and streets need fourteen feet. Selecting trees with these requirements in mind can prevent unnecessary pruning later.

The space available below ground could be restricted by utility lines, building easements, vertical retaining walls or heavily compacted soils. Considering the way that trees fill space helps to choose a tree that will fit the site when mature.

Climate may vary from one site to another. Sunlight, wind, moisture, air quality, and temperature extremes all influence the choice of tree species. The conditions in an urban park are similar to those in undeveloped areas. But cities also have concrete canyons with low levels of sunlight, and the urban equivalent of deserts-large paved areas with intense sunlight, high temperatures, and little moisture.

Soil conditions must be assessed to identify fertility, aeration and drainage. Matching these site factors to the needs of potential tree materials is critical to the success of any planting.

Design Considerations - Various elements in the landscape relate to each other on an aesthetic level as we] I as a physical level. Trees can be selected for the way their size, shape and color blends-or contrasts-with other elements. Trees may be selected for their texture or fragrance. The timing of leaf set and drop or the schedule of flowering may be important. The type of shade or the density of the crown could also affect the choice.

Figure 3: Overview of Eugene, Oregon, 1988.

A tree's above-ground branching system-or crown-determines the overall shape. Some tree shapes are more suitable than others for particular sites. And, shape affects the amount of volume that the tree will occupy. Common tree shapes include: oval, round, columnar, pyramidal, and vase.

Oval trees fit nicely into most street plantings, because they grow upward and don't interfere too much with adjacent vehicular traffic. Because of their somewhat upright growth habit, oval trees can generally be placed closer to buildings than more spreading trees. But, oval shaped trees may reach a lofty height, making them unsuitable for planting underneath power lines.

Round trees have a spreading growth habit, resulting from more than one terminal leader. Lateral branches growing from the leaders may ascend, extend horizontally or sweep downward. Trees with an ascending branch structure are more compatible with city streets where there are buildings nearby or large trucks that must pass underneath. Trees with a descending branch pattern may require more pruning while the tree is small to maintain visual or vehicular clearance. However, once they grow larger, downward sweeping branches are valued for the sense of enclosure they suggest. A horizontal branching structure demands a greater distance between adjacent trees or buildings, if it is to attain a natural form and require minimal pruning.

Columnar trees usually have tightly ascending branches with narrow branch angles and short branches. Trees with this shape are valued for their narrow width that enables them to be planted in tight spaces where there is not enough room for a tree with a spreading branch structure.

Pyramidal trees often start off as columnar. With age, the lower branches get bigger and droop under their own weight. Many trees in this category have a single dominant branch from which the side branches grow. While conifer trees, such as pines, best illustrate the pyramidal shape, there are also many broadleaf trees in this category.

Vase shaped trees may be the most appropriate for planting on city street trees, because they grow up and out. They also arch over streets and sidewalks to form a shady canopy, while maintaining visual clearance and vehicular access.

Design requirements usually grow from site characteristics. For example, street trees typically need clear trunks. They grow as individual specimens, widely spaced in streets, without competition from adjacent trees until nearly mature. By contrast, trees for parks often grow in large groups. Early in their lives, they protect each other. But long before reaching their mature size, they begin to compete.

Nurseries now offer a wide selection of crown shapes, especially from clones. For example, the red maple clones (Acer rubrum) range from the narrow upright Armstrong to the widespreading, rounded Schlesinger. Similar variation can be found in the Callery pear (Pyrus calleryana) and Norway maple(Acer platanoides) groups. This diversity of crown forms allows for more precise matching of trees to site conditions.

While clones are genetically predisposed to uniform growth, they do not guarantee uniform performance in the landscape. Site conditions vary considerably over relatively small distances. A series of clonal trees placed along a street could encounter a variety of site conditions that could affect tree performance and appearance.

Shoot development, or crown configuration, also affects the success of urban tree plantings. The development of a multi-trunked character, the height of the first branches and the presence of poor branch attachments may adversely impact the vigor of a tree, or require more intensive management to make the tree compatible with adjacent land uses.

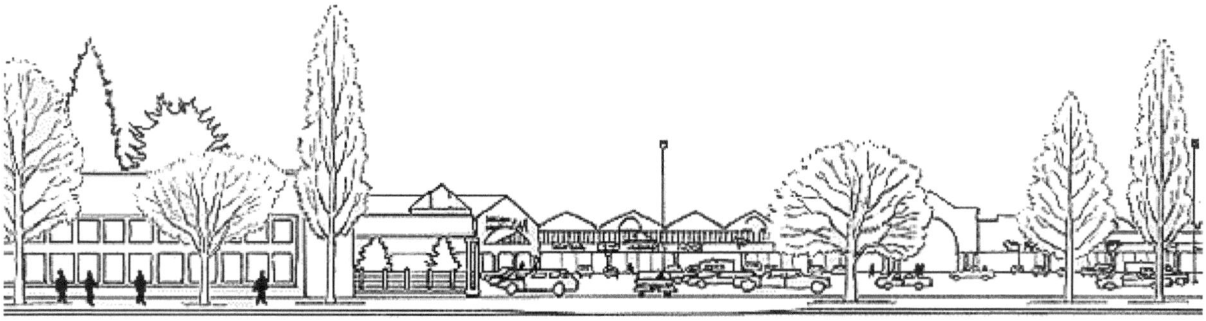

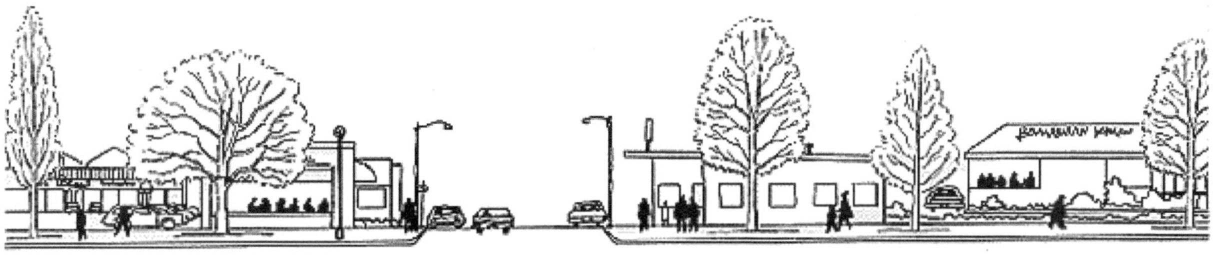

Maintenance considerations - Maintenance will re required of all tree species, but some will require far more than others. Given limited budgets, it makes sense to give maintenance requirements a relatively high priority among selection criteria. Pruning needs, fertilizer, water and pest management, and root control should all be anticipated.

Cities have long faced the problem of roots breaking sidewalks. Many people have searched for "deeprooted" species. However, the depth of rooting is more a function of site conditions, than an inherent characteristic of a species. Actually, most species are regarded as having shallow roots at maturity. Even so-called "taprooted" trees possess their vertically oriented roots for a short period of their lives. To reduce the importance of sidewalk damage as a criterion, use appropriately sized trees for the given space and planting techniques, such as large pits, root barriers, or deep planting.

Tolerance of stressful conditions is an important criterion for tree selection, and every site offers some type of stress. Generally, successful street tree species tolerate poor soil conditions, including fine texture and flooding. Tolerance to polluted soil may be needed. If salt is commonly used to clear the icy sidewalks, trees with a tolerance of salt should be selected. if the site has been disturbed, the pH level may influence the choice of plants.

Naturally, a species' susceptibility to harmful insects and disease should be considered, along with the cost of controlling them. For example, autumnal is flowering cherry (Prunus subhirtella) occurs widely, even though this genus is vulnerable to one of the most common diseases in the area: brown rot. Controlling brown rot requires an expensive multi-spray program. So, planting this species brings with it the burden of an intensive pest management effort.

Poor branch attachments, such as those with embedded bark, narrow angles of divergence or poorly spaced branches, become a major maintenance problem as trees mature. Some of these problems can be corrected at planting. Others can be eliminated by rejecting poorly structured trees from the nursery at planting.

Not every tree or every planting situation lasts for hundreds of years. Being dynamic, cities have spaces that are available for relatively short periods, making them suitable for short-lived trees or those that begin exhibiting undesirable traits as they mature. For example, big leaf maple (Acer macrophyllum) grows rapidly for 40-60 years. Then, branches begin to break out of the crown, and pruning requirements increase. In a similar manner, western white pine (Pinus monticola) reaches 60 to 80 feet in height after only 40 or 50 years. At this age and size, the trees become susceptible to white pine blister rust, which is always fatal. In situations where a tree is only needed for fifty years, both species would be excellent choices. While they develop problems as they grow older, in youth they offer a number of positive ornamental and management features.

Depending on the site, species' ability to withstand drought, freezing, wind, and air pollution should also be considered.

Urban areas offer a mosaic of environments and design possibilities, so there is tremendous opportunity to use a large range of trees in the urban forest. With such a great number of environments and design considerations, an equally large number of plants may be appropriate for use. This diversity reinforces the need for thorough site analysis and careful design for every project.

Providing Space For Trees In New Construction

Space for trees is readily available in open spaces around homes, in public parks, and school grounds. But in parking lots and along streets, adequate space for trees may only be available if it was specifically included during the design of the overall development.

During hot weather, parking lots can be as forbidding as great deserts. With proper planning, trees can minimize the heat. Spaces for trees should be scattered throughout the paved area, rather than only along the perimeter. These planting strips should be large enough to accommodate trees that will eventually grow large canopies and shade the parked cars. Areas surrounded with pavement should allow a square area at least six feet on a side. And, an area eight feet square would do much to prevent tree roots from eventually becoming a problem. Governmental bodies that regulate shopping center developments can require that the parking lots contain ample space for trees as well as for cars.

Street trees comprise about 85 percent of publicly-owned urban forests. They occupy an important and often overlooked-public land base. These unpaved ribbons along streets are commonly known as planting strips, parkways, or treelawns. Minimum widths identified by municipal codes are often inadequate for supporting mature trees, and some codes fail to mention trees at all. Since the size of existing street rights-of-way have remained the same for centuries, only future rights-of-way offer hope for adequate treelawns.

There are two practical ways to design treelawns of more appropriate location and width. The first approach combines the sidewalk and the curb as a single unit (monolithic design), so the treelawn occurs between the sidewalk and the right-of-way boundary. Called a boundary treelawn, it has been used widely on the West Coast since the middle of this century. The tree can be located off-center near the right-of-way boundary relatively far away from the sidewalk.

The more traditional arrangement, called a curbside treelawn, is located between the curb and the sidewalk. Besides using this space for street trees, curbside treelawns serve as storage for snow plowed from the street. Usually wide street rights-of-way make extra-wide curbside treelawns possible.

No standards exist for the width of treelawns. Prudence suggests that treelawns should be made extra-wide to accommodate large trees that will form a canopy over the streets. Wide treelawns also reduces the chance that roots will damage sidewalks. Preventing treelawns from cutting into the amount of land available for building is simply a matter of shortening the building setback.

Figure 4: Street trees may be planted within a curbside treelawn, median, behind the curbside sidewalk in the boundary treelawn, or outside the right-of-way, on private property.

Boundary treelawns make good use of space. For example, a tree planted one-half foot from the right-of-way boundary would be three and one-half feet from the sidewalk. While greater distance between trees and sidewalks may be better, three and one-half feet is reasonable. Since trees customarily grow midway between curb and sidewalk, achieving the same amount of space between trees and concrete in a curbside treelawn would require seven feet.

To implement any of the these concepts, a local government may have to amend its regulations concerning location of the treelawn, width of the right-of-way, and distance of the building setback.

Designing for Root Control

Tree roots are notorious for buckling the pavement of sidewalks and streets. Uneven surfaces may pose a hazard to pedestrians and motorists. Researchers are looking into several methods of preventing, or at least delaying, this type of damage. Planting time offers the best opportunity to prevent conflict between trees and pavement. Here are some techniques that may minimize the damage:

- Select species that best fit the available space (Fig. 5).
- Obtain trees produced on rootstocks with inherently deeper growth.
- Choose trees with columnar root balls.
- Install planting hole liners, deflection barriers (Fig. 6), pervious barriers, and chemicals barriers, to inhibit shallow root development.

Deflection barriers may be plastic sheeting that diverts roots away from a paved surface. Properly designed underground planters or planting hole liners can also block lateral growth, and force roots deeper. Pervious barriers could be fashioned from loosely woven plastic fabric or geotextile situated in the soil to prevent roots from penetrating into areas where they would cause damage. Chemical barriers, such as copper screen or geotextile impregnated with herbicide, inhibit growth in the tips of roots as they grow near the barrier.

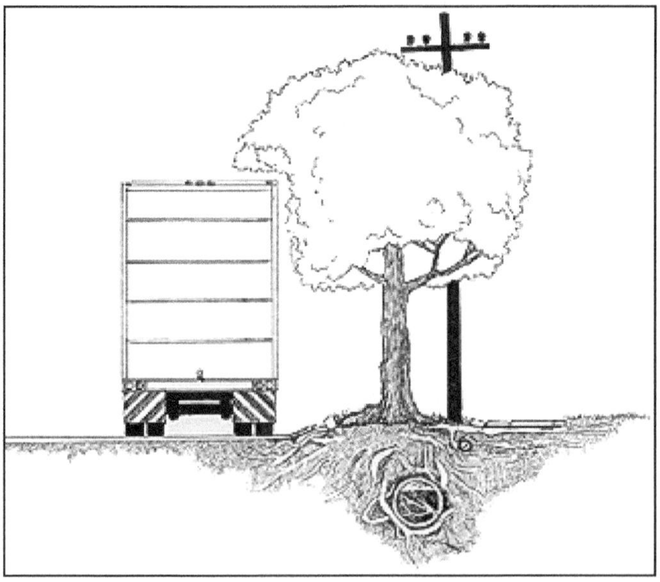

Figure 5: Careful attention must be given to site conditions, either existing or proposed, when selecting the trees to be planted.

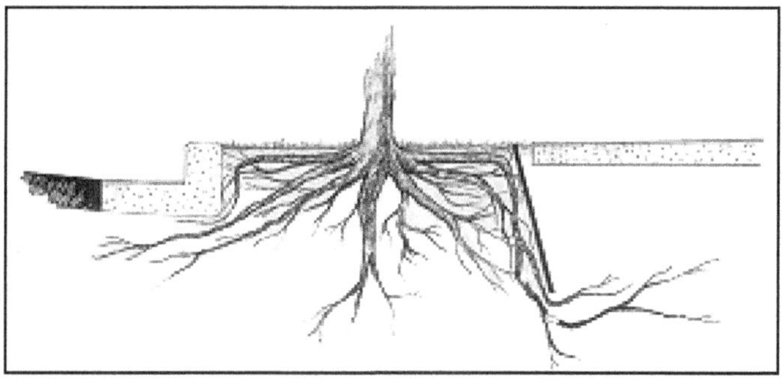

Figure 6: Deflection barriers direct the lateral root growth.

PLANTING AND EARLY CARE

Handling before Planting

Trees should be carefully inspected at the nursery or upon delivery to be sure they meet specifications for species, root quality, top conformation, and health. They may need to be acclimatized to their new site, a process known as hardening off. Both roots and tops must be kept moist and protected from temperature extremes.

It is best to plant bare-root deciduous trees and shrubs soon after being received. However, they can be held for a few weeks, if kept cool so neither roots nor buds become active. Balled-and-burlapped (B&B) and container grown trees can be planted almost any time, although it is wise to plant B&B plants soon after they are dug. Root balls and containers should be protected from the sun so that periphery roots are not injured.

Preparing the Planting Hole

In most soils, dig the planting hole one to two inches shallower than the depth of the root ball. In sandy soils, the hole can be as deep, or almost as deep, as the root ball. To prevent settling, make sure that the bottom of the hole is firm. The base of the trunk should be high enough to set slightly above the soil and to prevent water from collecting around it after rain or irrigation.

The hole should be about twice the diameter of the root ball, so backfill soil can be placed easily. For bare-root trees, the hole only needs to be large enough to take the roots without crowding. Roughen the sides and bottom of the hole with a shovel to enhance root penetration into the surrounding soil.

Preparing and Setting the Tree

Prune back the dead, diseased, broken, and twisted roots of bare-root plants to healthy tissue. For container grown plants, remove most of the roots that are matted at the bottom or are circling around the root ball.

Straighten the remaining ones. Roughen the outside of the root ball to improve contact with the fill soil. A number of factors affect how a tree should rest in the hole. Since some of these may conflict with one another, priorities will be needed. Set the plant so that it will be viewed to best advantage. Place the crook of the scion of budded (grafted) trees toward the afternoon sun to minimize bark injury. If this is not possible, shade the area or paint it with white exterior latex paint. Orient the side with low branches away from the areas of high activity. Point the side with the most branches into the direction of prevailing winds.

If none of the above are problems, orient the plant with the largest branch or most branches away from the afternoon sun. The less developed side will benefit from more sunlight. For greatest stability of bare-root trees, place the largest root in the direction away from prevailing winds. If the top of the tree is not vertical when the root ball or trunk base is straight, tip the root ball to bring the trunk more upright.

Backfilling

Backfill the hole with the original soil, unless it is in extremely poor physical condition. Amending backfill soil with organic or inorganic materials has seldom proven to be of physical benefit even during the first years. Backfill amendments may contain nutrients or organisms that are beneficial to a newly planted tree, but amending the backfill is not the easiest or cheapest way to obtain such benefits.

If backfill is amended, the final mix should be at least 40 to 50 percent amendment by volume and mixed thoroughly with the soil. Sand should not be used as an amendment, and fertilizer should not be added to the backfill or placed in the planting hole.

Set the roots or root ball on a firm base. Before putting the root ball of a B&B plant in the soil, cut the fabric covering in several places to make it easier for roots to penetrate. Work the soil in among the roots of bare-root and container plants so their roots are not compressed into a tight mass, but are spreading and supported by the soil. After adding each three to four inches of soil, firm it gently with one foot. Before adding the last layer of fill soil to a B&B plant, cut off any loose fabric that would remain exposed above the ball. Gently fold it down and cover it with the last layer of fill soil. Exposed fabric can act as a wick and dry the root ball. Do not cover the rootball with backfill soil, it could keep water from wetting the root ball.

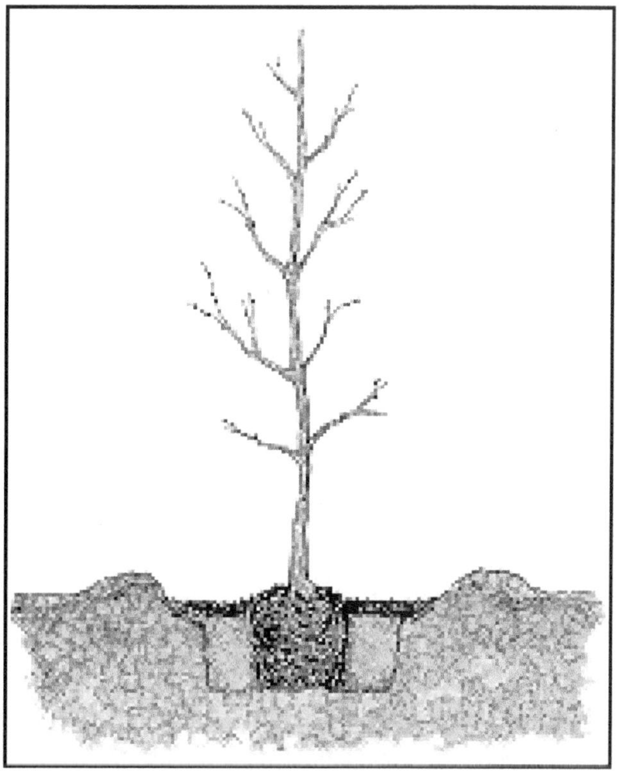

Figure 7: Planting detail.

Form a low berm around the edge of the planting hole and a higher one a foot farther out (Fig. 7). Water the inner basin to be sure the soil is moist and to help it settle. Double check the planting depth. If the original soil is at or below the fill soil, raise the tree. Gently lift bareroot trees by the trunk. Support B&B and container plants from below the root ball with a shovel while lifting gently on the trunk. Raise the tree two to three inches higher than finally wanted to allow for settling. Water the outer basin if the soil appears to be dry.

Staking and Guying

Trees that cannot stand upright on their own need to be braced with tall stakes or guy wires. Many nursery trees-particularly those in containers-are grown close together, tied to a single stake, and have their lower branches removed. As a result, they are unable to stand without support. The trunks of some trees are so spindly that they would bend below guys; these must be supported by stakes. if staking or guying is necessary, install the stakes or guys after planting but before watering the outer basin.

Two support stakes with one flexible tie near the top of each will hold a tree upright, provide flexibility, and minimize trunk injury and deformation (Fig. 9). Support stakes should be a little higher than needed to hold a tree upright in calm weather and allow the trunk to return to vertical after being deflected by hand or wind. Tie material should contact the trunk with a broad, smooth surface and have enough elasticity to minimize trunk abrasion and prevent girdling. Although widely used, wire inserted through hose can cause injury in windy areas. Support staking is temporary and should be removed soon after a tree can stand alone.

Most young trees that are exposed to moderate wind and weather concentrate their energy on growing strong enough to remain upright. They develop sturdy trunks that are tapered to bend without breaking, and strong root systems to hold trunks upright. It is wise to select such trees and to provide them with as much flexibility as possible when planted. Inspect stakes, ties, and the general condition of newly planted trees several times during the growing season to ensure tree health and well being.

If wind is not a problem, place the stakes in a way that provides maximum protection from traffic and equipment. In situations with moderate to strong winds, an imaginary line drawn between the support stakes should form a right angle to the most critical wind direction. The ability of stakes to withstand wind, particularly when the soil is wet, can be increased by connecting the two stakes with a wood cross tie partially or entirely below the soil surface (Fig. 9). Place the cross tie to the lee (downwind) of the stakes, being careful not to injure roots near the surface. If a single stake is used, it should be placed on the windward side of the tree to minimize trunk injury.

Avoid placing a tie on the trunk closer than 30 inches from the tip, especially on trees not previously staked. This arrangement will subject the trunk to maximum stress at the tie and increase the chances of it being deformed or broken by the wind. A flexible auxiliary stake of light spring steel or bamboo may be needed to support the trunk terminal. In a densely-headed tree, thinning out some of the branches will lighten the top and reduce its resistance to the wind.

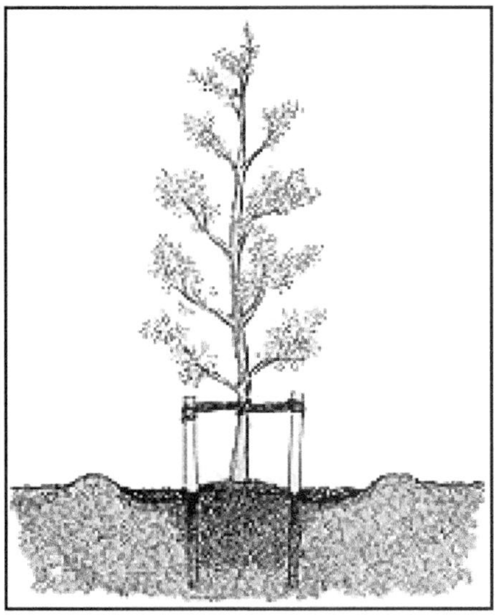

Figure 8: Protective staking.

Young trees are particularly subject to vandalism, and few staking devices have successfully prevented damage. Many agencies plant larger trees to reduce tree damage. Planting trees with fairly dense skirts of small twigs along the lower trunk obscures vulnerable trunks. Seattle has been fairly successful in protecting young trees in vandalism-prone areas by using five-eights-inch thick metal reinforcing bar ("rebar") driven four to five feet into the soil and extending five to six feet above ground. A two and one-half-inch or larger diameter tree is tied at three heights to the stake. The stake is removed after one year.

Even trees that can stand without support need protection from equipment operations and other activities. To protect the trunks, set two or three short stakes just outside the root ball with 18 to 24 inches exposed. Other trees may be poorly anchored by their roots, although the trunks seem sturdy enough to hold their tops upright. Anchor these trees by guying their trunks or tying their trunks to stakes similar in height to protective stakes (Fig. 8).

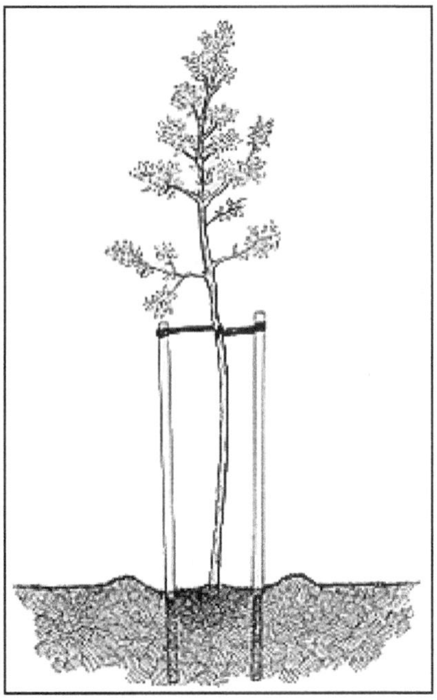

Figure 9: Support staking

Pruning and Training

In years past, severe pruning after planting was thought to be necessary. However, newly planted trees grow quite well if their tops are pruned lightly or not at all. On the other hand, up to 25 percent of the tops of newly planted trees can be removed without greatly affecting total future growth.

The key to pruning is to encourage the growth of several large permanent branches, called "scaffold branches," that will ultimately form the basic structure of the mature tree. At planting, remove branches growing close to potential scaffold branches, crossing one another, damaged, and those having included bark in their attachments. Also, remove or reduce the size of large branches that are too low.

Central-leader trees, conifers, and some hardwoods, require little pruning to grow strong and well shaped (Fig. 10). However, species that become round headed may need considerable pruning the first few years to ensure the desired height of branching and strong branch structure.

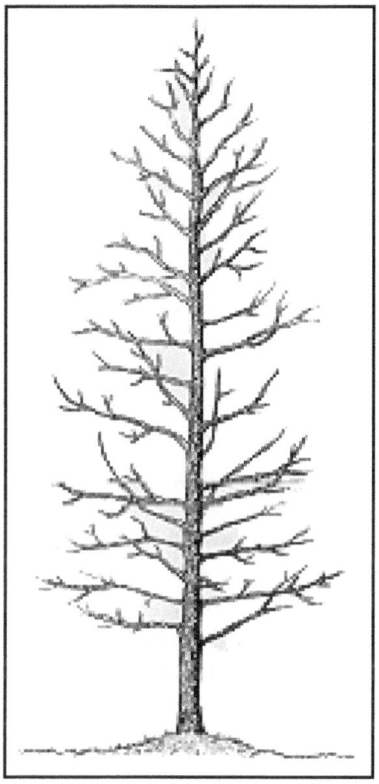

Figure 10:

During a tree's early years, frequent observation and directional pruning will guide the growth along a desirable course. This will improve the appearance and structural strength of the mature tree, and can greatly reduce the need for severe corrective pruning later. To do this, retard or remove large branches where they will not be wanted when the tree is mature. Temporary branches can protect the trunk from sun and vandals, as well as increase trunk growth and taper. Maintain temporary branches by encouraging twigs or keeping small branches cut back along the trunk below large permanent branches (Fig. 11). They should be kept small, and removed as the tree matures. This practice is in line with the observation that a tree is better able to withstand wind, snow, and ice if at least one-half of its foliage originates from branches along the lower two-thirds of the trunk.

The less a branch or tree is pruned the more total growth it will make. So, pruning should be used to control growth. For example, large branches that might be too low or that might compete or interfere with more desirable branches need to be pruned back or removed. Branches that will become the main scaffolds, particularly of large-growing trees, should be at least one-third smaller in diameter than the trunk where they arise, and should be vertically spaced at least 18 inches apart along the trunk (Fig. H5).

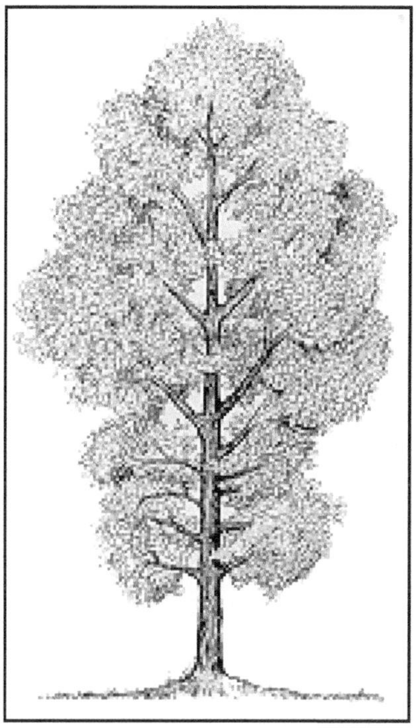

Figure 11.

If a branch is in a desirable location but is large compared to the trunk or other branches, it can be pruned to reduce its leaf area and slow its growth. This increases the strength of the branch attachment, because the trunk can grow around the branch and hold it more securely. The relative sizes of a branch and the trunk are more important than the angle of attachment in determining the strength of that attachment, although a sharp angle of attachment is a good indication that branch size may be a problem.

A tree will be more open and retain its natural form, if branches are removed completely (thinned) in contrast to being headed or stubbed back (Fig. 13). Heading concentrates subsequent growth just below the pruning cut and results in dense foliage with weakly attached branches

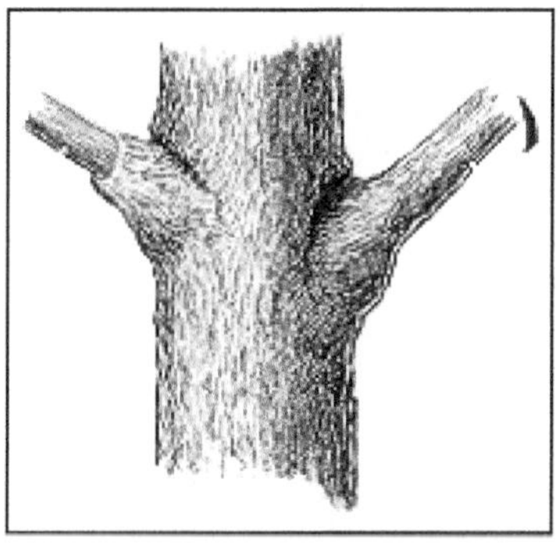

Figure 12

When removing a branch, make the final cut just outside the branch bark ridge in the crotch and the collar below (Fig. 12). This type of cut minimizes the size of a wound and the possibility that the trunk will decay. Painting pruning wounds has not been shown to reduce decay, and may even increase it.

Fertilization

In some tests, newly-planted trees have responded to fertilization following planting, and in others response occurred only the year after fertilizing. Unless experience indicates otherwise, fertilization after planting is recommended. The benefits may be well worth the modest cost.

In most soils, nitrogen is deficient and the only added nutrient to which trees will respond. Nitrogen fertilizers are or become water soluble, so they can be applied on the surface and be carried into the soil with water.

After the first application, apply fertilizer in the late summer or fall, so it is available for spring growth. Water after fertilization being careful not to wash the material to the bottom of the watering basin. This could create a toxic level of chemicals.

Slow-release forms of nitrogen may be advisable in sandy soils. In alkaline soils the availability of iron and manganese may be so low that certain plants exhibit the typical pale leaves with narrow (iron) or wide (manganese) darker green bands along the veins. Chelated forms of these nutrients can be applied to the soil or foliage according to the manufacturer's directions. If the trees respond to the chelated materials, a longer term and less expensive solution is to acidify the soil with sulfur.

Water Management

Young trees need adequate water to become established. Although rainfall may be adequate in some areas and in some seasons, additional water may be needed, particularly after planting when root systems are limited. After the initial watering at planting, deciduous trees do not need additional water until the first leaves have reached full size, or even later.

Pay attention to container and B&B plants that are in leaf. During the early part of their first growing season, they will need water more frequently than at the nursery. Container root balls hold less water when in the ground than in the container. And, the root system of B&B plants are greatly reduced when they leave the nursery.

At first, water the root ball and the backfill only enough to re-wet the soil. After the first few weeks, lengthen the irrigation interval. The area beyond the backfill needs to be watered only every two to three weeks. By the end of the season, the irrigation interval should be three to four weeks. Of course, rain must be taken into account. Overwatering can be as serious as under watering.

If winter rainfall has been adequate, trees should not need to be irrigated more than once a month during their second year. After that, most trees can survive with only one or no irrigations, although they would probably do better with monthly applications.

Soil Management

It is essential to maintain an area free of turf and weeds around tree trunks, because turf and weeds compete for water and nutrients, and some produce chemicals toxic to other plants. A small turf-free area around a tree also reduces the need for mowers to come close. This clear area must be at least one-foot in radius. Larger areas add little benefit. After four or five years, tree roots are extensive enough that other plants close to their trunks are not a problem, although mower operators should still exercise caution.

A three to four-inch-thick mulch, material placed on the soil surface, controls most weed seedlings. In addition, mulch protects the soil from compaction and erosion, conserves moisture, moderates soil temperatures, provides an all-weather surface for walking, and allows plants to root in the fertile and well-aerated surface soil. Keep mulches at least two inches away from the trunks of trees to minimize disease and rodent damage. A wide range of organic and inorganic materials can be used. Unfortunately, most bulky mulches are not satisfactory for trees in turf that is mown, but some plastic planting-basin covers will reduce turf and weed growth.

Power string trimmers are great for edging around borders and plants, but many trees have been fatally girdled by them. Trees are particularly vulnerable to girdling during the growing season when the cambium (the growing portion of the trunk just under the bark) is active. The cut weeds or turf return to their original more upright position obscuring the damage until the tree begins to weaken.

Figure 13: Thinning cuts (above) are much more desirable than heading cuts (at left).

Pest Management

Most trees get off to a good start, but serious problems can be avoided or minimized if the trees are periodically inspected. Inspect trees for the beginnings of insect and disease damage. At the same time, take care of any staking problems, do needed directive pruning, check on tree moisture status, and identify any other problems. The inspection should take only a few minutes per tree, but prompt action on any problems encountered would pay big dividends in healthier, stronger trees.

MAINTENANCE OF ESTABLISHED TREES

Maintenance of established trees differs significantly from that of newly planted trees. if a young tree fails, the tree itself is lost. But, improper maintenance or neglect of an established tree may result in damage to property or people. In addition to losing the benefits of a well-established tree, proper maintenance directly affects the safety, health, and welfare of urban inhabitants. Maintenance involves several categories of activities, including: scheduling, pruning, controlling insects and disease, fertilizing and aerating, as well as removing trees and stumps.

Tree Hazard Inspections

Most tree hazards can be prevented by regular checkups and proper treatments by tree professionals. To determine the scope of work, a field survey should answer thirteen questions. These questions were taken from Alex Shigo's Tree Hazards: Your Trees Can Kill:

1. Target: If the tree falls will it hit cars, houses, power lines or people? If so, the need for immediate action becomes much greater.
2. Architecture: Has the tree grown beyond its normal form into a dangerous form?
3. History: Has the tree lost large branches recently?
4. Edge Tree: Were neighboring trees cut away recently leaving tall trees at the edge?
5. Dead Branches: Are there dead tops or branches? Is the tree dead?
6. Cracks: Are there deep, open cracks in the trunk and branches? Cracks are major starting points for trunk and branch failures. Crack drying is just as important a factor leading to failures as over loading and decaying wood.
7. Crotch Cracks: Are there deep, open cracks below joining stems?
8. Living Branches: Do living branches bend abruptly upward or downward where tips of large branches were cut off? Living branches may pull out of trunks that are weakened by rot or cracks. Long periods of hot, dry weather may dry out the rot or cracks and weaken the union of the branch on the trunk. Beware of large branches on rotten or cracked trunks.
9. Topping: Are large branches growing rapidly from topping cuts on big trees? Sprouts that lean away from topping cuts have weak attachments. Sprouts near the edge of a cut may roll inward as it grows and further weaken the attachment.
10. Storm Injury: Are there broken branches, split trunks, or injured roots? Are branches close to power lines?
11. Root Rot: Are there fungus fruit bodies (mushrooms) on roots? Were roots injured by construction?
12. Rots and Cankers: Are there hollows or cankers (dead spots), some with fungus fruit bodies? is the tree leaning?
13. Construction Injury: Have roots, trunk, or branches been injured? Is there a new lawn or garden over injured roots? Water and fertilizer applied to new lawns over injured roots are absorbed by the tree's smaller, non-woody roots. The water and fertilizer stimulate the growth of the fungi that are rotting the injured woody roots. While the woody support roots grow weaker, the tree top gets larger and heavier. Once the root structure is sufficiently weakened, a moderate storm could cause the tree to fall.

Scheduling

Most public agencies use two approaches for deciding when to do tree maintenance. They respond to crisis situations that demand immediate attention, and they schedule maintenance according to a well planned program. While crises will surely occur, urban forestry programs should strive to maximize the amount of scheduled work. Usually, well managed and adequately funded programs are most efficient about scheduling.

Many urban forestry organizations in the U.S. have become victims of crisis management, responding to one unplanned situation after another. A crisis in might include: a broken limb, an uprooted tree, an obstructed stop sign, or angry people-citizens and civic leaders. Often these situations must be resolved before other maintenance can be performed. This creates an unpredictable drain on already limited resources.

Scheduled or programmed tree management allows field crews to work in predetermined zones of the city. This approach reduces transit time and increases productivity. Crews remain in a zone until all necessary work has been completed. Because they inspect each tree and perform the appropriate maintenance, the chances of a safety problem developing are significantly reduced. Street tree inventory information for a particular zone can also be gathered and updated while the crew is there.

Information from Santa Maria, California and other western cities shows that the cost per unit of maintenance is generally twice as high with crisis management as it is with the scheduled approach.

Closely related to work schedules are specifications. Because many communities have increased their use of contractors, there is a growing need for a clearly identified scope of work. Some situations lend themselves to a per unit bid and others to an hourly time and material estimate. Specifications communicate needs, form the basis of bids and serve as a standard for evaluating the quality of the completed work. In-house crews also require clearly defined expectations and objective evaluation of their work. Specifications will help them, too. When developing tree work specifications, consider the following material.

Pruning Guidelines

In writing a work plan for in-house staff or contractual specifications, the purpose and scope of pruning needs to be clearly identified.

Trees respond in predictable ways to pruning. By studying these responses arborists have developed pruning practices that preserve or enhance the beauty, structure, and function of trees.

The Western Chapter of the International Society of Arboriculture (ISA) has developed standards for pruning that provide general guidelines. Of course, each tree has a unique form and structure, so pruning needs may not always fit strict rules. However, it is the arborist that must take responsibility for special pruning practices that vary greatly from the standards. Pruning Standards is available from the Western Chapter, International Society of Arboriculture, P.O. Box 424, St. Helena, California, 94574.

The following descriptions of the various types of pruning mature trees were taken from these:

Crown cleaning or cleaning out is the removal of dead, dying, diseased, crowded, weakly attached, and low vigor branches and watersprouts for the tree crown.

Crown thinning includes crown cleaning and the selective removal of branches to increase light penetration and air movement into the crown. Greater light and air movement stimulates and maintains interior foliage, which improves branch taper and strength. Thinning reduces the wind-sail effect of the crown and the weight of heavy limbs. Thinning the crown can emphasize the structural beauty of the trunk and branches, as well as improve the growth of plants beneath the tree by increasing the light penetration.

When thinning the crown of mature trees, no more than one-quarter of the foliage should be removed. At least one-half of the remaining foliage should grow from branches that originate in the lower two-thirds of the tree. Removing laterals from a branch requires a similar approach. Try to retain inner laterals and leave the same distribution of foliage along the branch. Trees and branches pruned in this way have stress more evenly distributed.

Removing the inside lateral branches also produces an effect known as "lion's-tailing." By removing all the inner foliage, weight is moved to the ends of the branches, which may cause the branch structure to weaken and limbs to break. Greater light penetration may cause sunburned branches and stimulate watersprouts.

Crown reduction, also known as drop-crotching, decreases the height and spread of a tree. Thinning cuts will maintain the structural integrity and natural form of a tree, and delay the time when it will need to be pruned again. To make this type of cut, prune the branch back to its point of attachment or to a lateral that is at least one-half the diameter of the cut being made.

Crown restoration improves the structure and appearance of trees that have been topped or severely pruned using heading cuts. Select one to three main branch stubs that will grow to reform a more natural looking crown. Thinning or even heading may be required to match the weight of the new branches with the strength of their attachment. Restoration may require several prunings over a number of years.

Crown raising provides clearance for buildings, vehicles, pedestrians, and vistas by removing lower branches. It is important to maintain at least one-half of the tree's foliage on branches that originate in the lower two-thirds of the crown. This ensures a well-formed, tapered structure and uniformly distributed stress. When pruning for view, it's better to open "windows" through the foliage of the tree, rather than severely raising or reducing the crown.

Traffic Control

Take measures to expedite public passage through or around the work area and to prevent accidents, damage and injury. It is also important to give adequate warning to vehicles and pedestrians of any dangerous conditions that may be encountered.

Restricting traffic flow should be done only with the consent of the appropriate municipal official, such as city arborist, public works director or traffic engineer.

Tree removal should be conducted in such a manner as to insure continuous traffic flow in the street at all times. Restrict tree operations to one side of the street. Try to maintain two lanes for traffic. In situations where it possible to maintain only one lane of traffic, station flaggers at both ends of the operation area to control the flow in both directions.

Be sure to identify the agency or contractor responsible to providing traffic control personnel and equipment, such as hats and signs.

Traffic control must comply with the requirements of the American National Standards Z133 "Safety Requirements for Tree Pruning, Trimming, Repair or Removal," published by the American National Standards Institute, Inc., 1430 Broadway, New York, New York, 10018.

Insect and Disease Control

Because trees planted in urban areas are frequently subjected to hostile growing conditions, they are not always in the best of health. Trees under stress are much more susceptible to attack by harmful insects and diseases and are more likely to exhibit symptoms of the adverse environment. The type, severity and duration of a particular insect, disease or environmental problem will vary greatly, depending upon the tree's location, climate and other environmental factors. Many excellent reference texts and highly qualified local authorities can be consulted regarding specific problems. Here are some ideas to keep in mind when identifying problems and seeking advice on corrective measures.

It is a good idea to examine trees and other vegetation on a regular basis-whether or not a significant problem currently exists. These examinations help identify problems sooner, by providing a history of the plants that serves as a point of comparison when a serious problem develops.

Several types of local authorities are available for technical assistance. For free advice call on the county cooperative extension service, agricultural commissioners office and municipal parks or forestry offices. Information is also available for a nominal fee from local retail nurseries, tree services, pest control contractors or arboricultural and horticultural consultants.

These experts will need information in order to diagnose the problem. Look closely at the leaves, twigs, trunk, and the ground around the tree. Are there any deformed or dead parts of the plant? Overall, how do the trees appear compared to those close by, a little farther down the street and in a completely different part of town? If the plants seem similar, chances are that any irregularities are typical of the species or locality, and may not be serious enough to treat.

If one tree is significantly different from others in the immediate area, it will be wise to check with someone. Before contacting an expert, document as many of the symptoms and as much of the tree's history as possible. Include a description of any changes in the tree's environment, such as construction, new walks, utilities, and any special treatments of the turf, ground cover or nearby vegetation.

Figure 14: Regular tree inspections and follow-up maintenance or removal are necessary to minimize potential hazards.

Next, call an authority and describe any symptoms in detail. It is usually helpful to have a sample in hand when describing the problem over the telephone. If it is not possible to get a clear identification of the problem and recommendation for treatment over the telephone, the expert will have to see a fresh sample. Ask to bring a sample to the expert or ask the expert to examine the site. Expect a charge for site visits.

In preparing a sample for examination, consider several factors. The sample should be as fresh as possible and a manageable size. Label the sample with name, address and telephone number, description of the plant and its history. Try to cut the samples into pieces under a foot long and place them in a sealed plastic bag. If possible, bring normal appearing shoots as well as affected shoots. Place insects in a small bottle of rubbing alcohol, and bring along a sample of the plant they were found on. When it comes to samples, different experts may have different needs, so it will pay to ask for specific instructions and follow them closely.

If the problem sounds more complex than you expected or involves several problems simultaneously, it may be worth spending some money to have a specialist develop an assessment and annual treatment plan. This can be very cost effective, particularly when construction injury is involved, because a less detailed analysis may yield recommendations on treating the symptoms rather than the real problem.

Fertilization and Aeration

Many street trees grow in an extremely hostile environment both above and below ground. The physical, chemical, and biological conditions of the soil may need to be managed just as elements of the trees' above-ground environment are managed. Two activities-fertilization and aeration-are the primary methods of managing this area below the soil surface. Both are vital to the health and longevity of the urban forest.

Trees need adequate amounts of nutrients, water and air in the soil. Without these elements trees will grow with less vigor and will be more susceptible to secondary problems, such as attack from insects and diseases. Properly nourished trees will be more able to withstand the attack of insects and diseases and tolerate the adverse growing environment afforded by most urban situations.

Fertilizers - Plants require at least sixteen chemical elements for proper growth and development. Three of these elements-carbon, hydrogen, and oxygen-are provided by air and water. The other essential elements are obtained by the roots from soil.

The specific fertilizers applied should be based on need. Every site is different, so soil tests should be conducted every two to three years. Cooperative Extension Services can analyze samples of soil taken from each tree site and make specific recommendations.

Trees most commonly require large amounts of nitrogen. Because it is easily leached and often volatile, if necessary it may be necessary to apply nitrogen once or twice a year. Other chemical elements, such as calcium, phosphorous, potassium, and magnesium, seldom need to be added.

Acidity and alkalinity are measured on a pH scale, with lower pH indicating greater acidity. Soil pH affects the availability of certain nutrients, particularly iron and manganese, so it's important to maintain the pH between 5.5 and 7.0 for most plants.

Trees may show signs of a specific nutrient deficiency, such as interveinal chlorosis, even though soil tests indicate the presence of that nutrient. in this case, the nutrient may be present in the soil in a form that is not available to the tree. Or, nutrient uptake by trees may be inhibited by soil compaction, poor drainage or poor aeration. An analysis of the foliage may be required to determine the plants needs.

These guidelines were taken from the National Arborist Association Standard for Fertilizing Shade and Ornamental Trees. Be sure to study the complete NAA standards before proceeding with a fertilizer program.

Generally, the ideal time of the year to fertilize is in the late summer or fall. Fertilizer can also be applied after leaves open fully until early July. Avoid treatment with readily available inorganic nitrogen in heavy doses between July and September, because it could cause a late flush of growth that would not harden off before freezing weather.

There are a variety of methods for fertilizing trees. Fertilizers can be broadcast over the surface of the ground, poured into holes drilled into the soil, injected as a liquid into the soil, sprayed onto the leaves, or injected directly into the trunk.

Surface applications are most commonly used for fertilizers that contain only nitrogen. A properly calibrated, mechanical spreader broadcasts fertilizer over the ground in a pattern of concentric circles or linear strips beginning two or three feet from the trunk and extending five or ten feet beyond the drip line. Care should be taken to avoid excessive overlapping.

Apply fertilizer when grass is dry and then wash the grass off thoroughly with a lawn sprinkler or irrigation system. Grass blades could be burned by fertilizer that becomes slightly wet after a light rain or dew.

Drill holes allow relatively insoluble fertilizers, such as phosphorous, to be placed close to roots. Drill holes also provide valuable aeration in compacted soils. Arrange drill holes in concentric circles beginning about two to three feet from the tree trunk and extending five or ten feet beyond the drip line. Space the holes two to three feet apart. Drill eight to twelve inches deep, depending on the species of tree, pattern of root growth and type of soil. In compacted soil situations, it makes sense to drill deeper holes. Distribute the fertilizer uniformly among the holes after mixing it with peat moss, calcine clay, perlite, small crushed stone, sand or other appropriate soil amendment. If the area under the canopy is restricted, reduce the amount of fertilizer in proportion to the area fertilized or the number of holes drilled. Keep the fertilizer at least four inches below the surface.

If shrubs live within the root area of the tree, take care not to drill holes closer than six inches to the crown of the shrubs.

In turfed areas, the hole can be closed by pressing from different angles with the heel of a shoe, or by filling it with a plug of sod. Irrigation will help prevent injury to turf. Be careful not to flood the area, because it might bring dissolved fertilizer to the surface, injuring the turf.

Liquid injection also puts fertilizers into direct contact with tree roots. Drive the soil probe or lance eight to twelve inches into the soil. Make injections about two and one-half feet apart beginning about two to three feet from the tree trunk and working out to five or ten feet beyond the drip line. The soil lance should have three or four horizontal discharge holes in its point. Use a hydraulic sprayer that generates 100-200 pounds of pressure. inject one-half gallon of fertilizer into each hole.

Foliar spray cannot provide an adequate amount of all necessary minerals. However, spraying liquid fertilizer onto the foliage can correct minor deficiencies, especially iron. Micronutrient applications should be applied when first leaves reach full size. Within two to eight weeks, the tree should respond with greening foliage and normal bud growth on affected shoots. Species, age, time of year, soil conditions, and the severity of the deficiency will all affect the response time. One or two applications per year will generally control deficiencies, but under some conditions, maintaining healthy growth may require several treatments each year.

Trunk injections and implants are used for deficiencies in specific elements, such as iron and manganese. Trees growing where there is limited soil surface under the drip line may also need injections or implants. This treatment requires a clean, sharp drill to penetrate the bark as low as possible on the trunk. For implants, the penetration should extend into the xylem tissue. The best time for this type of fertilization is early in spring before growth starts. Wounds made at this time of year close rapidly. Trees under four inches in diameter should not receive injections or implants, Be careful not to administer injections or implants when soil moisture is low, because it may cause severe foliar burning.

Aeration - Pore spaces in the soil supply trees with air, as well as water. Some soils are naturally tight. Other soils may start out well aerated, but through the actions of pedestrians, vehicles, and even water, pore spaces are compressed. Soil compaction is far easier to prevent than it is to remedy. Preventive measures include limiting pedestrian and vehicular access, and mulching exposed soil to minimize compaction caused by the impact of rain or irrigation water.

If planting into a site with poor aeration, first rip or deep plow the soil, replace the soil with a suitable medium, or try to select trees tolerant of low soil aeration.

Mechanical aeration is possible, but at best, it's a temporary solution. Physically separating soil particles merely buys time until the pore spaces cave in again. Physical aeration is similar to the sub-surface fertilizing methods mentioned previously. An auger or high pressure water probe breaks up soil particles that have been pressed tightly together. Surface penetrations should be eight to ten inches deep two to three feet apart and extend up to one-fourth of the radius beyond the drip line of the tree. Openings should be made away from the trunk and main roots to avoid injury to the tree. For the benefit of aeration to continue, the procedure must be repeated on a regular basis Aeration may also be improved by adding an organic mulch on the soil.

Tree and Stump Removal

Trees and stumps are removed for several reasons. Tree inventory information can supply a set of criteria to guide tree removal decisions. Safety is perhaps the most important reason, and considering public liability for injuries, this becomes ample justification for removal. Trees could be removed to prevent the spread of harmful insects, disease or vegetation that may be harmful to the environment. Finally, aesthetics may dictate removing a particular tree.

As opposed to pruning, removal refers to disposing of the entire tree, including the leaves, branches, trunk, stump and major roots. When removing trees and stumps consider the following guidelines.

Stumps are defined as the lower portion of a tree-up to a maximum height of four feet-that remains after the foliage, limbs, branches, and the upper part of the trunk have been cut off.

Tree removal.

The stump removal area is generally between the sidewalk and the curb, or between the curb and another curb if growing in a median strip. For trees planted in an open space, the area for stump removal is that which causes the surface of the ground to be higher than the adjacent grade. Roots within the stump removal area should be taken out as deep as 24 inches below the finished grade. This is especially true of roots that are exposed at grade and those adjacent to or growing over a curb or sidewalk. All exposed surface roots beyond the stump area should be removed to a depth of twelve inches below grade. Soil that has been displaced by deeper roots shall be leveled to the existing grade.

The hole or depression resulting from the removal work should be filled with topsoil. Do not use chips, leaves, brush, sawdust or tree debris as filler. This organic matter would eventually decompose allowing the soil to settle. Because any type of soil will settle a little over the first several months additional soil of the same quality should be added. The entire area should be made level with the existing grade.

Stump removal

Cleanup involves removing all soil, leaves, twigs, or trash resulting from the work. Remove debris daily to approved disposal sites. Sewer systems, landfills and ocean dumping sites are not appropriate destinations for tree debris. Depending on the nature of the material, it could be used for lumber, arts and crafts, firewood, mulch or compost.

Damage to property resulting from this work should be repaired within a reasonable time. Before starting work at a new site, survey the condition of the area, including adjacent proper ties. During clean up, survey the area again to identify damage caused by the tree work.

Figure 15: Cleanup activity may include brush chipping.

Overhead power lines pose a potential danger during removal operations. Exercise extreme caution to avoid damage to the lines or workers. If damage does occur, report promptly to the City Arborist, designated inspector, utility company or police department.

PRESERVING TREES AFFECTED BY DEVELOPMENT

Mature trees provide many benefits to development sites. They enhance the aesthetic character of the area, give scale to the new buildings, provide shade, give a look of maturity to the landscape, and provide habitat for wildlife. it may be possible to retain trees with minimum forethought. However, preserving specimens that will survive and perform well in the new environment requires thorough planning, careful implementation, and adequate maintenance.

Effective tree protection is a long process. First, evaluate the trees to determine which specimens are suitable for preservation. Next, work with the planners and engineers to design improvements so that suitable trees are preserved. Then, monitor construction around the trees to see that the trees are not injured. Finally, routinely evaluate the trees to identify maintenance needs.

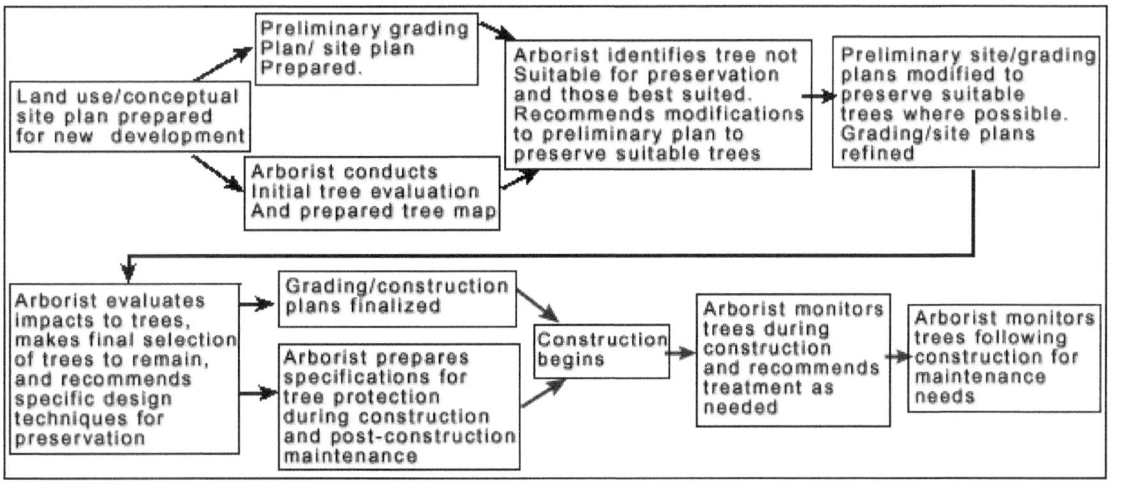

Identifying Trees to be Protected

Deciding which trees to preserve, and designing development around them often seems like a 'chicken or the egg' problem. The trees suitable for preservation must be identified before designing around them. On the other hand, the design strongly influences the selection of trees to be preserved.

One way out of this dilemma involves a preliminary evaluation of the trees to determine those that are suitable for preservation. Compare that information with conceptual site plans. Identify the trees suitable for saving, and modify the site plan to accommodate the desirable trees. Finally, work out the details of grading and preservation for trees that will remain (Fig. 14).

In most cases the costs to preserve trees are significant. Costs accrue because the land dedicated for tree preservation is unavailable for building, structures, and techniques to minimize damage to trees require extra design and construction attention, and affected trees must be maintained to support long-term health. These costs will be borne by the public, whether through increased taxes to pay for trees in public rights-of-way, higher homeowner association dues, greater commercial rents, higher home prices or direct payment for maintenance. Trees should be selected carefully, keeping their value and contribution to the new environment in mind.

In determining which trees are to be preserved, four factors must be considered:

- Suitability of the species to the new land use;
- Tree health and structural stability;
- Species tolerance to changes in environment; and
- Level of maintenance that will be provided following impact.

Species Suitability - The suitability of the species to the location and use of the site are important factors in determining whether to save a tree or not. For example, species that normally grow along streams, such as willow and alder, would perform poorly in a parking lot.

The tree should survive in the new landscape for a considerable length of time to warrant the cost and effort of protection. An oak that can be expected to survive for many decades has a greater value than a mature acacia that may only live for a few years.

Tree Health - Young, vigorous, healthy trees are the best candidates for protection, because they grow new tissue quickly and adapt readily to new environments. However, it is large, old trees that are most often the focus of preservation. Of course, it is possible to preserve old trees as long as they are healthy, but younger ones may give the best return on investment.

Vigorous trees usually have full canopies and healthy leaves. Three conditions indicate poor tree health. First, the leaves are small and pale for the species. Second, some of the branches are dead. Finally, most of the foliage arises from short twigs along the major limbs, known as epicormic growth. Trees with large cavities or other structural weaknesses are not good candidates for preservation, unless the problems can be alleviated by pruning, cabling or bracing.

Tolerance To Changes In Environment - The ability of the tree to tolerate injury and changes in the environment is another important factor to consider. Trees that regenerate roots quickly and have adaptations to control water loss seem to be better able to tolerate construction impacts.

For example, coast live oak (Quercus agrifolia), which has a stiff, cupped leaf with a thick cuticle, tolerates greater root injury than California black walnut (Juglans hindsii), which has a smooth, thin leaf. On the other hand, willows, which also have smooth, thin leaves, typically adapt well to new environments, because they have the ability to regenerate new roots quickly.

Environmental changes can cause tree decline even when the tree is not directly injured. Significant changes in water table levels or rechannelization of streams or runoff can seriously weaken or kill trees. Road fills placed over stream beds can raise water levels upstream. It is important to consider large scale alterations in the overall ecology of the area, as well as the specific changes that will occur next to trees.

Level Of Maintenance That Will Be Provided - Trees stressed by construction require considerable maintenance to prevent loss of vigor and attack by harmful insects and diseases. Irrigation, fertilization, pest and disease control, and pruning are all aspects of this maintenance. For example, Monterey pine trees (Pinus radiata) stressed by root injury may require regular irrigation and spraying for bark beetles or other insects. If those treatments cannot be provided, it may be better to remove the trees rather than trying to preserve them.

Responsibilities for maintenance of affected trees should be established early in the planning phase. If the trees will not be maintained, it is important to minimize stresses to the trees that would affect tree health.

Preparing a Tree Location Map

A tree location map is critical to successful preservation. Use a topographic map showing vegetation lines as a base. Early in the design process locate the trunks approximately. Later, the engineer's survey can pinpoint the locations and show the base elevation of the trunks. The engineer should project to the center line of the tree when plotting locations. An indication of the dripline for each tree would also be helpful. On 20-scale or larger maps, plot the circumference of the trunk for larger trees.

Be sure that tree locations show on all building site plans. Number the trees on the map so that all discussions can address specific trees. It is helpful to have a corresponding numbered tag attached to the tree, as well.

Assessing Potential Impacts to Trees

Evaluate the development plans around the trees to estimate the potential impact. This requires skill in reading plans and an understanding of construction procedures. The following plans and specifications should be reviewed:

- Topography and tree survey plan;
- Grading plan and specifications;
- Plot or development plan;
- Utilities placement and depth;
- Drainage plan; and
- Geotechnical survey.

The plans will provide information on the depth of cuts, fills, utilities, subdrains, and other excavations, as well as their distance from trees. The geotechnical survey describes the soil profile. It also specifies the compaction and over-excavation required under structures, pavement, and fills.

To better visualize the changes that will occur next to trees, draw cross-sectional views showing existing and future grades, and all improvements (Fig. 15). Based on the geotechnical requirements, the amount of soil work could also be estimated and recorded on the drawing.

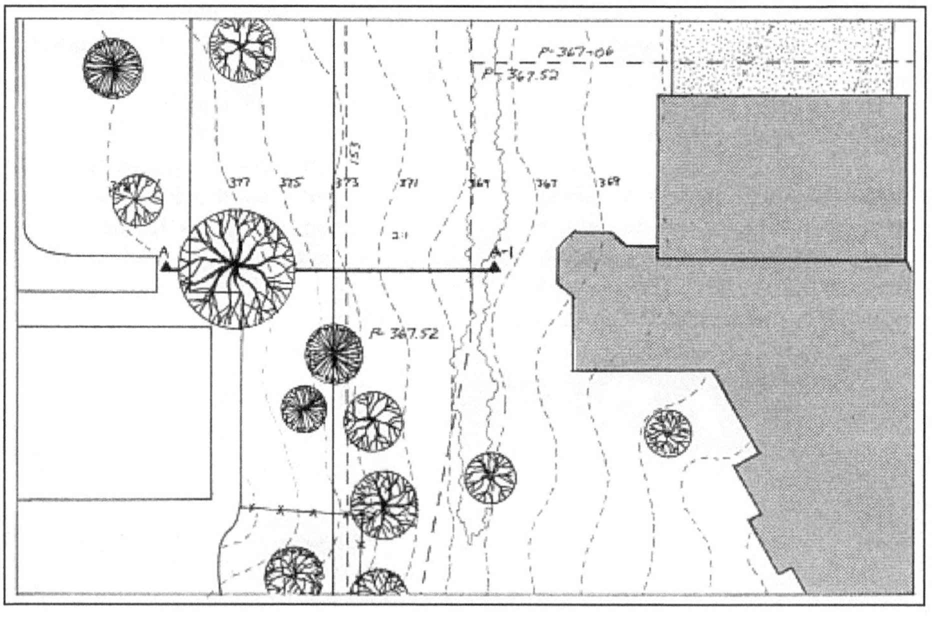

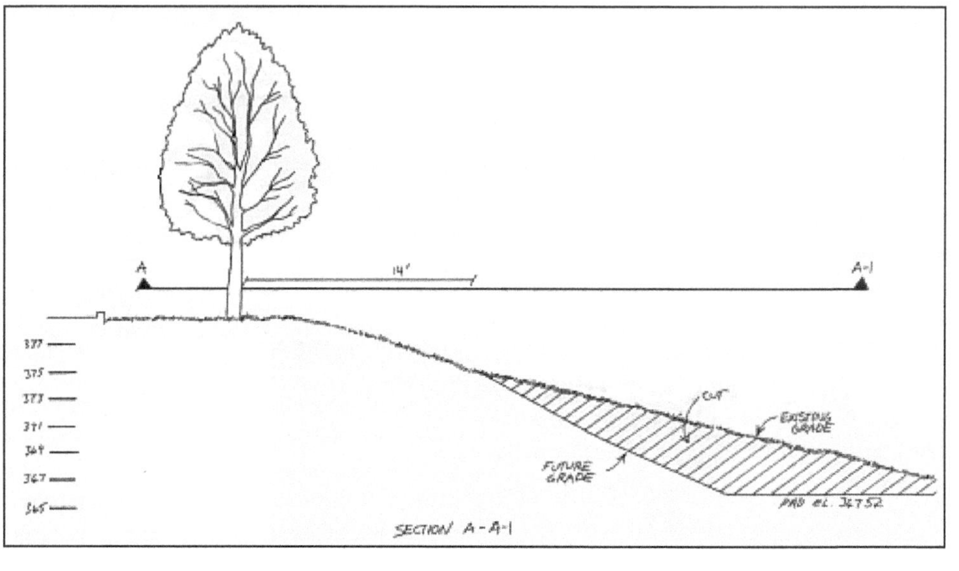

Soil Alterations During Construction

Soil favorable for root growth is a mixture of mineral particles, organic matter, air and water. About half of the volume of the soil is pore space containing varying amounts of water and air. To soil engineers pore spaces are known as voids. Although an asset to plant growth, voids reduce a soil's value as structural material. To stabilize the soil, engineers compact it to remove as many of the voids as possible. Compacting soils for construction involves removing the soil to a specified depth, mixing the soil with water, and replacing the soil in thin lifts (usually six to eight inches thick). Heavy equipment, designed to maximize the compaction, drives repeatedly over each lift. This activity hurts tree roots by first cutting them, and then by breaking down the soil structure.

Location of Tree Roots

To protect tree roots from damage, it is important to know where to look for them. Although root systems are often depicted as mirror images of the tops, they usually cover a much larger area. Roots can extend far beyond the dripline, as much as two to three times the diameter of the crown. The major portion of the absorbing roots system of a mature tree is within the top three feet of soil, and most of the fine roots active in water and nutrient absorption are in the top twelve inches (Fig. 16). Many trees form vertical sinker roots that arise from larger horizontal roots near the trunk of the tree. Sinker roots aid in water and nutrient absorption from deeper layers of aerated soil during times of drought.

Root patterns also are affected by topography and characteristics of the soil or substrate. Trees on slopes tend to have more roots on the downhill side. Roots of trees along streams will parallel the bank.

Developing a site is seldom possible without hurting tree roots to some extent. Even preliminary grading, stripping the site of debris and organic laden topsoil, can cause significant root damage. It is commonly thought that a healthy tree tolerates removal of one-third of its roots. Most guidelines for tree preservation advise holding construction and grading outside of the dripline. However, based on a typical root structure, even that restriction could lead to removal of over half the tree's roots. As land values rise, there is more pressure to encroach within the dripline to gain usable space.

Several variables affect a tree's ability to tolerate encroachment, including the health, species, root structure and environmental factors. The type of construction that will occur and how it will be executed are equally important factors. For example, construction of a concrete sidewalk on natural grade requires about six inches of excavation. This would cause less injury than an asphalt road requiring at least twelve inches of excavation. A road with water and sewer improvements along the curb-usually in trenches four to six feet deep- must be held farther away from trees than one with no improvements.

If work must be done close to trees, extra care can minimize the damage. For example, lowering the grade two feet at a distance fifteen feet from the trunk of a tree that forms sinker roots may be possible. But it would require that the excavation be done by hand and the roots cut cleanly with a saw, rather than with equipment that rips and shatters roots.

Evaluating the potential of trees to survive and adapt to the new environment must be made on a tree-by-tree basis. The evaluation involves estimating the percentage of root loss, the potential of the tree to tolerate immediate water stress, and the ability of the tree to produce new roots.

Figure 16

Tree Preservation Techniques

Major construction impacts and techniques to minimize them are described in Table 1. Following is a discussion of the major points to consider in design and construction.

Preserving Trees in Groups -Whenever possible, preserve trees in groups so that relatively large areas can be set aside for native vegetation. These areas serve as green belts or open space. In forest situations the amount of thinning the trees will tolerate depends on the size, age and species present. Mixed forest stands, which contain conifers and hardwoods of different ages and sizes, should be preserved in large blocks (at least 60 feet wide).

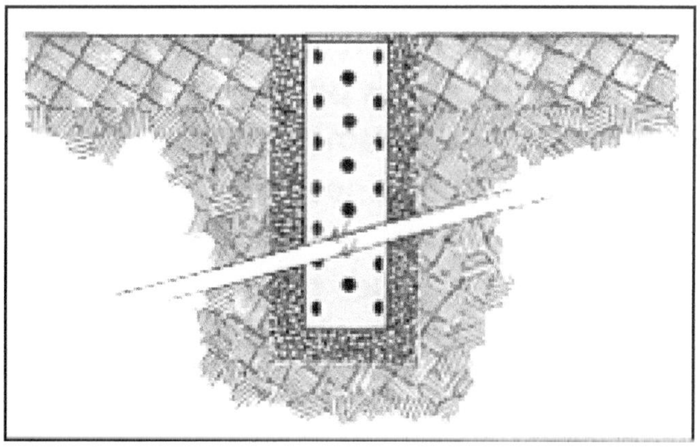

Figure 17.

On the other hand, single oaks growing in the open can often be preserved effectively.

If only a small amount of land can be dedicated to tree preservation, it may be better to allot a larger area around the few best specimens than to attempt saving a greater number of trees that would sustain heavy damage.

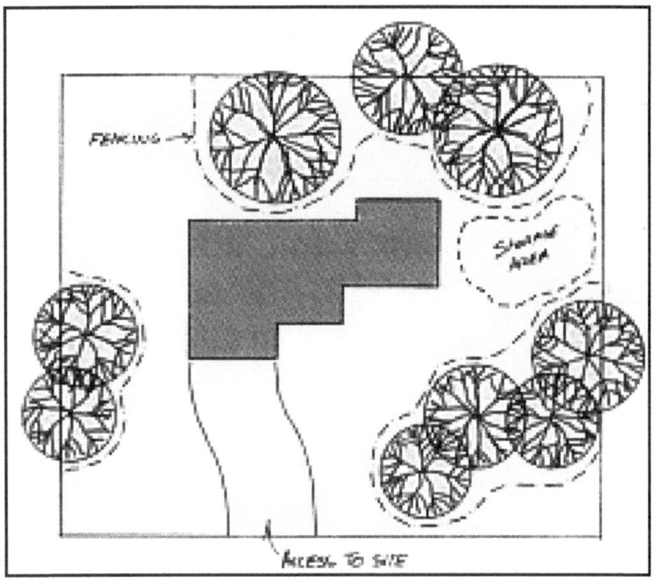

Figure 18.

Site Preparation and Clearing Specifications - Clearly identify the trees to be preserved on all plans. Flag and fence them in the field. When next to trees that will remain, trees to be removed should be cut rather than pushed over with equipment. This avoids possible root damage to remaining trees.

Specify no stripping of top soil or grubbing of understory in tree preservation zones.

Locate storage for equipment and materials, access roads to the site, and traffic patterns within the site well away from trees to avoid unnecessary root injury and soil compaction. These areas should be identified on the site plans (Fig. 18).

Protecting Trees from Fill - Few trees can tolerate having fill soil placed over their trunks. Furthermore, the soil work involved in placing a fill causes root damage and creates an environment unfavorable for new root development (see Table 1).

Figure 19.

Figure 20.

The following treatments can minimize the adverse effects of fill:

— Hold fill away from the tree with a retaining wall designed with a discontinuous footing (Fig. 19 & 20).

— Unless the fill will support a structure, ask the soil engineer to specify the minimum soil compaction (usually about 85 percent).

— If fill will be imported, provide specifications for the chemical characteristics of the soil, so that trees will not be harmed by toxic conditions, such as high salts. Engineering specifications usually cover only the physical properties of the soil

— For fill covering large portions of the roots, consider installing an aeration system on natural grade prior to placing the fill. If the potential for soil subsidence is a problem, air vents can be installed through the fill into natural grade after construction is complete (Fig. 17).

Table 1: major construction impacts and methods to minimize damage

Table 1: Major Construction Impacts and Methods to Minimize Damage

IMPACT TO TREE	CONSTRUCTION ACTIVITY	METHODS/TREATMENTS TO MINIMIZE DAMAGE
Root loss	Stripping site of organic surface soil during mass grading	Restrict stripping of topsoil around trees. Any woody vegetation to be removed adjacent to trees to remain should be cut at ground level and not pulled out by equipment, or root injury to remaining trees may result.
	Lowering grade, scarifying, preparing subgrade for fills, structures	Use retaining walls with discontinuous footings to maintain natural grade as far as possible from trees (Fig. 21 & 22). Excavate to finish grade by hand and cut exposed roots with a saw to avoid root wrenching and shattering by equipment, or cut with root pruning equipment. Spoil beyond cut face can be removed by equipment sitting outside the dripline of the tree.
	Subgrade preparation for pavement	Use paving materials requiring a minimum amount of excavation (e.g. reinforced concrete instead of asphalt). Design traffic patterns to avoid heavy loads adjacent to trees (heavy load bearing pavement require thicker base material and subgrade compaction). Specify minimum subgrade compaction under pavement within dripline (extra reinforcement in concrete or geotextile under asphalt may be needed).
	Excavation for footings, walls, foundations	Design walls/structures with discontinuous footings, pier foundations (Fig. 22). Excavate by hand. Avoid slab foundations, post and beam footings.
	Trenching for utilities, drainage	Coordinate utility trench locations with installation contractors. Consolidate utility trenches. Excavate trenches by hand in areas with roots larger than 1" diameter. Tunnel under woody roots rather than cutting them.
Wounding top of tree	Injury from equipment	Fence trees to enclose low branches and protect trunk. Report all damage promptly so arborists can treat appropriately.
	Pruning for vertical clearance for building, traffic, and construction equipment	Prune to minimum height required prior to construction. Consider minimum height requirements of construction equipment and emergency vehicles over roads. All pruning should be performed by an arborist, not by construction personnel.
Unfavorable conditions for root growth; chronic stress from reduced root systems	Compacted soils	Fence trees to keep traffic and storage out of root area. In areas of engineered fills, specify minimum compaction (usually 85%) if fill will not support a structure. Provide a storage yard and traffic areas for construction activity well away from trees. Protect soil surface from traffic compaction with thick mulch. Following construction, vertical mulch compacted areas. Install aeration vents (Fig. 17).
	Spills, waste disposal (e.g. paint, oil, fuel)	Post notices on fences prohibiting dumping and disposal of waste around trees. Require immediate cleanup of accidental spills.
	Soil sterilants (herbicides) applied under pavement	Use herbicides safe for use around existing vegetation and follow directions on the label.
	Impervious pavement over soil surface	Utilize pervious paving materials (e.g. interlocking blocks set on sand, Fig. 23). Install aeration vents in impervious paving (Fig. 24).
Inadequate soil moisture	Rechannelization of stream flow; redirecting runnoff; lowering water table; lower grade	In some cases it may be possible to design systems to allow low flows through normal stream alignments and provide bypass into storm drains for peak flow conditions. (Usually flood control and engineering specifications are not flexible where the possibility of flooding occurs.) Provide supplemental irrigation in similar volumes and seasonal distribution as would normally occur.
Excess soil moisture	Underground flow backup; raising water table	Fills placed across drainage courses must have culverts placed at the bottom of the low flow so that water is not backed up before rising to the elevation of the culvert. Study the geotechnical report for ground water characteristics to see that walls and fills will not intercept underground flow.
	Lack of surface drainage away from tree	Where surface grades are to be modified, make sure that water will flow away from the trunk, i.e. that the trunk is not at the lowest point. If the tree is placed in a well, drainage must be provided from the bottom of the well.
	Compacted soils; irrigation of exotic landscapes	Compacted soils have few macropores and many micropores. Core vent to improve drainage. Some species cannot tolerate frequent irrigation required to maintain lawns, flowers, and other shallow-rooted plants. Avoid landscaping under those trees, or utilize plants that do not require irrigation.
Increased exposure	Thinning stands, removal of undergrowth	Preserve species that perform poorly in single stands as groups or clusters of trees. Maintain the natural undergrowth.
	Reflected heat from surrounding hard surfaces	Minimize use of hard surfaces around trees. Monitor soil moisture needs where water use is expected to increase.
	Pruning	Avoid severe pruning where previously shaded bark would be exposed to sun. Where pruning is unavoidable, provide protection to bark from sun.

Lowering the Grade - Obviously, root damage occurs when soil is stripped away to lower the grade. Here are a few ways to minimize the damage that occurs:

— Keep cuts as far from trees as possible by installing retaining walls. If the cut is greater than three feet, a continuous footing can be used because few roots are encountered below that depth. For shallow cuts, use discontinuous footings to minimize root injury (Fig. 21 & 22).

— Excavate by hand at the cut face, cutting the exposed roots cleanly with a saw. Once the trench has been dug to the depth of the finish grade a backhoe can be used to pull away the soil. The backhoe should sit outside the dripline and remove the soil to finish grade.

Figure 21.

Figure 22.

Paving - Paving inflicts more extensive root damage than might be expected. Because it is often considered a structure, pavement requires soil compaction similar to buildings. Excavation must be deep enough to accommodate the compacted sub grade, the base material, and the pavement itself. To minimize damage caused by paving:

— Maintain an area of several feet around the base of the trunk free of all pavement. Mulch the soil surface.

— Use paving materials that allow water to penetrate, such as inter locking bricks on sand, where possible over the root zone of trees.

— Avoid excavation into the root zone by placing base material and pavement on natural grade, making the level of the pavement higher than the tree flair (Fig. 23). Make sure water will not collect in the well surrounding the tree.

— Install aeration vents in impervious pavement (Fig. 24).

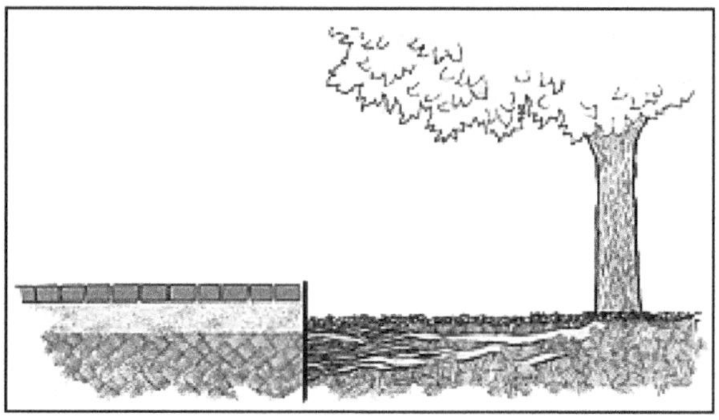

Figure 23.

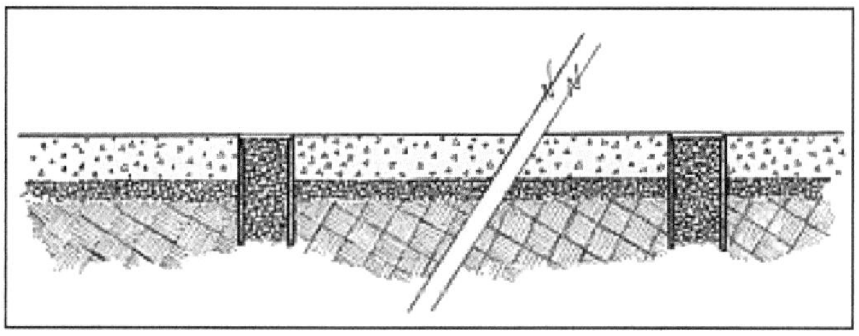

Figure 24.

Utilities and Street Improvements - Underground utilities are often overlooked as a possible cause of root injury. Water and sewer main lines are usually placed just inside the curb and gutter during road construction. Depth requirements vary, but four to six feet deep is typical. Contractors connect buildings to utility lines and take the shortest and straightest path, unless otherwise instructed. These connections are usually placed in trenches three feet deep. Consider the following methods to minimize damage from installation of utilities:

— Place notices on tree protection fences informing workers to prevent encroachment by routing all trenches around trees.

— Trench by hand when digging close to trees, so the woody roots can be bridged, or tunnel under tree roots.

— Curve trenches for street improvements towards the middle of the road or move them to the other side of the street.

Post-Construction Site Re-Assessment

Once construction is complete and the site clean, the tree care professional begins to direct maintenance. The remaining trees were chosen by others in the development process, and the reasons for selecting them may not be known. If protection measures were applied, they may be hidden. So, before post-construction maintenance can begin, the tree care professional must re-assess the condition of the site.

Begin by determining the condition of the trees and sites. Safety, value to the property, and the chances of remaining vigorous are all factors with considering. The maintenance prescription should compensate fore construction damage, including: disruptions to drainage, restricting surfaces, impacts of fill, and interference with the trees.

Trees stressed by construction must be carefully maintained to avoid loss of vigor or attack by harmful insects and disease. They require special attention to irrigation, fertilization, pruning, and pest control. In extreme cases, improper preservation techniques or severe construction damage could prompt heavy pruning—or even removal of whole tree. To determine the need for and the practicality of restoring a site severely changed by construction, the tree manager should draw on information from building plans as well as the designers and contractors involved in construction

Damage or modification to the above-ground portion of the trees is easy to detect. Impacts to the roots and soil are more common, but generally difficult to uncover. These impacts are too often overlooked. Be sure to consider all factors affecting tree health, before developing a maintenance plan.

Table 2: typical grading and construction specification and terms

Table 2: Typical Grading and Construction Specifications and Terms

TERM	DESCRIPTION
Stripping	Removal of all surface vegetation, organic laden topsoil, and debris, usually 6-12" depth. Often results in removal of soil through the major rootzone.
Scarification	Subsurface preparation under fills and slabs by loosening soil, usually to a depth of 6-12", then recompacting to 90-95%.
Percent compaction	The ratio of the fill material as compacted in the field to the maximum dry density of the same material, expressed as a percentage of the dry density.
Fill	Soil or rock material placed to raise the natural grade of the site or to backfill excavations (including trenches).
Engineered fill or compacted fill	Fill subjected to geotechnical engineering tests for proper compaction. Required percentage of compaction is created through the depth of the fill by spreading the fill material in uniform lifts not exceeding 6-8" in uncompacted thickness. Each layer (lift) is brought to a water content at or near the laboratory optimum and compacted with heavy equipment. 90-97% compaction is usually required.
Pavement	A typical pavement section for roads includes 2-5" of asphalt pavement, underlain with 8-20" of compacted aggregate base, placed on 6-12" of compacted subgrade. Concrete pavements usually require less excavation.
Slopes	The engineer determines the maximum steepness of cut and fill slopes (e.g. 2:1, 3:1). Fills on or against existing slopes are placed on horizontal benches cut into the slope. The toe of the slope may be keyed into native soils by cutting into the native grade 3 to 5 feet.
Foundations	Soils with potential for expansion can exert significant uplift pressures on shallow foundations and pavements. Foundations and slab-on-grade floors may need to be supported on sections of engineered fill, requiring excavation into the native grade.
Footings	Footings anchor walls. There are several types of footings, but most are continuous and are about 12-18" deep and 18-24" wide. Discontinuous footings can be designed for most walls and foundations. These rely on deep pillar type supports in holes drilled several feet deep and several feet apart. Minimum root damage will occur if the major woody roots are located and the pillars spaced to avoid them.
Street improvements	These include sewer, storm drains, and water lines. They are usually placed in trenches 4 to 6 feet deep along the edges of roads. Sewer and water lines must be placed some minimum distance apart according to local requirements.

UTILITY ARBORICULTURE

Local utilities generally prune more trees in public rights-of-way than all other parties combined. And, underground services have a similar impact. Trees and overhead lines have been getting in each other's way ever since the first utility pole was erected along a tree lined street. Initially, trees were considered obstructions to the electrification of a nation. Lines were built and the public got electric power, often at the expense of trees.

It should come as no surprise, that the public has long complained about the seeming insensitivity to trees displayed by utility pruning programs. At the same time, utility arborists have legitimate objections about common tree planting practices in utility rights-of-way. Both trees and overhead lines have a legitimate place in human communities. For this reason, it is prudent for every community to take a close look at the practices of local utility companies and the way those practices affect the urban forest. To develop a constructive working relationship, it's important to understand the opportunities for cooperation as well as conflict. Minimizing conflicts requires an examination of the goals and objectives of each function.

Purpose

The local utility company is responsible for maintaining safe and reliable electrical service to its customers. Trees growing near overhead lines can pose a threat to reliability and public safety, so trimming and removing trees along the "utility strip" is one important utility function. Simply put, utilities manage the portion of the urban forest growing near overhead power lines. However, the community is responsible for maintaining and preserving all components of the urban forest, including parks, greenways, and street trees growing in the "utility strip."

Attitude

A municipal arborist views the urban tree as a generalist; each tree is a valuable and integral part of the entire urban forest. Condition, vigor, size, and maintenance needs are items that require attention for the good of the community. On the other hand, a utility arborist focuses on trees as they relate to overhead lines. The goal of a utilities tree trimming efforts is that clearance be obtained between the power lines and the trees. When the need for line clearance work is assessed, the decision to trim or remove is made from that perspective. Work closely with your local utility to broaden priorities and benefit the urban forest as well as the utility system.

Approach

Some utilities have developed vegetation management plans, which are similar to municipal street tree management plans. Before scheduled maintenance is performed, a crew conducts a pre-job survey of the tree work along the line. This inventory identifies the amount of trimming and tree removal work required. Inventory information also helps allocate labor and equipment resources, and establish crew makeup and work practices

"Round-over" or "topping" techniques of pruning are not acceptable standards for tree trimming.

Planting inappropriate tree species under power lines creates a difficult situation for both the utility and the community. For example, trying to control the height of trees with the genetic drive to grow eighty feet tall poses a real maintenance challenge. As long as the wrong trees are planted in the wrong place, utilities will be obligated to prune them or remove them altogether. To minimize this, many utilities have developed tree lists suggesting alternatives to such favorites as oaks, maples, London plane trees, and other tall-growing species. The recommended species are selected for low maintenance, insect and disease resistance, availability, and beauty. Mature height must be 25 feet or less.

Normally, the utility's vegetation management budget is significantly larger than the community's urban forestry budget. Cooperation allows both organizations to extend their resources. In large and small communities alike, utility efforts represent the most significant force for change in the urban forest. Where a community lacks resources, a cooperative effort may represent the only way to manage the urban forest.

In many communities, a history of cooperation exists between the local utilities and government agencies. For example, cooperation was common during efforts to remove elm trees during the peak of the Dutch elm disease in the late 1960's and early 1970's. If such a relationship does not exist, one should be initiated immediately.

Cooperation occurs in situations where both parties benefit by working together. In many cities, the continual decline of "old growth, even age" trees along streets and in parks poses a hazard recognized by both utilities and communities. But, this situation also presents an opportunity to improve the composition of the urban forest through a cooperative program of tree removal and replacement. Selectively removing the overlymature trees and replanting with more desirable species-those more compatible with urban conditions-creates a new urban forest.

Conflicting ideas will not disappear overnight. For example, it may be difficult to reconcile the public's desire for streets shaded by majestic trees with the utility's need to eliminate interference between trees and overhead lines. Where appropriate small scale street trees can be integrated into the community's planting plan. In other cases, pruning practices that appeal to community aesthetics may be needed. Overall, a joint effort will lead to a compromise that comes closest to meeting the needs of both groups.

Common Concerns

In many communities, trees are caught between utilities trimming from the top down and communities trimming from the bottom up. This creates serious problems for the trees. Maintaining reliable and safe service is the main mission of utilities, and it requires clearance between trees and distribution lines.

While not universally practiced, many utility companies continue the practice of "topping" or "round-over" pruning. This technique describes a method of pruning where cuts are made at arbitrary points on a branch along a uniform plane within the crown. The resulting crown form is artificially uniform following pruning. Unfortunately, topping causes more problems than it solves. Because the tree sprouts grow rapidly, the tree must be pruned frequently.

Two commonly accepted pruning alternatives are "lateral drop crotch pruning," also known as "natural pruning," and "side-pruning." Lateral or natural pruning requires the tree trimmer to pick out the branches growing toward the lines and remove them where they attach to the next lateral limb. Properly placed cuts produce few sprouts and natural growth rates. The next cycle of trimming requires fewer cuts than the topping approach, because the problem branches were removed during the previous cycle and sprout growth has been slower. This method is used both on trees growing beneath or to the side of utility lines.

When the wrong species has been planted directly underneath the power lines, there may be no other alternative short of removal and replanting.

Increasingly, utilities install underground distribution lines that require trenches. These trenches frequently damage tree roots, which slows tree growth, and leads to decline of the crown and root system. Branches may die, increasing the chances of wind damage and invasion of wood decaying fungi or insects. Augering, tunneling, or boring through the root zone of the tree will cause less damage to the root system.

Safety and liability are issues of concern to everyone. Dead or dying trees in a park or along a street will eventually fall, possibly causing injuries or fatalities. If such trees are close to overhead power lines, their failure could tear down lines and poles, causing outages and other damage. Trees with low branches may tempt climbers and pose a safety concern. Someone climbing into the crown of the tree, particularly a child, risks touching a high voltage wire and suffering serious injury or death. Vehicular and pedestrian traffic is another concern. Street trees must be managed to ensure visibility and clearance for street lights and signs

Height reduction pruning by "topping" causes more problems than it solves

COORDINATION

The utility and the community are both responsible to the public. By working together they can obtain better results than by working alone. Communication is the key. Coordinating work plans and schedules increases efficiency. And, cooperation reduces the need for restrictive municipal regulations.

Through cooperation the utility will become more aware of the impacts that their operations have on the urban forest. Utilities with a negative image-whether justified or not- may be eager to demonstrate their helpful attitude.

Height reduction pruning using this "natural" or "drop crotch" technique is the correct way to "top" a tree.

Through cooperative efforts, utilities are becoming more sensitive to the value of incorporating sound, arboriculture techniques into their vegetation management programs. And, municipal arborists are beginning to recognize utilities as an arboriculturally.

THE POLITICS OF TREES

The purpose of this guide is primarily to provide both professional and nonprofessionals with easily understood technical information on proper tree management. While information is important, having the opportunity to apply this knowledge is even more important. Successful urban forestry programs around the nation have something in common. They exercise the political process to attain their goals.

Urban forestry professionals feel much more comfortable dealing with technical subjects based on clear and widely accepted principles. However, the political process often fails to respond to clear principles, making it difficult to understand and hard to control. It's no surprise that technically competent people sometimes choose to avoid politics. But without some political involvement, it is unlikely that tree programs will receive the support they need. The following section highlights the basic elements of decision making that affect trees in urban areas.

Decision-Making Process

Local decision making is seldom a transparent process, so it is important to be prepared. Begin by identifying a goal: to save a group of trees threatened by new construction; to provide additional funding for a tree planting program; or, to implement a hazardous tree removal program. Reconnaissance is next. Identify key players and their roles. The roles may be technical, financial, or administrative. Determine the person or people who will make the final decision. Match peoples' roles with the goal, so that energy and resources can be focused on people who have influence in the appropriate area. Devise a strategy. Decide how best to persuade the decision makers.

A successful strategy usually includes gathering and distributing information. Identify technical experts who are credible to all concerned parties. They provide the objective technical information. Early in the process, give this information to decision makers. This allows everyone a chance to review the information and ask questions. When the time comes, the decision will have been made after careful consideration, and not on the spur of the moment. Communication is one of the keys to success.

In local governments, the staff who implement policies are frequently consulted by decision-makers. The staff is also a good source of advice for citizens. They can point out major obstacles or opponents. Overcoming the obstacles may mean doing research, answering questions, or meeting with people. But, once this is done there is a much greater chance of receiving approval. Luckily, trees fall into the "motherhood and apple pie category," and people have difficulty not supporting them.

Some goals could be achieved administratively, saving the effort to develop and implement a legislative solution. Local government staff, especially planners, can be allies. They are generally supportive of actions that improve the liveability of the community, particularly trees and other types of landscaping. For example, planners often include street tree requirements in zoning codes. Let them know that a vocal, citizens group favors this approach and will support them, with elected officials, if necessary.

On the other hand, staff may see problems implementing the proposal and try to discourage the effort. Thank them for their help and advice, but inform them that the goal will be pursued anyway. The appeal could then be taken to a higher authority. And, if the system does not work, work outside the system. Some wonderful accomplishments have been achieved by citizens who diligently refused to take "no" for an answer.

Local officials are often swayed by organized movements, even though they may actually be quite small. Encourage distinguished citizens, such as doctors, lawyers, and businessmen to endorse the cause, so the organization represents the entire community. Identify well-known people with important titles, such as Ph.D., M.D. and J.D., as supporters of the proposal. Stationery listing the Board of Governors in the margin communicates the organization's seriousness and commitment. Looking as professional as possible always helps to establish credibility.

When influential people support the tree proposal, they should be thanked publicly. The organization could issue a news release identifying the good works of the individual or present an award. Whatever the method, make sure the community knows about those people that helped the effort. Most policy decisions are made at public meetings, and a good showing of support is critical to achieving a favorable outcome. Encourage friends, relatives, neighbors, and the general public to attend-the more the better.

Select several articulate and well-dressed people to endorse the proposal on behalf of the group. And, remember the most credible and influential people: "voters who care about what is going to happen to their trees." Occasionally, more than one appearance may be necessary.

Even though it may seem that the odds are long, there are many examples of successful campaigns that operated on a shoestring. Well directed efforts taking a few days and less than $100 have launched tree programs costing hundreds of thousands of dollars. A small effort could result in substantial long term benefits to the community.

Once the policy decision has been made, the easy part is over. Implementing the policy will require continued involvement from the group that promoted it. They must remain visible to support and monitor implementation. Support of the staff who will be implementing the policy will be helpful.

Technical vs. Value Decisions

Tree management frequently offers several alternatives for action, and weighing them can be difficult. The dilemma often arises from trying to balance two types of considerations. Some factors can be measured, such as size, shape, condition, health, and cost. Others are intangible values, such as beauty, history, attitude, and aesthetics. Urban forestry professionals often find themselves balancing tangible factors against intangible ones. For example, preserving existing trees during development must often be compared to the value of the trees and their chances of surviving construction. Communities often require a developer to spend hundreds of dollars preserving a tree that becomes damaged. It is worth less than the cost to preserve it, and then it dies a year or two after the project is completed. There is no formula for arriving at the best solution. Only professional judgement, honed by experience and tempered with public input, can make these decisions.

Liability

Trees pose some unique liability problems. While most trees cause few problems, there are occasions where problems can be significant and should be avoided if possible. Liability problems can best be avoided by clearly assigning responsibility for a tree's care. Visual clearance for traffic signals must be maintained. Dead wood or hazardous trees on public property must be removed. And, property damage resulting from trees, such as broken sidewalks, must be repaired.

Scheduled tree maintenance programs limit liability to some extent by minimizing the number of dangerous situations that develop. Such a program can also maximize the benefits from trees by educating people, regulating tree care activities, and providing some central organization to undertake this task.

A number of communities have attempted to pass liability for trees to adjacent property owners, while retaining regulatory authority over anything done to the trees. While the stated purpose of this action is to eliminate the community's liability for trees, it cannot be accomplished in this manner. At most, the property owner shares liability with the local government.

Coordination

As a community resource, trees may come under the jurisdictions of various groups. Although often overlooked, one task for urban forestry programs is coordinating the activities of others. Coordination is needed in three situations: public/public, public/ private and private/private.

Public/Public - Any public tree care agency is regularly faced with adverse effects from other public agencies who have responsibilities for things such as sewer systems, street rights-of-way and overhead utility lines. This is also true of different departments within the same public agency. Lack of attention to the work of others can be so damaging that it may outweigh the good work done by the urban forestry program. It is essential that one department coordinate all tree-related activities on public property, if the tree resources are to be protected. Since the tree care programs in most areas suffer from a lack of funds, it helps to establish good informal working relationships with other groups. In one city, the department of public works, which removed a number of trees in order to widen a city street, decided to replant new trees as part of the project. They did this based solely on the encouragement of the forestry department. The neighborhood was delighted with the result.

Public/Public - This is probably the most typical situation in many municipalities. Coordination often comes in the form of a permit system. Any tree work, such as pruning, planting, and applying pesticides, requires a permit. Such a system coordinates the activities of homeowners, businesses, tree care contractors, utility companies, and developers. Some cities even extend this regulation to private property by requiring a permit to remove any tree over a certain size, usually four to six inches diameter. Since it affects the largest percentage of the urban forest, public/private coordination will have the greatest benefits.

Public/Public - This type of effort addresses what private property owners do to their trees. It is difficult to play as direct a role as in the previous two examples, but it is still worth considering. Private/private coordination fills the remaining pieces of the urban forest, and allows a comprehensive approach to management.

Public education campaigns can stress the value of trees to the community, their contribution to property values, environmental benefits, and economic contributions. Other topics could be more technical: proper tree care procedures, choosing a reputable tree care company, and minimizing the liability.

This role is more "facilitation" than the direct regulation or coordination mentioned earlier. This role could be played by either a public agency or a private non-profit group, such as Tree People in Los Angeles, California. A number of communities in British Columbia, Oregon, Washington, and California cultivate community spirit to encourage private individuals to preserve historic and otherwise significant trees.

Remember that maintaining and preserving the urban forest for future generations is a big job-more than one person can do alone. Success will require cooperation and coordination among many people. The more people involved the more successful the effort will be.

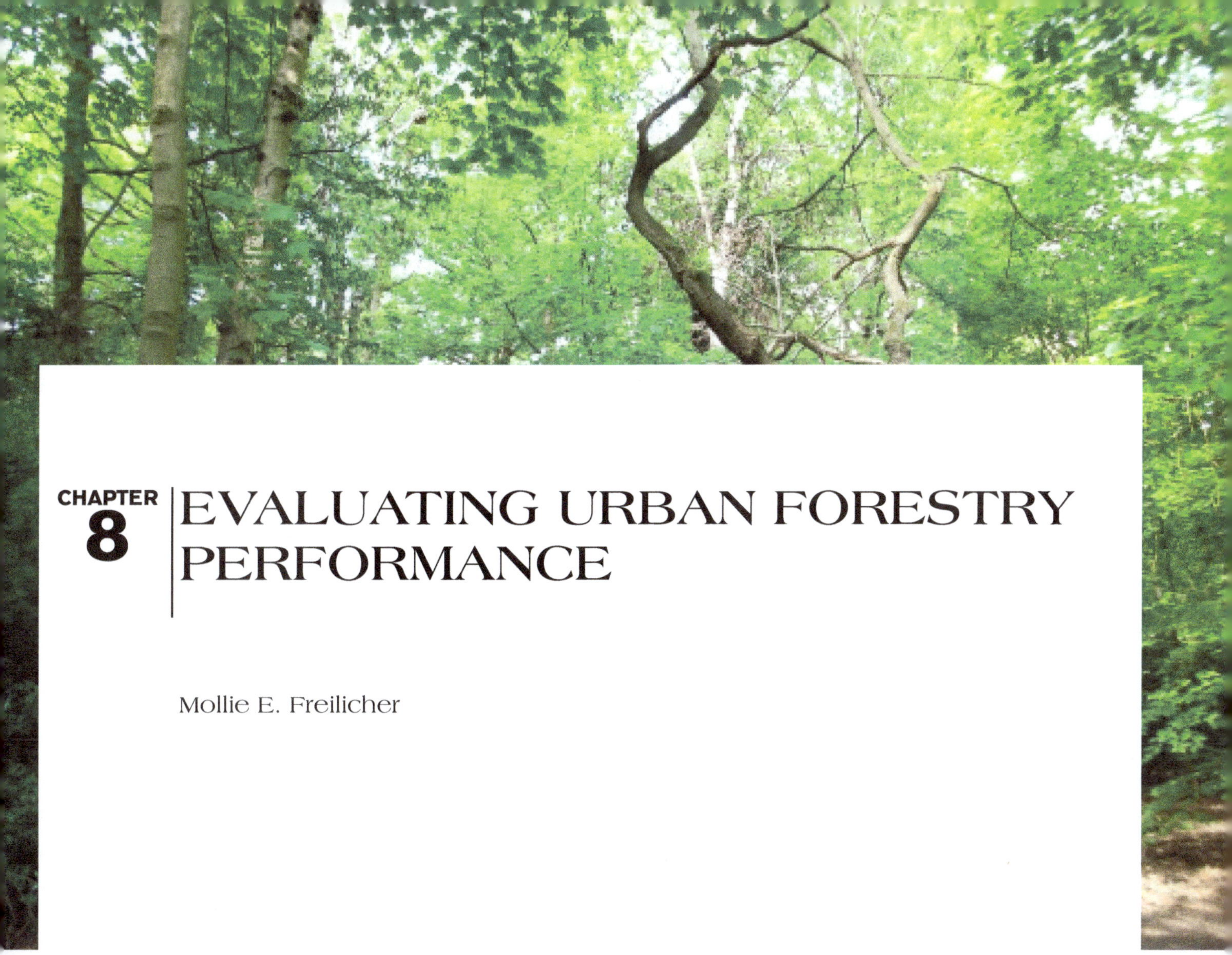

CHAPTER 8 EVALUATING URBAN FORESTRY PERFORMANCE

Mollie E. Freilicher

INTRODUCTION AND LITERATURE REVIEW

Introduction

Urban and community forest managers can measure urban forest performance at least two ways: by measuring characteristics of trees or by measuring the management of the resource. In 2006, the United States Department of Agriculture, Forest Service (USDA-FS or Forest Service) selected the latter and implemented performance measures to document management performance in communities across the United States (USDA Forest Service n.d.)

Forest Service involvement in urban and community forestry is a relatively recent phenomenon, but the Forest Service has quickly become an important source of funding for urban and community forest managers. A 2002 survey of state urban foresters found that 100% of respondents (n=41) received some funding from the Forest Service (Hauer and Johnson 2008). In fiscal year 2008, the Forest Service budget allocated $17.4 million to its Urban and Community Forestry (U&CF) program, a 38% decrease from 2007 (USDA-FS 2007). This was amended to nearly $27.7 million, but shows that threats to urban forestry funding exist (USDA Forest Service 2009). In the face of shrinking budgets for urban and community forestry, accountability is an important tool for ensuring that existing funds are being used to achieve Forest Service goals.

The United States Congress enabled the Forest Service to work with states and local communities under the Cooperative Forestry Assistance Act of 1978. This permitted the Forest Service to "provide financial, technical, and related assistance to State foresters [...]" and to work with local government to encourage urban forestry programs (16 U.S.C. § 2105). This act led to the creation of the Urban and Community Forestry program, which oversees performance measurement of urban forest programs. A 2004 congressional report noted that the Forest Service lacked the ability to track state performance and progress related to U&CF programs (U.S House of Representatives 2004). The Forest Service introduced a new system, the Community Accomplishment Reporting System, (CARS) in 2005, replacing the Performance Measurement Accountability System (PMAS) (USDA-FS 2005a). A Forest Service official, in a 2004 congressional report, cited PMAS as inadequate for meaningfully assessing accomplishments (U.S. House of Representatives 2004). Another report the Forest Service commissioned in 2004 on the U&CF program pointed out that PMAS assumes "that if communities create and strengthen programs related to urban forests, then healthy and sustainable urban forests will result" (Hortscience and Aslan

Citation: Freilicher, Mollie E., Evaluating Urban Forestry Performance, http://scholarworks.umass.edu/theses/509

2004). The same assumption also lies at the base of CARS, but CARS is organized at the community level, rather than the state level, facilitating further inquiry. Meeting the CARS measures and reporting them to the Forest Service, however, does not address the condition of urban forests across the United States. There is no accounting for diversity, age structure, condition, or for quantifiable environmental and monetary benefits trees provide.

The CARS performance measures the Forest Service adopted included four aspects of urban forestry programs that pertain to management (Table 1.1). These are part of what the Forest Service describes as a "family" of performance measures that includes non-management measures as well, such as volunteer community service hours in urban forestry activities (USDA-FS Service n.d.). These other non-management measures were not considered in this study. Throughout this thesis, the four CARS measures that pertain to management will be referred to as "CARS measures." The Forest Service tracks attainment of CARS measures through a database that state urban forest coordinators access to update information about state programs. The CARS database includes information for urban and community forestry programs across the United States. The Forest Service allocates part of its funding for state urban and community forestry programs according to attainment of the CARS measures (USDA Forest Service 2005). The Forest Service also takes into account other factors such as how much funding a state has received. It is unclear how the Forest Service selected the four CARS measures, as opposed to other measures that might be equally or more informative, such as tree budget, per capita spending on tree care, the number of staff devoted to urban forest management, or measures concerning urban forest structure. Many studies have investigated the prevalence of the CARS measures or similar measures in urban forestry programs across the country (Kielbaso 1990; Davis 1993; Elmendorf et al. 2003; Schroeder et al. 2003; Treiman and Gartner 2004; Ries et al. 2007, Rines et al. In Review). Other studies have described the structure of urban forests (McPherson and Rowntree 1989; Gartner et al. 2002), while some have evaluated environmental functions of urban forests (McPherson et al. 1994; Scott et al. 1998; Xiao et al. 1998; McPherson et al. 2005), and condition of urban forests. [For a comprehensive review of the urban forest of Sacramento, California see Journal of Arboriculture 24(2) and 24(4)]. From these studies, however, it is difficult to draw the relative importance of each measure or the aggregate result of using all these measures on an urban forest. A holistic study of these measures and urban forest performance is lacking. Additionally, the CARS measures evaluate management of urban forests, but do not take into account the condition of urban forests. Without known improvements in urban forest resources in communities meeting CARS measures, it is difficult to justify the importance of the measures. It is vital to know if there is a relationship between the CARS measures and urban forest condition.

In 2006, the University of Massachusetts, and the Massachusetts Department of Conservation and Recreation (DCR), conducted a survey to determine the performance of urban forestry programs in Massachusetts. The survey was directed to Tree Wardens. Every community in Massachusetts must have an individual, the Tree Warden, who is responsible for the care of public trees (M.G.L. Ch. 87; M.G.L. Ch. 41). Tree wardens from 143 communities in Massachusetts responded to the survey (a 41% response rate). The survey consisted of twenty questions regarding the position, training, staff, management priorities, and attitudes of tree wardens. From the tree wardens' responses, investigators determined whether a community met one, two, three, or all four CARS measures. Following completion of the study, more communities responded, increasing the overall response rate.

The purpose of the current study is to assess what it means to meet the four CARS measures. The Forest Service refers to communities that meet four CARS measures as "managing communities" which is adopted in this study. This study will address five research questions related to the CARS measures and urban forest performance:

1. Does investing in an urban forestry program provide financial benefit to a community?
2. Are there community demographic variables that are associated with CARS measures?
3. Do managing communities plant at least as many or more trees than they remove?
4. What is the role of arborist certification in tree planting and removal activities?
5. Do managing communities remove dead (and thus hazardous) trees in a timely fashion?

Judicious urban forest management can help maximize environmental, economic, and social benefits of urban and community forests, in addition to reducing tree-related hazards. As communities become more densely populated these benefits increase as does the importance of maintaining a safe urban tree population.

LITERATURE REVIEW

Urban Forests

Urban forests consist of all trees and woody vegetation in an urbanized area, including vegetation on public and private land (Miller 1997). The urban forest is not necessarily a connected grouping of trees, although this does occur, but the agglomeration of trees over an area of relatively dense human settlement (Miller 1997). Dwyer et al. (2003) identified three characteristics of the urban forest: diversity, connectedness, and dynamics, all with management implications. They identified diversity among trees, microclimates, soils, urban infrastructure, and landscape uses. Urban forests connect people with natural resources, buildings with the landscape, and wildlife with urbanized areas. Trees are also connected to a number of urban issues: water quality, air pollution, recreation, and aesthetics. They identified urban forests as a dynamic resource, changing over time. A successful, sustainable management program addresses all of these characteristics. Traditional definitions of sustainability stem from the United Nations Brundtland Commission Report of 1987 that defined sustainable development as "meet[ing] the needs of the present without compromising the ability of future generations to meet their own needs." (World Commission on Environment and Development 1987). Researchers have refined this to fit forestry and to include other connected elements besides the tree resource. Environmental, economic, and community factors were part of Salwasser's 1993 definition (Quoted in

Clark and Matheny 1997). This is especially important in urban areas where these elements are directly and closely connected with the resource. Dwyer et al. (2003) describe urban forest sustainability as "maintaining healthy and functional vegetation and associated systems that provide long-term benefits desired by the community."

Clark and Matheny (1997) described sustainability for urban forests. Forming the foundation of their model are three elements. Two of these elements are reflected in the CARS measures: the community framework and resource management. The third element, the vegetation resource, is not reflected in the CARS measures. An urban forest is only sustainable if the resource itself is diverse and provides a meaningful level of benefits (Clark and Matheny 1997; Dwyer et al. 2003). From the CARS measures alone, the state of the urban forest is unknown.

In 1987, the Forest Service and the National Association of State Foresters published guidelines for Effective Tree Care Programs (Hanson et al. 1987, as cited in Davis 1993 p. 201-202). These guidelines included numerous provisions, some of which the Forest Service integrated into the CARS program. These guidelines, however, go beyond the CARS measures and include, among other concepts, outreach to the community, a dedicated budget that is funded by more than one source, a tree planting program, use of local media, inclusion of developing areas in tree management programs, and local citizen participation in tree care activities (cited in Davis 1993). These guidelines include many elements of Clark and Matheny's (1997) model for sustainability, with emphasis placed on the community facet of urban forestry and the management of the resource.

While many studies focus on management programs or elements of management (Kielbaso 1990; Reeder and Gerhold 1993; Vitosh and Thompson 2000; Dickerson et al. 2001; Elmendorf et al. 2003; Schroeder et al. 2003; Treiman and Gartner 2004; Kuhns et al. 2005; Hauer and Johnson 2008; Lewis and Boulahanis 2008; Stevenson et al. 2008) fewer focus on the structure and condition of urban forests (McPherson and Rowntree 1989; Davis 1993; Galvin 1999; Gartner et al. 2002; Xiao 2002; Galvin and Bleil 2004; McPherson et al. 2005). In a survey of urban forestry in Pennsylvania communities, Reeder and Gerhold (1993) asked respondents about the condition of the urban forest and the results give some idea of the perceived and/or actual quality of the urban forest. Elmendorf et al. (2003) also asked about the occurrence of systematic tree care and the importance of systematic tree care in Pennsylvania communities. They found that often actions did not live up to attitudes. Davis (1993) also examined urban trees through assessment of tree inventories for communities in Kansas. Through his assessment he evaluated urban forest composition and trends toward increasing diversity, and also trends in planting activities to identify communities that were increasing the urban forest resource. Davis found mixed results among communities with inventories. In some areas, the street tree population was improving, and in others, it was deteriorating. Galvin and Bleil (2004) examined the urban forest resource in their assessment of canopy cover and its relation to demographic measures in Maryland.

BENEFITS OF URBAN TREES

Urban trees help mitigate many environmental problems common to urban areas. [For a broad review of benefits, see Nowak and Dwyer (2007).] Two major environmental benefits of urban trees are reduction of stormwater runoff and air pollution. Urban trees intercept rainfall on leaves, branches, and bark, reducing subsequent runoff by slowing the flow rate. This service has economic value as well as environmental value. In Santa Monica, California the reduction in flood control costs and the avoided costs for treating intercepted stormwater totaled $110,980 (Xiao and McPherson 2003), and in Modesto, California, stormwater benefits were greater than $600,000 (McPherson and Simpson 2002). Trees mitigate urban air pollution directly, by temporarily retaining pollutants on leaf surfaces and reducing peak flow, and indirectly, by reducing air temperatures and hampering the formation of ozone. Trees can also reduce heating and cooling needs for buildings, which reduces pollutant emissions at the point source of electricity generation (Nowak et al. 2006).

Urban and community forests also provide social and economic benefits (Dwyer et al. 1992; McPherson et al. 2005). The environmental services described above have economic value through avoided costs that a community would otherwise incur. Urban trees also provide economic benefits through increased residential property values (Morales 1980; Martin et al. 1989). This aesthetic benefit also positively impacts for business owners in well-planted business districts where consumers prefer to spend more time (Dwyer 1992; Wolf 2003; Wolf 2004).

Researchers have been developing methods to quantify benefits of urban forests. Many studies have quantified the value of the urban forest as a whole using methods from economics such as the public's willingness to pay for tree care programs. Such valuation indirectly assesses the value of the resource, but complicating factors such as differences between perceived willingness to pay and actual behavior or misstating willingness to pay in circumstances where that value may be used to levy new taxes or otherwise pay to manage a resource limit the method's rigor (Field and Field 2002). More recently, methods have focused on putting a dollar value on the environmental services urban forests provide, either through costs of controlling pollution or costs of mitigating damage ("damage value"). Wang and Santini (1995) describe these two methods in detail for quantifying the cost of air pollution. For air pollution, calculating control costs is a two-step process that first begins with identifying the marginal control measures. This method assumes that the marginal control measures represent the ideal for air quality standards. The second step is to calculate the cost of those measures. The underlying assumption is that the marginal control cost equals the marginal damage value for the pollutant. This assumption, however, is not realistic based on all that goes into government-established emissions standards. This method is simpler than calculating the costs for mitigating damages which is a lengthy process that involves many steps: recognizing emissions sources, estimating emissions, and estimating health effects and costs of treatment. It is difficult to take into account all assumptions, effects, and costs, making this a difficult mode for estimating costs (Wang and Santini 1995; Field and Field 2002).

For several years, researchers at the Forest Service and cooperating agencies have been developing software to determine environmental benefits provided by trees of different species growing throughout the United States. One program is the Street Tree Resource Analysis Tool for Urban Forest Managers (STRATUM), which is currently bundled in the i-Tree software suite. STRATUM calculates environmental and monetary benefits of trees. From species and diameter at breast height (dbh measured 1.37 m from the ground), STRATUM estimates environmental services and concomitant economic value provided by a particular tree. STRATUM then sums the services and values from all trees in the community. [For more detailed information on STRATUM, see McPherson et al. (2007).]

The Forest Service intended the STRATUM program to be accessible for urban forest managers. With that in mind, there are only two urban forest inputs that managers have to enter into the program: species and dbh. At the most general level, STRATUM calculates environmental information including benefit prices based on regional default values. Users select their region and the program defines prices for various pollutants: CO_2, PM_{10}, NO_2, SO_2, volatile organic compounds, and stormwater. STRATUM also calculates avoided electricity and heating costs based on local average electricity and natural gas prices.

The ability of urban forests to provide environmental benefits is increasing in importance as more areas across the United States become urbanized. Massachusetts is the fourth most urbanized state, with 34.2% of its land described as urban (Nowak et al. 2005). With increasing urbanization comes the conversion of land from agriculture or forest to housing, roads, and other infrastructure for people, buildings, and vehicles— and subsequently, increased pollution. Even with these pressures and the potential benefits of a well-maintained urban forest, many communities are lacking adequate urban forest management programs. In 2003, 90% did not have programs performing at the sustained level, the highest level of performance, under the Forest Service PMAS reporting scheme. The sustained level designated programs that functioned with community support and municipal activity to systematically manage urban forests and natural resources in communities (Hauer and Johnson 2008).

URBAN FOREST PERFORMANCE MEASURES AND COMMUNITY DEMOGRAPHICS

The nature of urban forest management programs is often associated with community demographics. Across the United States, communities with greater population often have more elements of a comprehensive tree program (Dickerson et al. 2001; Schroeder et al. 2003; Kuhns et al. 2005). This association also held true in Massachusetts (Rines 2007). In addition to population, population density has been correlated with urban forest management performance. Reeder and Gerhold (1993) showed that cities in Pennsylvania were more likely to have tree programs than less populated townships and boroughs. Larger communities in Missouri were also more likely to have urban foresters with a degree in forestry or a related field (Treiman and Gartner 2003).

Among small southern towns, the most important predictor of urban forest performance was the presence of a tree department with an urban forester and a dedicated budget, not affluence and education level of residents (Lewis and Boulahanis 2008). In New Jersey, Tate (1984) showed that communities felt that they could not provide the most favorable tree care because of insufficient funding. Research has also shown associations between community demographics and presence of professional staff in tree management programs. In Maryland, canopy cover, as a measure of urban forest quality, was correlated not with population, but with the area of a municipality (Galvin and Bleil 2004). Researchers detected these and other patterns as they relate to the CARS measures that the Forest Service adopted.

TREE INVENTORIES AND MANAGEMENT PLANS

A tree inventory can be a tool for justifying tree care funding and increasing work efficiency in communities (Tate 1985), but is frequently lacking in municipal management programs (Kielbaso et al. 1982; Reeder and Gerhold 1993; Elmendorf et al. 2003; Schroeder et al. 2003). Recent evidence, however, indicates that more communities are conducting inventories. In Utah, the number of communities with inventories has doubled since the 1990s (Kuhns et al. 2005), while in California, the percentage of communities with inventories has increased to over 80% (Thompson 2006), and in Oregon, to 56% (Ries et al. 2007). In Massachusetts, 62% of communities reported that they had tree inventories (Rines 2007). In Utah, the use of inventories was more common in communities with populations greater than 10,000 (Kuhns et al. 2005).

In Massachusetts, management plans were the least likely CARS measure for communities to meet. Approximately 1/3 of responding communities reported having a management plan in the 2006 survey and this was positively associated with population (Rines 2007). Utah showed similar levels of attainment for management plans (Kuhns et al. 2005). There have been associations between population and adoption of management plans elsewhere in the United States. In Pennsylvania, communities with more than 20,000 inhabitants were more likely to have adopted a management plan than communities with fewer than 2,500 people. For communities in between, however, the trend was not clear with medium-sized communities (between 5,001-10,000) more likely than communities in the next highest population tier (10,001-20,000) to have a management plan (Stevenson et al. 2008). Among cities in Oregon, only 9% had a management plan (Ries et al. 2007). Sometimes municipal officials view tree management plans as important (Elmendorf et al. 2003), but in Massachusetts, only 56% of tree wardens agreed or strongly agreed that management plans were important (Rines 2007).

PROFESSIONAL STAFF

In larger communities in Illinois, the person in charge of managing urban trees was more likely to be certified or have a degree in urban forestry or a related field. In small communities, however, the principle urban forest manager did not have any formal training in the care and maintenance of trees (Schroeder et al. 2003). Trends in Oregon were similar. There, as community size increased, so did the likelihood of ISACertified Arborists on staff. Only 26% of small communities had an ISA-Certified

Arborist, either as a manager, on staff, or on contract, but 100% of large cities did (Ries et al. 2007).

TREE ORDINANCES

Less populous communities in Illinois were less likely to have comprehensive tree ordinances (Dickerson et al. 2001; Schroeder et al. 2003). Dickerson et al. (2001) also showed that some provisions of tree ordinances were associated with affluence and education. In Oregon, 62% of cities had a tree ordinance (Ries et al. 2007). While 73% of communities in Pennsylvania had a tree ordinance, only 28% had systematic tree programs, raising some doubts about the ability of communities to enforce tree ordinances (Reeder and Gerhold 1993). Ordinances were correlated with population in Pennsylvania. Communities of greater than 20,000 people were almost twice as likely to have ordinances than communities of fewer than 2,500 (Stevenson et al. 2008). The same was true in Utah (Kuhns et al. 2005), where the percentage of communities having an ordinance increased with community size.

ADVOCACY AND ADVISORY ORGANIZATIONS

Usage of advisory or advocacy organizations varied across the United States. In Utah, 23% of communities had a shade tree commission or tree board (Kuhns et al. 2005). While the authors did not correlate this with population, they speculated that smaller communities might not be able effectively to take advantage of support. In California, however, approximately 80% of incorporated cities and counties reported having a tree board, but this figure dropped to 25% when focusing on advocacy organizations (Thompson 2006). In Pennsylvania 57% of communities indicated that they had a shade tree commission and that nearly 90% met at least once a year (Elmendorf et al. 2003). In Oregon, cities were 20% behind with 38% of cities in 2004 reporting existence of an advisory organization (Ries et al. 2007).

Table 1: USDA Forest Service, Urban & Community Forestry Program Performance Measures (USDA-FS, CARS Glossary)

Management Plans: A detailed document or set of documents, developed from professionally-based inventories/resource assessments, which outline(s) the future management of the community's trees and forests. The plan must be active (i.e., in use by the community and updated as needed to incorporate new information).

Professional Staff: Individuals who have one or more of the following credentials, and who the community directly employs or retains through written agreement to advise and/or assist in the development or management of their urban and community forestry program: 1) a degree in urban forestry or a closely related field (e.g., forestry, horticulture, arboriculture, etc.), and/or; 2) International Society of Arboriculture Certified Arborist (ISA) or equivalent professional certification.

Ordinances/Policies: Statutes or regulations that direct citizens and local governments in the planting, protection and maintenance of urban and community trees and forests.

Advocacy/Advisory Organization: Organizations that are formalized or chartered to advise (organizations established by the local government) or advocate or act (non-governmental organizations active in the community) for the planting, protection and maintenance of urban and community trees and forests.

REFERENCES

1. Cappaert, D., D.G. McCullough, T.M. Poland, and N.W. Siegert. 2005c. Emerald ash borer in North America: A research and regulatory challenge. American Entomologist 51:152–165.
2. Clark, James, R. Nelda P. Matheny, Genni Cross and Victoria Wake. 1997. A model of urban forest sustainability. Jour of Arboriculture. 23(1): 17-30.
3. D'Agostino, R.B. 1970. Transformation to normality of the null distribution of g1. Biometrika. 57(3):679-681.
4. Davis, R. L. 1993. Street tree trends in Kansas and the influence of community factors. Journal of Arboriculture 19(4): 201-208.
5. Dwyer, J.F., G.L. Peterson, and A.J. Darragh. 1983. Estimating the value of urban trees and forests using the travel cost method. Journal of Arboriculture 9(7):182-185.
6. Dwyer, J. F., E. G. McPherson, and H. W. Schroeder. 1992. Assessing the benefits and costs of the urban forest. Journal of Arboriculture 18, (09): 227-34.
7. Dwyer, J. F., D. J. Nowak, and M. H. Noble. 2003. Sustaining urban forests. Journal of Arboriculture 29, (1) (01): 49.
8. Elmendorf, W. F., V. J. Cotrone, J. T. Mullen. 2003. Trends in urban forestry practices, programs, and sustainability: contrasting a Pennsylvania, U.S. study. Journal of Arboriculture 29(4): 237-248.
9. Field, B.C and M.K. Field. 2002. Environmental economics: an introduction. 3rd edition. Boston: McGraw Hill Irwin.
10. Gartner, J.T., T. Treiman, and T. Frevert. 2002. Missouri urban forest—a ten-year comparison. Journal of Arboriculture 28(2):76-83.
11. Grado, S.C., D.L. Grebner, M.K. Measells, and A.L. Husak. 2006. Status, needs, and knowledge levels of

Mississippi's communities relative to urban forestry. Arboriculture and Urban Forestry. 32(1): 24-32.

12. Harrell, J.B. and P.P. Gornicki. 1981. Urban forestry in the sunshine state. Journal of Forestry. 79(8):506-508.
13. i-Tree. 2008 i-Tree Software Suite v2.0: User's Manual.
14. i-Tree. www.itreetools.org
15. Kielbaso, J. J., B. Beauchamp, K. Larison and C. Randall. 1988. Trends in urban forestry management. Baseline Data Report. 20(1), International City Management Association. Washington, DC.
16. Kielbaso, J. J. 1990. Trends and issues in city forests. Journal of Arboriculture 16(3): 69-76.
17. Kuhns, M. R., B. Lee, and D. K. Reiter. 2005. Characteristics of urban forestry programs in Utah, U.S. Journal of Arboriculture 31(6):285-95.
18. Lewis, B. L. and J.G. Boulahanis. 2008. Keeping up the urban forest: predictors of tree maintenance in small southern towns in the United States. Arboriculture and Urban Forestry 34(1):41-46.
19. Magurran, A.E. 1988. Ecological diversity and its measurement. Princeton, N.J.: Princeton University Press.
20. Martin, C. W., R. C. Maggio, and D. N. Appel. 1989. The contributory value of trees to residential property in the Austin, Texas metropolitan area. Journal of Arboriculture 15, (3; 3) (03):72-6.
21. Massachusetts Department of Agricultural Resources. n.d. Massachusetts Prohibited Plant List. Boston, MA. http://www.mass.gov/agr/farmproducts/ proposed_prohibited_plant_list_v12-12-05.htm. Accessed 10 January 2010.
22. Massachusetts State Data Center/Donahue Institute, University of Massachusetts President's Office. 2009. Boston, MA. http://www.massbenchmarks.org/statedata/data.htm. Accessed 28 May 2009.
23. McCullough, D.G., N.F. Schneeberger, S.A. Katovich. 2008. Emerald Ash Borer. Pest Alert. USDA Forest Service, State and Private Forestry, Northeastern Area, NAPR-02-04.
24. McPherson, E. G. 2006. Urban forestry in North America. Renewable Resources Journal 24(3): 8.
25. McPherson, G., and R. Rowntree. 1989. Using structural measures to compare twentytwo U.S. street tree populations. Landscape Journal 8:13-23.
26. McPherson, G., J. R. Simpson, P. J. Peper, S. E. Maco, and Q. Xiao. 2005. Municipal forest benefits and costs in five US cities. Journal of Forestry 103(8) (12):411-6.
27. McPherson, G.E.; J.R. Simpson, P.J. Peper, S.L. Gardner, K.E. Vargas, and Q. Xiao. 2007. Northeast community tree guide: benefits, costs, and strategic planting.
28. Gen. Tech. Rep. PSW-GTR-202. Albany, CA: U.S. Department of Agriculture, Forest Service, Pacific Southwest Research Station.
29. Mendenhall, W., R J. Beaver, and B. M. Beaver. 2006. Introduction to probability and statistics. 12th ed. Belmont, CA: Thomson.
30. Miller, R.W. 1997. Urban forestry. 2nd edition. Upper Saddle River, NJ: Prentice Hall. The National Arbor Day Foundation. n.d. Tree city standards: the four standards of aTree City USA. http://www.arborday.org/programs/treeCityUSA/standards.cfm. Accessed 3 June 2009.
31. Nowak, D. J., D. E. Crane, and J. C. Stevens. 2006. Air pollution removal by urban trees and shrubs in the United States. Urban Forestry & Urban Greening 4(3): 115.
32. Nowak, D.J. and J.F. Dwyer. 2007. Understanding the benefits and costs of urban forest ecosystems. In Urban and Community Forestry in the Northeast. 2nd ed. John E. Kuser, ed. New York: Springer.
33. Nowak, D. J., J. T. Walton, J. F. Dwyer, L. G. Kaya, and S. Myeong. 2005. The increasing influence of urban environments on US forest management. Journal of Forestry 103, (12):377-82.
34. Phillips, L. 2006. Understanding i -TREE. 2. street tree managment tool for urban forest managers (STRATUM). Arbor Age 26, (09):28-31.
35. Reeder, E.C., and H.D. Gerhold. 1993. Municipal tree programs in Pennsylvania. Journal. Arboriculture 19(1):12–19.
36. Ricard. R.M. 1994. 1994 urban and community forestry survey results. University of Connecticut Cooperative Extension System.
37. Ricard, R. M., and D.V. Bloniarz. 2006. Learning preferences, job satisfaction, community interactions, and urban forestry practices of New England (USA) tree wardens. Urban Forestry & Urban Greening 5(1):1-14.
38. Ries, P.D., A.S. Reed, and S.J. Kresse. 2007. The impact of statewide urban forestry programs: a survey of cities in Oregon, U.S. Arboriculture and Urban Forestry 33(3):168-175.
39. Rines, D. M. 2007. A survey of tree wardens to assess urban and community forestry performance in Massachusetts (U.S.A.). Unpublished master's thesis. University of Massachusetts, Amherst, Department of Natural Resource Conservation.
40. Schroeder, H. W., T. L. Green, and T. J. Howe. 2003. Community tree programs in Illinois, U.S. A: statewide survey and assessment. Journal of Arboriculture 29(4):218-224. Simpson, E.H. 1949. Measurement of diversity. Nature. 163:688.
41. Stovin, V. R., A. Jorgensen, and A. Clayden. 2008. Street trees and stormwater management. Arboricultural Journal: The International Journal of Urban Forestry 30, (4) (03):297-300.
42. Tate, R.L. 1984. Tree management in New Jersey. Journal of Arboriculture 10(8):229- 233.
43. Tate, R.L. 1985. Uses of street tree inventory data. Journal of Arboriculture 11(7):211- 214.
44. Thompson, J.R., D.J. Nowak, D.E. Crane, and J.A. Hunkins. 2004. Iowa, U.S., communities benefit from a tree-planting program: characteristics of recently planted trees. Journal of Arboriculture 30(1):1-10.
45. Thompson, R. P. 2006. The state of urban and community forestry in California: status in 2003 and trends since

1998. Urban Forest Ecosystems Institute. Technical Report No. 13. San Luis Obispo, CA.

46. Treiman, T., and J. Gartner. 2004. Community forestry in Missouri, U.S.: Attitudes and knowledge of local officials. Journal of Arboriculture 30(4):205-213.

47. Treiman, T., and J. Gartner. 2005. What do people want from their community forests? Results of a public attitude survey in Missouri, U.S. Journal of Arboriculture 31(5):243-250.

48. U.S. Census Bureau. 2009. American Factfinder. http://factfinder.census.gov/home/ saff/main.html?_lang=en. USDA Forest Service. 2004. FY 2004 Budget Justification. http://www.fs.fed.us/ budget_2004/documents/fy2004_overview.pdf. Accessed 10 June 2009.

49. USDA Forest Service. n.d. Urban and community forestry: family of performance measures for the urban and community forestry program. http://www.fs.fed.us/ucf/performance.html. Accessed 30 Oct 2008.

50. USDA Forest Service. 2005. Performance-based methodology for allocating Urban and Community Forestry program funds.

51. USDA Forest Service. 2005a. Northeastern area news notes. October 26, 2005. http://www.na.fs.fed.us/newsnotes/2005/oct05/oct05.pdf. Accessed 30 September 2008.

52. USDA Forest Service. 2005b. Fiscal Year 2006 President's Budget Overview. http://www.fs.fed.us/publications/budget-2006/fy2006-forest-service-budgetoverview.pdf.Accessed 10 June 2009.

53. USDA Forest Service. 2006. CARS community accomplishments reporting system: for the urban and community forestry program. User's Guide. National Information Center for State and Private Forestry. Release 1.2.

54. USDA Forest Service. 2008. Asian Longhorned Beetle. Pest Alert. State and Private Forestry, Northeastern Area, NA-PR-01-99GEN.

55. USDA Forest Service. 2009. Overview of FY 2010 Budget Justification. http://www.fs.fed.us/publications/budget-2010/overview-fy-2010-budgetrequest.pdf. Accessed 10 June 2009.

56. U.S. House of Representatives. 2004. A Report to the Committee on Appropriations U.S. House of Representatives on the U.S. Forest Service Urban and Community Forestry Program. Surveys and Investigations Staff, Committee on Appropriations, The U.S. House of Representatives, March 2004. http://www.treelink.org/docs/National_UCF_Partners/ForestServiceUrbanForest rypgmreport.pdf. Accessed 1 February 2007.

57. Wang, M., Santini, D., 1995. Monetary Values of Air Pollution Emissions in Various US Areas, Transportation Research Board Paper no. 951046, Washington DC.

58. Warren Group. 2008. Single-family home sales by community, 2006-2007. http://www.boston.com/realestate/specials/06_07_sales_single/. Accessed 5 May 2008.

59. Wolf, K. L. 2004. Trees and business district preferences: A case study of Athens, Georgia, U.S. Journal of Arboriculture 30, (6) (11):336-46.

60. World Commission on Environment and Development. 1987. Our Common Future. New York: Oxford University Press.

61. Xiao, Qingfu, and E. Gregory McPherson. 2002. Rainfall interception by Santa Monica's municipal urban forest. Urban Ecosystems. 6(4) (12):291-302.

62. Zar, J.H. 1999. Biostatistical Analysis. 4th Ed. Prentice Hall: Upper Saddle River, NJ.

CHAPTER 9

SUSTAINABILITY FOR URBAN FORESTS

Camilo Ordóñez and Peter N. Duinker

Keywords: Sustainability; urban forest; urban-forest sustainability; urban forest management

INTRODUCTION

Much has been written about how urban forests are valued positively by citizens and how they help mitigate some daunting environmental problems [1]. However, little has been said about the way sustainability is interpreted for urban forest systems. Most authors explore urban-forest sustainability through fairly narrow definitions of a sustainable urban forest, and thus lead to singular interpretations of the concept as the maintenance of tree health. These definitions lack consideration of the wealth of knowledge about sustainability in general and sustainable forest management (SFM) in particular.

The purpose of this paper is to further the thinking around urban-forest sustainability and to reveal alternative interpretations of the concept. To do so, three relevant conceptual domains are explored: concepts behind urban forest systems, interpretations of sustainability through the concept of sustainable development, and SFM. An analysis of interpretations to date is delivered and an argument is advanced on how sustainability could be understood better for urban forests.

Citation: Camilo Ordóñez and Peter N. Duinker, Sustainability for urban forests, doi: 10.3390/su2061510

THREE CONCEPTUAL DOMAINS

The Urban Forest

Urban forests can be broadly defined as the natural and planted trees in urban areas. Urban forests have several characteristics that make them different from hinterland forests. These characteristics have been thoroughly documented and are summarized below.

Structure

Urban forest structure refers to the biophysical and geographical characteristics of the urban forest. Urban forest structure may include the arrangement of trees in relation to each other and to objects, the forest composition and diversity, as well as age classes and health status. It may also include other geographic characteristics that imply nature-society interactions, such as the location of trees in different property categories. Urban forest structure determines the way urban forests function and the services they provide [2].

Connectedness

Urban forests do not display the biophysical connectivity of hinterland forests and are highly fragmented. However, they have intricate connectedness to their wider physical and social environment. The trees benefit or take a toll on people's lives, and they are in turn affected in myriad ways by human decisions and actions [3].

Dynamics & Driving Forces

Urban forests have a complex dynamic in the sense that they couple natural development processes and human processes, both of which operate at a variety of rates. The following forces can alter urban forests: (1) anthropogenic direct actions (e.g., removal of vegetation cover); (2) anthropogenic indirect actions (e.g., budget cuts); and (3) natural forces (e.g., fires, storms, insects) [4].

Values

The terms goods, benefits, functions, services and values have all been used interchangeably to mean what is important to sustain in and by urban forests. A discussion on how these concepts might be differentiated will aid our exploration of urban-forest sustainability as a concept.

Goods are economically constrained, as they can be broadly defined as something for which you pay to increase utility [5]. This term does little to encompass all the non-economically-valued characteristics of ecosystems. As for benefits, these are usually associated with costs (as in the well-known technique called benefit-cost analysis), and an economic overtone is also implicit with this term. The typology of ecosystem functions is diverse and sometimes contradictory. In a broad interpretation, functions may mean processes and structures by which ecosystems influence nature and/or people [6]. Many uses of the term inherently imply that ecosystems function for people, and not for themselves or for nature, therefore entailing a utilitarian, and usually economically-oriented, interpretation [7]. However, many services and/or values associated with the urban forest are not physical and cannot be understood through typologies that refer to physical structures or processes. The narrow perspective implied by the terms good, benefit and function is not inclusive of all what urban forests provide and do for society.

Services and values are, then, more appropriate terms. However, the widest definition of services is what is provided by society to better people's lives. In that sense, services say nothing about the social or even psychological underpinning behind ecosystem management and the values humans put to nature [8]. As for values, these can be defined in many ways. By values people refer to an adapted abstraction that emerges from social dialogue [9], or to fundamental orientations, life goals, or guiding principles, which serve as the basis for organizing individuals' beliefs and attitudes and guiding their respective behaviours [10]. However, these definitions refer to broader socio-cultural and moral categorization, and not to particular human-nature relationships.

In the forest literature, a value can be a characteristic, component, or quality considered by an interested party to be important in relation to a forest [11]. The concept of value behind a human-nature system can be broad, but it is also flexible, and can be furthered categorized as assigned value (a value for humans) or existence value (a value in itself) [12]. This further categorization reflects a semantic flexibility not possible with the use of other terms, such as goods, benefits, functions and services, and an ability to reason our way through their meaning without changing typologies. Because of this flexibility and broadness, the term values conveys the most comprehensive account of what must be sustained in urban forests.

The values associated with the urban forest may fall within three broad categories: ecosystem, social and economic. Ecosystem values include air pollution removal [13], carbon capture [6], regulation of the hydrological cycle, *i.e.*, rainfall interception [14], regulation of the micro-climatic environment [15], harbouring wildlife [16], among others. Social values include positive psychological effects [17], aesthetic quality [18], and emotional and spiritual benefits [19], among others. Economic values include increased real-estate values [20], recreational values [1], savings due to carbon dioxide sequestration and air pollutant removal [21], among others.

Sustainable Development

An interpretation of sustainability for urban forests implies an interpretation of sustainability itself. Sustainability may be broadly defined as the persistence or continuation of a system over time. In trying to capture what this means for natural and human systems, scholars have found it easier to interpret sustainability by defining a series of associated concepts, such as sustainable development. This term is used as a synonym for sustainability in some instances, though it has completely different interpretations in other instances depending on the field of knowledge and/or the management regime being addressed. Because of this, and because it is the most extensively defined term in the sustainability literature, a brief, but by no means exhaustive, review of its most useful definitions is important in our examination of sustainability (see Table 1)

Table 1: Some useful definitions of sustainable development.

Definition of Sustainable Development	Source
The capacity to maintain ecological, social and economic stability in the transformation of biospheric services into human ones	[22]
Meeting or optimizing present and future generations' needs	[23]
The persistence of a necessary and desired system (socio-political or natural) over an indefinite future	[24]
The integration of ethical, economic, social and ecological aspects of human life into a coherent view of the world so that current and future generations of people and other living organisms may continue indefinitely	[25]
Meeting needs and fulfilling aspirations under environmental, social and technological constraints	[26]
Living in harmony with nature and with one another	[27]
Keeping quality conditions in the relationship between humans and nature	[28]

Before the most useful interpretations of the above definitions can be explored, it is worthwhile to note that any interpretation of sustainability implies adhering to a specific view of sustainability. For example, sustainability can be viewed mechanically. This is favoured by technical people, who assign ideal quantitative measures to criteria-and-indicator frameworks that help characterize ideal states of a system. Examples of this view in the urban forest sector abound, as it would be discussed later. Because the possibility of such sustainable states is doubtful, as our understanding of what is sustainable may change over time, the mechanical view is challenged by another view of sustainability, the evolutionary view. In this view, one admits that a sustainable state can never ultimately be achieved because the benchmarks are ever-changing. Thus, sustainability is a process or journey because it can never be fully achieved [29], and no stable specification of a particular future sustainable system is possible.

Others see sustainability impressionistically: what does one think of when the word sustainability is brought up? The World Commission on Environment and Development's [23] definition serves this view. This view has considerable value in garnering participation in sustainability discussions by non-technical people. Another view of sustainability can be termed empirical—whatever people say is sustainable at a specific time and place is what is indeed sustainable.

There can also be a comprehensive view of sustainability which calls on managers to sustain everything that is valued. Criteria-and-indicator frameworks are also embodied in this view, trying to cover all possible grounds for pursuing sustainability [30]. Advocates of this view may also strive for an ideal sustainable state, as with the mechanic view, but both quantitative and qualitative indicators may be fitted into such a framework, making it possible to develop non-mechanical interpretations.

Another view of sustainability is dogmatic where people hold strongly to specific principles, such as the precautionary principle. Such normative approaches constrain sustainability strictly to one or a few management principles. Finally, a rule/tool view of sustainability is recognized, in which one simplifies matters to the sole application of certain rules or tools. In the forest-management context, for example, people's penchants to call for the abolition of clearcutting, chemical pesticides and genetically modified organisms represent this view.

The seven basic views of sustainability above are not mutually exclusive. Many views may be interplaying at one time, yet, usually, we attach ourselves to one or another depending on the system being studied and our motivation for addressing sustainability at the moment. However, no view holds the definitive answer about what sustainability really is or how it is to be achieved. The views that bound the most useful interpretations of sustainability, including our own, will be discussed below.

As evident in Table 1, early definitions of sustainable development [22,23,25] were tailored to agree with a classic definition of development as the process of economic growth and the alleviation of poverty. These definitions usually lead to mechanical interpretations. For its advocates, achieving sustainability requires a mere shift of the elements involved in the development process. This leads many technical people to try and fit new technologies and economic measures around a defined sustainable state. However, merely modifying the economic and technological elements of development can not capture the whole spectrum of ideas behind sustainability [27]. Indeed, many definitions of sustainable development force the term to be used lightly for mechanistic purposes [31] and try and fit an interpretation that does not do true justice to what sustainable development is about.

Mechanical interpretations are premature, as it is already known that no particular definition of what sustainability is at a given moment may satisfy all fields of knowledge [32]. To foster a broader interpretation of sustainability, later definitions of sustainable development [24,26,27,28] tried to adhere to a broader conceptual framework of sustainability. In this approach, sustainable development starts to be synonymous with sustainability as a complex body of theory. However, it is important to recognize that definitions of sustainable development do not define sustainability by themselves.

It is then in our interpretation of sustainability where the real meaning of the concept lies. In interpreting the definitions listed in Table 1, a few essential qualities of sustainability stand out: the need or inherent goodness of the continuation of the natural world; the ecological constraints on economic growth; the promotion of intergenerational equity; and sustaining social, economic and environmental values. The first qualities described can be seen more as principles of sustainability, since they do not imply operational strategies but rather philosophical assumptions behind sustainable human-nature systems. The last quality, however, has operational utility. Because value is a superior concept than service, as discussed earlier, this interpretation implies seeing sustainability, in its broadest form, as the process of maintaining the values associated with human-nature systems in the future. This interpretation is determined by both a comprehensive and evolutionary view of sustainability, as explained above, as they try to refer to the action of sustaining the myriad environmental, social and economic values people have, and recognizing this as a process that changes over time. This interpretation is perhaps the broadest and most useful interpretation of the concept, and may apply to the majority of human-nature systems, one of which is the urban forest system.

Sustainable Forest Management

The concepts used to explain SFM have had a rich history and provide further insights for interpreting sustainability for urban forests. The general idea of forest sustainability was conceived early in the environmental movement [33]. However, until the early 1990s, when the term SFM was coined, the early understanding of SFM emphasized sustained yield. This referred to the maintenance of forest resource productivity and the maintenance or enhancement of outputs of a narrow range of objectively measurable products and services. The first move to broaden the SFM concept was by coupling sustained yield with a broader sense of integrated forest management and ecological integrity [34]. This process of conceptual development was furthered with time (seeTable 2).

Table 2: Some useful definitions of Sustainable Forest Management.

Definition of Sustainable Forest Management	Source
The maintenance of all specific natural processes or ecological features of a forest site to meet human needs	[35]
The conservation of the ecosystem base of forests equated with a management that is ecologically sound, economically viable, and socially responsible	[36]
The maintenance of the biodiversity, productivity, regenerative capacity, vitality, and the potential to fulfil relevant ecological, economic, and social functions	[37]
The recognition, association and sustainability of values associated to forests	[38]
The process in which forests are used to produce goods and environmental services that increase, or at least maintain, the welfare of people living today, while protecting for future generations the environment and natural resource base on which future production will depend	[39]

Some believed that a fullsome statement about the fundamental principles behind SFM and a true interpretation of SFM could only be accomplished through the development of criteria and indicators [40]. Some mid-1990s criteria-and-indicator reports were delivered, but not until the Montreal (for boreal and temperate forests) and Helsinki (for European forests) processes did a real momentum start for developing a common framework. Eventually, many such frameworks and SFM models were developed, but a few nations remain exemplary in practice, e.g., Canada, [41].

The SFM criteria-and-indicator approach reflects many of the issues discussed above for interpreting sustainability. First of all, modern SFM is not bound to a definition by this approach but rather an interpretation. A so-called "constructive ambiguity" is seen as the more useful way to interpret certain concepts such as SFM [42]. Second, some SFM criteria-and-indicator frameworks reflect many views of sustainability. For example, some reflect a comprehensive view of the concept, as they adhere to the general idea that SFM should be as broad and complex a concept as possible so as to help converge the full amplitude of values and elements that should be part of it [43]. Key qualities in recent SFM criteria-and-indicator frameworks reflect an advanced version of this notion. For example, many criteria-and-indicator frameworks today embrace an extremely wide range of forest values, much beyond timber exploitation, and include recreation uses, water and air quality, carbon storage, special forest products, scenic quality, traditional land uses of Aboriginal peoples, and others. Terms such as integration and consideration of multiple values are commonplace in exemplary SFM criteria-and-indicator frameworks [44].

Some SFM criteria-and-indicator frameworks also reflect a mechanical view of sustainability, as they focus on quantitative indicators. Many scholars recognize that this notion is troublesome, as the number, ranking and weighing of indicators would always need consensus on technical ordinance [45]. In fact, many recent criteria-and-indicator frameworks respond to this limitation by adopting an evolutionary view and redefining the criteria-and-indicator approach as an ever-ongoing process, or by considering qualitative indicators to assess unquantifiable sustainability issues. Other qualities are also important to consider in the SFM criteria-and-indicator approach. For example, time and space frames are explicitly considered in modern SFM criteria-and-indicator frameworks, with both having been dramatically expanded. For example, time horizons for forest planning now extend to several centuries in many jurisdictions [11].

In many ways, interpreting SFM through the key qualities above enriches a now outdated and mechanical, quantitative-only SFM definition and is clearer in expressing how SFM is conceptualized today [46]. This interpretation echoes the most useful interpretation of sustainability drawn above, that of sustaining the values associated with human-nature systems, though it adds the inclusion of expanded time and space frames. As with the interpretation of sustainability, this interpretation may apply to the majority of human-nature systems, one of which is the urban forest system.

INTERPRETING SUSTAINABILITY FOR URBAN FORESTS

The concepts of urban forests, SFM and sustainability discussed above guide our quest to find strong interpretations of sustainability for urban forests. Our launching pad rests on the several attempts that have already been made to interpret urban-forest sustainability.

Interpretations of Urban-forest Sustainability

Most of the urban forest literature tries to give meaning to sustainability as applied to urban forests through definitions of the concept of a sustainable urban forest. This ideal sustainable urban forest model is seen in the early literature as one where species selection and diversity, inventory and landscape planning, tree care and wood utilization, and public relations and support are the main elements to consider [47]. Others favoured comprehensiveness in their definition of sustainable urban forests, and define the concept as the process of maximizing energy efficiency, minimizing waste production and pollution, capturing the benefits of natural processes while minimizing damage from natural events, and meeting the economic and social needs of all people in a manner that does not degrade or destroy the productivity or health of its natural systems [48]. Later, a sustainable urban forest came to mean healthy and functional vegetation that provides the long-term benefits desired by the community [49]. Such interpretations are useful in that they try to embrace some of the most pressing issues behind sustainability, but are focused on interpreting the concept through a definition of an ideal state, and do so by addressing the more operationally relevant considerations. Indeed, such an approach is only possible if the concept is combined with a technical backdrop of tree maintenance, mainly standards such as pruning, species diversity, and so on; standards against which the sustainability of the urban forests is to be assessed.

As with SFM, many saw that a useful interpretation of sustainability as applied to urban forests could only be achieved through a criteria-and-indicator framework that described a sustainable urban forest model. One of the best efforts to present such models, in that it has been extensively referenced in many urban-forest management plans in North America, is that of Clark and Matheny [50]. In this model, issues related to the urban forest are associated with quantitative performance indicators. Though some performance indicators can be measured in the field, other measures of sustainability rely on the judgement of urban foresters [51]. All in all, though, the empirical measurement is arbitrarily designed according to an ideal state of the urban forest. Other models of this kind exist [52].

Analysis of Urban-Forest Sustainability Interpretations

Most of the interpretations of urban-forest sustainability discussed above come up short in two significant ways. First, the concept is given meaning through a definition of a sustainable urban forest, an ideal state of the system, focusing solely on trees and their caretaking. Models of sustainable urban forest management that attach to this view usually have a strong technical component, and indeed, in many it is the only component. For example, urban forest management plans in North America include goals that define the sustainability of their urban forest by striving for the general health of trees, canopy cover goals, a maximum representation for a single tree species, a standard of planting native species, and so on [53,54]. Our exploration of promising interpretations of sustainability suggests a process of sustaining all values associated with that system. Keeping trees healthy does not automatically satisfy all urban forest values, even if people's values respond to their maintenance. Indeed, although what people value in the urban forest is not independent from its biophysical structure, other value considerations may be also important in progressing towards sustainability.

Second, a deeper gap in such an interpretation of urban-forest sustainability is that the concept has been viewed for a long time as a mechanical problem, a technical-ecological issue. As the discussion about sustainable development and SFM demonstrated, interpreting sustainability through a mechanical view brings it down to the level of a biophysical process. Allocating an arbitrary empirical measurement to urban-forest sustainability according to an ideal state of urban forests, defined via standards of tree health, canopy cover, species mix, and other urban tree maintenance yardsticks, is too restrictive. Relying heavily on technical and numeric standards is not only a sterile way to describing the sustainability of a system but disregards unquantifiable indicators. Indeed, social issues associated with urban forests also contribute to the definition of a sustainable urban forest but do so in a non-quantitative way. Indeed, agreement among experts and stakeholders in the ranking of value-laden issues may be impossible to achieve.

Nonetheless, certain aspects about mechanical approaches are practical for certain constituencies. Coupling urban-forest sustainability to canopy cover, for example, is practical for an urban forester and tree care-takers because the concept conforms to an easily operationalized measure of progress. This point is illustrated by how people who deal daily with urban trees define a sustainable urban forest, stressing the importance of the biophysical health of trees. Others define the concept in terms of the interest and commitment of groups of people who deal with urban trees. Urban forest stakeholders, experts and non-experts alike, have widely differing views of sustainability [55].

The complexity of defining a sustainable forest portends considerable difficulty in serving society's needs well if one grounds an urban-forest sustainability framework in the above interpretations. Urban-forest sustainability is a broader concept and includes the concepts, principles, operational framework and applications associated with the urban forest system and how to get to any desired state. Thus, though not entirely devoid of mechanical techniques, urban-forest sustainability is much more complex an issue that requires a more inclusive interpretation.

How Sustainability Could Be Interpreted Better for Urban Forests

Our examination of the sustainable development and SFM literature reveals useful interpretations of sustainability for human-nature systems, such as the urban forest system. A vision of sustainability for urban forests may well be that at any point in the foreseeable future (a viable timeframe of, say, 100 years), the management of urban forests applied today should be able to provide a strong array of values at all times, with acceptable spatial distribution, except when natural or human catastrophes prevent such.

This interpretation reflects many of the key qualities that are valuable in any interpretation of sustainability as our discussion on SFM showed. Firstly, a holistic understanding of urban forest values gives meaning to the concept of urban-forest sustainability in a deeper and broader way. Leading urban forest scholars remind us that the look, function and management of a sustainable urban forest depend on which ecological functions and social benefits are desired, who chooses them, and at what space and time scales these elements are to be sustained [56]. Exploring what it means to frame urban-forest sustainability as a values-based enterprise may require, as with SFM, constructing a comprehensive criteria-and-indicator framework. How this framework may be structured according to the values associated to the urban forest, and how it can be made operational, is a matter of future exploration. Secondly, because sustainability is a vision for the future within a spatial context, following short-term or space-limited conceptions of the urban forests could too easily end in urban forest unsustainability. The time-frame of 100 years chosen here echoes the discussion about SFM in general, which reflects both the long-term vision that managers should have and the far-reaching implications of deciding how to manage urban forests today with a values-based approach.

It is interesting to realize how giving meaning to urban-forest sustainability in such a broad way opens the door for many other considerations in urban forest management. For example, ecosystem values are defined in many ways by our contact with nature and the way people see the good in nature [38]. We hypothesize that people's values associated with urban trees is determined partly on tree-ownership patterns, and urban forest management must address these patterns more deeply. In some countries, urban trees are mainly owned privately, e.g., US [57], yet much is to be said about ownership patterns elsewhere in the world. As with SFM, sustainable urban forest management should accommodate ownership fragmentation, as this feature defines different management objectives. Moreover, this interpretation is open to address the integration of urban-forest sustainability with the wider concept of overall urban sustainability. This has been recognized as an issue for future research [58]. Indeed, many urban forest studies have focused on the technical backdrop to determine policy measures related to tree maintenance [59], but not on the inclusion of urban-forest sustainability in the wider urban planning and development context. Finally, this interpretation is open to addressing climate change. Projected future climates across the globe may significantly affect and change urban forest structures and the ways we value them. Thus, actions associated with mitigation of and adaptation to climate change must be accommodated in urban forest management programs.

CONCLUSIONS

The concepts explored in the three conceptual domains in this paper provide a valuable backdrop to help interpret urban-forest sustainability more broadly and deeply than we have found evidence for so far. The applicability of this conceptual framework may have wider relevance than just in relation to urban forest systems. Moreover, the interpretation of urban-forest sustainability elucidated here reflects the myriad of values associated with the urban forest, helps engage people with interdisciplinary interests in the urban forest, and may provide the conceptual foundation for addressing key issues in urban forestry such as ownership patterns, integration and coordination with other management regimes, and climate change. How the interpretations of urban-forest sustainability found here may be framed and made operational in sustainable urban forest management is a matter for future thought. Enriched views of urban-forest sustainability are necessary for people to be able to care for their urban forests in a future characterized by growing human populations and a changing climate.

REFERENCES

1. Nowak, D.J.; Noble, M.H.; Sisinni, S.M.; Dwyer, J.F. People and trees: Assessing the US urban forest resource. *J. Forest* **2001**, *99*, 37–42.
2. Nowak, D.J.; Dwyer, J.F. Understanding the benefits and costs of urban forest ecosystems. In*Urban and Community Forestry in the Northeast*; Kuser, J.E., Ed.; Springer: New York, NY, USA, 2007; pp. 25–46.
3. Dwyer, J.F.; Nowak, D.J. A national assessment of the urban forest: An overview. In Proceedings of the Society of American Foresters 1999 National Convention, Portland, OR, USA, 11–15 September 1999; US Forest Service: Washington, DC; USA, 2000; pp. 157–162.
4. Nowak, D.J. Historical vegetation change in Oakland and its implications for urban forest management. *J. Arboric.* **1993**, *19*, 313–319.
5. Turner, R.K.; Pearce, D.; Bateman, I. *Environmental Economics: An Elementary Introduction*; Harvester Wheatsheaf: Hertfordshire, UK, 1994.
6. de Groot, R.S.; Wilson, M.A.; Boumansb, R.M.J. A typology for the classification, description and valuation of ecosystem functions, goods and services. *Eclo. Econ.* **2002**, *41*, 393–408.
7. Turner, R.K.; Paavola, J.; Cooper, P.; Farber, S.; Jessamy, V.; Georgiou, S. Valuing nature: Lessons learned and future research directions. *Ecol. Econ.* **2003**, *46*, 493–510.
8. Dwyer, J.F.; Schroeder, H.W.; Gobster, P.H. The significance of urban trees and forests: Toward a deeper understanding of values. *J. Arboric.* **1991**, *17*, 276–284.
9. Homer, P.M.; Kahle, L. A structural equation test of the value-attitude-behaviour hierarchy. *J. Pers. Soc. Psychol.* **1998**, *54*, 638–646.
10. Dutcher, D.D.; Finley, J.C.; Luloff, A.E.; Johnson, J.B. Connectivity with nature as a measure of environmental values. *Environ. Behav.* **2007**, *39*, 474–493.
11. *Sustainable Forest Management: Requirements and Guidance*; Canadian Standards Association (CSA): Ottawa, ON, Canada, 2003; pp. 1–78.
12. Moyer, J.N.; Owen, R.J.; Duinker, P.N. Forest values: A framework for old-growth forest with implications for other forest conditions. *Open For. Sci. J.* **2008**, *1*, 27–36.
13. Nowak, D.J.; Crane, D.E.; Stevens, J.C. Air pollution removal by urban trees and shrubs in the United States. *Urban For. Urban Greening* **2006**, *4*, 115–123.
14. Xiao, Q.; McPherson, E.G.; Ustin, S.L.; Grismer, M.E.; Simpson, J.R. Winter rainfall interception by two mature open-grown trees in Davis, California. *Hydro. Proc.* **2000**, *14*, 763–784.
15. Heisler, G.M.; Grimmond, S.; Grant, R.H.; Souch, C. Investigation of the influence of Chicago's urban forests on wind and air temperature within residential neighbourhoods. In*Chicago's Urban Forest Ecosystem: Results of the Chicago Urban Forest Climate Project*; Gen. Tech. Rep. NE-186; McPherson, E.G., Nowak, D.J., Rowntree, R.A., Eds.; U.S. Department of Agriculture, Forest Service, Northeastern Forest Experiment Station: Radnor, PA, USA, 1994; pp. 1–201.
16. Adams, L.W. In our own backyard: Conserving urban wildlife. *J. Forest* **1994**, *92*, 24–25.
17. Ulrich, R. View through a window may influence recovery from surgery. *Science* **1984**, *224*, 420–421. [PubMed]
18. Smardon, R.C. Perception and aesthetics of the urban environment: Review of the role of vegetation. *Landscape Urban Plan.* **1988**, *15*, 85–106.
19. Chiesura, A. The role of urban parks for the sustainable city. *Landscape Urban Plan.* **2004**, *68*, 129–138.
20. Tyrväinen, L.; Miettinen, A. Property prices and urban forest amenities. *J. Environ. Econ. Manag.* **2000**, *39*, 205–223.
21. McPherson, E.G.; Simpson, J.R.; Peper, P.J.; Xiao, Q. Benefit-cost analysis of Modesto's municipal urban forest. *J. Arboric.* **1999**, *25*, 235–248.
22. *World Conservation Strategy: Living Resource Conservation for Sustainable Development*; International

Union for Conservation of Nature (IUCN): Gland, Switzerland, 1980; pp. 1–77.

23. WCED. *Our Common Future: A Report to the World Commission on Environment and Development of the United Nations, A/42/427*; Oxford University Press: Oxford, UK, 1987; pp. 1–375.
24. Robinson, J. *Defining a Sustainable Society: Values, Principles and Definitions*; Sustainable Society Project, Department of Environment and Resource Studies; University of Waterloo: Waterloo, ON, Canada, 1990; pp. 36–46.
25. Ekins, P.; Max-Neef, M.A. *Real-life Economics: Understanding Wealth Creation*; Routledge: London, UK, 1992.
26. Harte, M.J. *The Many Perspectives of Sustainable Development, CSRD-R3*; Centre for Sustainable Regional Development: Victoria, BC, Canada, 1993.
27. Mebratu, D. Sustainability and sustainable development: Historical and conceptual review.*Environ. Impact Assess. Rev.* **1998**, *18*, 493–520.
28. Martinez-Alier, J.; Roca, J. *Economía ecológica y política ambiental*; (Spanish); Fondo de Cultura Económica: Mexico DF, Mexico, 2001.
29. Costanza, R.; Patten, B.C. Defining and predicting sustainability. *Ecol. Econ.* **1995**, *15*, 193–196.
30. *Criteria and Indicators for the Conservation and Sustainable Management of Temperate and Boreal Forests*; Montreal Process Liason Office: Santiago, Chile, 1995; pp. 1–25.
31. Costanza, R.; Low, B.S.; Ostrom, E.; Wilson, J. *Institutions, Ecosystems and Sustainability*; CRC Press: Maryland, NY, USA, 2001; p. 270.
32. Pezzoli, K. Sustainable development: A transdisciplinary overview of the literature. *J. Environ. Planning Manage.* **1997**, *40*, 549–574.
33. Clawson, M. *Forests for Whom and for What?* John Hopkins University Press: Washington, DC, USA, 1975; p. 175.
34. Bormann, B.T.; Brookes, M.H.; Ford, E.D.; Kiester, A.R.; Oliver, C.D.; Weigand, J.F. *A Broad, Strategic Framework for Sustainable-ecosystem Management*; USDA National Forest System, Forest Service Research: Washington, DC, USA, 1993; pp. 1–10.
35. Webster, H. Some thoughts on sustainable development as a concept, and as applied to forests. *Forest Chron.* **1993**, *69*, 531–533.
36. Salwasser, H.; MacCleery, D.W.; Snellgrove, T.A. An ecosystem perspective on sustainable forestry and new directions for the US national forest system. In *Defining Sustainable Forestry*; Aolet, G.H., Johnson, N., Olson, J.T., Sample, V.A., Eds.; Island Press: Washington, DC, USA, 1993; p. 44.
37. Wiersum, K.F. 200 years of sustainability in forestry: Lessons from history. *Environ. Manage. 1995,* **1995**, *19*, 321–329.
38. Bengston, D.N.; Xu, Z. *Changing National Forest Values: A Content Analysis*; Research Paper NC-323; USDA Forest Service, North Central Forest Experiment Station: St. Paul, MN, USA, 1995; pp. 1–28.
39. Gregersen, H.; Lundgren, A.; Byron, N. Forestry for sustainable development: Making it happen. *J. Forest* **1998**, *96*, 6–10.
40. Mosseler, A.; Bowers, W.W. Criteria and indicators of sustainable forest management—from concept to reality. *IUFRO News* **1998**, *27*, 9–11.
41. *Defining Sustainable Forest Management in Canada: Criteria and Indicators*; Canadian Council of Forest Ministers (CCFM): Ottawa, ON, Canada, 2003; p. 27.
42. van Damme, L. Forest Sustainability in Ontario. *Forest. Chron.* **2009**, *85*, 415.
43. Maser, C. *Sustainable Forestry: Philosophy, Science, and Economics*; St. Lucie Press: Delray Beach, FL, USA, 1994; p. 373.
44. Adamowicz, W.L.; Burton, P.J. Sustainability and sustainable forest management. In *Towards Sustainable Management of the Boreal Forest*; Burton, P.J., Messier, C., Smith, D.W., Adamowicz, W.L., Eds.; NRC-CNRC Research Press: Ottawa, ON, Canada, 2003; pp. 41–65.
45. Lowe, P.D. The limits to the use of criteria and indicators for sustainable forest management.*Comm. For. Rev.* **1995**, *74*, 343–349.
46. Duinker, P.N.; Bull, G.Q.; Shindler, B.A. Sustainable forestry in Canada and the United States: Developments and prospects. In *Two Paths toward Sustainable Forests: Public Values in Canada and the United States*; Shindler, B.A., Beckley, T.M., Finley, M.C., Eds.; Oregon State University Press: Corvallis, OR, USA, 2003; pp. 35–59.
47. Thomson, R.; Pillabury, N.; Hanna, R. *The Elements of Sustainability in Urban Forestry*; California Department of Forestry and Fire Protection: Riverside, CA, USA, 1994.
48. Gangloff, D. The sustainable city. *Am. Forests* **1995**, *101*, 30–34.
49. Dwyer, J.F.; Nowak, D.J.; Noble, M.H. Sustaining urban forests. *J. Arboric.* **2003**, *29*, 49–55.
50. Clark, J.R.; Matheny, N.P.; Cross, G.; Wake, V. A model of urban forest sustainability. *J. Arboric.* **1997**, *23*, 17–30.
51. Clark, J.R.; Matheny, N.P. A model or urban forest sustainability: Application to cities in the United States. *J. Arboric.* **1998**, *24*, 112–120.
52. McPherson, E.G. Structure and sustainability of Sacramento's urban forest. *J. Arboric.* **1998**,*24*, 174–190.
53. Clark, J.R. *Urban Forest Plan: City & County of San Francisco*; San Francisco Government (SFGov): San Francisco, CA, USA, 2006.
54. Kenney, W.A. *Urban Forest Strategic Management Plan, Town of Oakville: 2008–2027*; Urban Forest Innovations (UFII): Oakville, ON, Canada, 2008; p. 70.
55. Elmendorf, W.F.; Cotrone, V.J.; Mullen, J.T. Trends in urban forestry practices, programs, and sustainability: Contrasting a Pennsylvannia, US study. *J. Arboric.* **2003**, *29*, 237–248.

56. Dwyer, J.F.; Nowak, D.J.; Watson, G. Future directions for urban forestry research in the United States. *J. Arboric.* **2002**, *28*, 231–236.

57. *Urban and Community Forestry in the Northeast*; Kuser, J.E., Ed.; Springer: New York, NY, USA, 2007.

58. James, P.; Tzoulas, K.; Adams, M.D.; Barber, A.; Box, J.; Breuste, J.; Elmqvist, T.; Frith, M.; Gordon, C.; Greening, K.L.; Handley, J.; Haworthk, S.; Kazmierczaka, A.E.; Johnston, M.; Korpelam, K.; Morettin, M.; Niemelão, J.; Pauleitp, J.; Roeq, M.H.; Sadler, J.P.; Thompson, C.W. Towards an integrated understanding of green space in the European built environment. *Urban For. Urban Greening* **2009**, *8*, 65–75.

59. Escobedo, F.J.; Wagner, J.E.; Nowak, D.J.; De la Maza, C.L.; Rodriguez, M.; Crane, D.E. Analyzing the cost effectiveness of Santiago, Chile's policy of using urban forests to improve air quality. *J. Environ. Manage.* **2008**, *86*, 148–157

CHAPTER 10 URBAN TREE CANOPY COVER AND ENVIRONMENTAL JUSTICE

Kirsten Schwarz, Michail Fragkias, Christopher G. Boone, Weiqi Zhou, Melissa McHale, J. Morgan Grove, Jarlath O'Neil-Dunne, Joseph P. McFadden, Geoffrey L. Buckley, Dan Childers, Laura Ogden, Stephanie Pincetl, Diane Pataki, Ali Whitmer, Mary L. Cadenasso

INTRODUCTION

The environmental justice (EJ) community, including activists, academics, and policy makers, has long been concerned with recognized patterns of low-income and minority communities disproportionately burdened by environmental hazards [1,2]. Recently, the scope of environmental equity has been broadened from a sole focus on burdens to include the distribution of environmental goods or amenities [3]. To date, studies on the allocation of amenities across urbanizing landscapes have focused mostly on the placement of parks and open spaces [4–7], as well as vegetation cover [8–10]. Generally these studies report uneven distribution of environmental amenities that disfavor racial and ethnic minority and low income neighborhoods.

Environmental justice studies typically focus on single case studies. This reflects the grassroots and local nature of environmental justice struggles that have informed analytical approaches [11]. However, comparative analyses of multiple cities using the same methodology can reveal whether there are consistent and significant patterns across cities of different sizes, morphologies, biophysical contexts, ages of development, and histories. Although comparative ecological methodologies have been promoted in the field of urban ecology, comparative urban environmental justice studies are relatively few [12–15] and tend to focus on measuring environmental burdens rather than benefits [16,17]. Such analyses on generalizable patterns of environmental inequity should ideally lead to investigations of processes responsible for common and differing patterns observed, including the role of historic, place-specific drivers of urbanization and environmental change [11,18]. Before such process-based studies can be undertaken, however, investigators should carefully examine the distributive equity of benefits or burdens.

Urban Tree Canopy (UTC) cover is widely regarded as an environmental good or amenity. UTC cover as an environmental amenity stems from the direct flow of perceived benefits, or ecosystem services, to people and neighborhoods where UTC cover is found. UTC cover has been linked to the provisioning of multiple ecosystem services, including regulation of regional climate and water cycles [19–26]. In addition to UTC, "greenness", as an indicator of vegetation cover, has been associated with reductions in childhood obesity rates [27]. Presence of vegetation can also decrease cognitive fatigue, improve worker attitudes on the job, and reduce stress as well as feelings of anger, depression, or anxiety [28,29]. UTC cover has been associated with improved aesthetics, noise reduction, and stronger social cohesion and community empowerment [19,30–32]. It is important to note, however,

Citation: Schwarz K, Fragkias M, Boone CG, Zhou W, McHale M, Grove JM, et al, Urban Tree Canopy Cover and Environmental Justice, PLoS ONE 10(4): e0122051. doi:10.1371/journal.pone.0122051

that UTC can also create disamenities such as increased water demand, maintenance costs, allergies, and perceived safety concerns [33,34]. Some street tree campaigns have been met with resistance from residents who simply do not want trees in front of their houses. The potential costs or burdens of UTC cover depend on a number of factors, including climate, vulnerability and price of water supply in a changing climate, socio-demographic preferences, built environment characteristics, and fiscal capacity to maintain UTC cover [35]. While benefits from and support for UTC are not universally positive, for the purposes of this analysis, we treat UTC as an environmental amenity.

Fairness of public investment in the distribution, delivery, and maintenance of services derived from urban tree canopies is a basic environmental justice concern [19,27,30,31,33,36]. Equity assessments must take into account existing distributions of benefits as they relate to social groups, especially ethnic/racial minorities and lower-income groups who have been traditionally disadvantaged and marginalized, or lack the resources or capacity to overcome a scarcity of environmental benefits [37]. These considerations point to the need for UTC assessments at a scale smaller than municipal jurisdictions because distributional patterns likely vary depending on the scale of analysis [38]. Research must match the scale of analysis to the social processes that drives the distribution of amenities in order to examine patterns at scales meaningful to environmental justice inquiries. Important social processes may occur at the scale of a neighborhood, a defined territory that reflects and reinforces group belonging. In general, studies relating UTC cover to demographics and race use census tracts as a proxy for neighborhood. For this paper, we use the census block group (CBG), a subset of the census tract, as the unit of analysis. We tested for scale differences using the census tract and census block group as analytical units but did not find meaningful differences between the two.

The spatial resolution of biogeophysical datasets, similar to census data, is predicted to affect UTC assessments, a phenomenon that is referred in spatial sciences as the Modifiable Areal Unit Problem [39]. Individual trees or lines of street trees—both common in urban areas—are not captured by moderate-resolution imagery [40]. Historically, moderate-resolution data derived from 30 meter resolution Landsat Thematic Mapper imagery has been used to classify tree canopy cover. These moderate resolution datasets, such as the National Land Cover Dataset (NLCD) [41], have allowed for nation-wide comparisons, but their usefulness is limited in urban systems that exhibit fine spatial heterogeneity [42]. The advancement of high resolution imagery and adoption of geographic information systems (GIS) by government entities has resulted in an abundance of high resolution geospatial data that can be used to derive very high (< 1m) spatial resolution vegetation data [43] including individual tree canopies. Accurate land cover data from high resolution imagery may help uncover patterns in UTC cover masked by the NLCD. Still, even with higher spatial resolution, important issues such as tree species appropriateness for different climate zones remains difficult to ascertain [44].

High spatial resolution of social and biogeophysical datasets matter little if the appropriate statistical analyses are not used. Some statistical methodologies used in ecology do not account for the spatial structure of data. For example, ordinary least squares (OLS) regression is a common technique to assess relationships between dependent and predictor variables, yet it fails to account for spatial autocorrelation, which may violate the assumption of independence of the errors in the regression. Failure to account for spatial autocorrelation can result in biased regression estimates due to a lack of accounting of spatial dependence in variables; or it can lead to higher standard errors in the regression estimators affecting statistical inference [45]. Alternatively, spatial autoregression (SAR) models account for spatial autocorrelation and therefore correct for violations of the assumptions in the classic linear regression models. The importance of SAR models has been widely demonstrated in various social and natural science fields and at their intersection [46,47].

In this study we examine potential inequities associated with the distribution of urban tree cover in relationship to race/ethnicity and income, in seven cities across the US: Baltimore, MD, Los Angeles, CA, New York, NY, Philadelphia, PA, Raleigh, NC, Sacramento, CA, and Washington, DC. This work is the product of a NCEAS (National Center for Ecological Analysis and Synthesis) working group on the urban ecology of environmental justice. NCEAS working groups focus on discovering novel patterns in existing datasets rather than creation of new datasets. The NCEAS participants chose these cities based on their expert knowledge and availability of high spatial resolution UTC cover. We hypothesized that urban tree cover would be positively correlated with increasing income and negatively correlated with minority populations in all cities. In particular, we anticipated that bivariate analyses would indicate a statistically significant and strong negative relationship between tree cover and minority populations, but the use of multivariate regressions incorporating control variables such as population density, housing age, education, and income would change that relationship. As all of our results can also be influenced by moderate and fine resolution biogeophysical and social data, we analyzed these relationships using high resolution land cover classifications (1 m) as well as census tracts and census block groups. We hypothesized higher spatial resolution data would indicate stronger relationships among tree cover, income, and race/ethnicity due to increased accuracy in describing the heterogeneity of urban systems. Many studies have examined the relationships associated with UTC and demographics, using ordinary least squares regression (OLS), but because these variables are not randomly distributed across the landscape, we hypothesized that spatial autoregression (SAR) would provide more robust results.

MATERIALS AND METHODS

City Site Descriptions

The seven case study cities vary in social (Table 1) and biogeophysical (Table 2) characteristics. Although high spatial resolution remotely sensed imagery is now freely available nationwide through the National Agriculture Imagery Program (NAIP), land cover classifications for cities using this imagery are still rare because they are expensive and time consuming to complete. The seven cities in our study were not randomly selected but based on the pooling of data from the NCEAS working group. Even though the cities are not stratified by social or biogeophysical characteristics, there are important differences among cities. Average annual temperatures fall within a band of 12.6 to 19°C (Table 2), and Los Angeles and Sacramento experience higher mean temperatures and receive less rainfall compared to the other cities. The arid cities of Los Angeles and Sacramento also have longer growing seasons and were not forested historically. In these regions, trees grew in and around riparian corridors along with oak forests in the surrounding foothills. For these California cities, water is a more important limiting factor to plant growth than the length of the growing season. In addition, with a predicted drying climate, Sacramento and Los Angeles are expected to experience declines in water supply. Although the East Coast cities (Baltimore, Raleigh, Philadelphia, New York City and Washington, D.C.) have longer winters and more freeze days annually, growing season increases slightly at the more southern locations, permitting higher growth rates and faster regrowth after removal of UTC. Existing UTC cover varies considerably across the case study cities, with the highest mean tree canopy percentage by CBG in Raleigh, NC (55%) and the lowest in Philadelphia, PA (13%) (Table 2).

Table 1: Site Descriptions—Social variables for the seven study cities. http://dx.doi.org/10.1371/journal.pone.0122051.t001

	Population (2000)	Percent Asian (2000)	Percent Black (2000)	Percent White (2000)	Percent Hispanic (2000)	Median Income (1999)	No HS Diploma (%)	BA Degree or Higher (%)
Baltimore, MD	651,154	1.5	64.3	31.6	1.7	30,078	31.6	19.1
Los Angeles, CA	3,694,820	10.0	11.2	46.9	46.5	36,687	33.4	25.5
New York, NY	8,008,278	9.8	26.6	44.7	27.0	38,293	27.7	27.4
Philadelphia, PA	1,517,550	4.5	43.2	45.0	8.5	30,746	28.8	17.8
Raleigh, NC	276,093	3.4	27.8	63.3	7.0	46,612	11.5	44.8
Sacramento, CA	407,018	16.6	15.5	48.3	21.6	37,049	22.7	23.9
Washington, D.C.	572,059	2.7	60.0	30.8	7.9	40,127	22.2	39.1

Please note that not all race categories are included and that respondents can select more than one race for the 2000 Census. Race and Hispanic origin are considered separate. Median income refers to household income.

Table 2: Site Descriptions—Biogeophysical variables for the seven study cities. http://dx.doi.org/10.1371/journal.pone.0122051.t002

	Mean Tree Canopy % by CBG	Mean Annual Precip	Mean Annual Temp (°C)	Cooling Degree Days	Heating Degree Days	Median Spring Freeze Day (-2.2°C)	Median Fall Freeze Day (-2.2°C)	Growing Season Days*
Baltimore, MD	22.34	41.94	12.6	4720	1147	3/30	11/10	226
Los Angeles, CA	17.61	15.14	19	1506	928	0/00	0/00	365
New York, NY	16.35	49.69	12.6	4754	1151	3/25	11/28	249
Philadelphia, PA	12.65	42.05	12.9	4759	1235	3/26	11/19	239
Raleigh, NC	54.64	46.49	15.3	3431	1456	3/15	11/22	253
Sacramento, CA	23.66	17.93	16.2	2666	1248	1/7	12/24	352
Washington, D.C.	27.52	41.94	12.6	4720	1147	3/30	11/10	226

The cities represent a range of sizes, from New York City and Los Angeles (population 8,008,278 and 3,694,820 in 2000) to much smaller cities such as Raleigh, North Carolina (population of 276,093 in 2000) (Table 1). The racial and ethnic makeup of the cities varies (Table 1). Over 88% of the population in Baltimore, Philadelphia, Raleigh, and Washington D.C. self-identify as either Black or White, and not Hispanic. In contrast, Los Angeles, Sacramento and New York City are more racially and ethnically diverse. Differences in median household income range from $30,078 in Baltimore to $46,612 in Raleigh. Populations vary in educational attainment (Table 1). For example, 33.4% of residents in the city of Los Angeles do not have a high school diploma, while only 11.5% do not in Raleigh. Raleigh has the highest percentage (44.8%) of residents with a Bachelor's degree or higher compared to Philadelphia, which has the lowest percentage of residents with a Bachelor's degree or higher (17.8%).

Social Data

GIS data layers of census block groups (CBG) for the seven cities were derived from the 2000 US Census TIGER (Topographically Integrated Geographic Encoding and Referencing System) dataset. Only census block groups that were completely within the boundary of the cities were used in this study (census boundaries do not always align with municipal boundaries). The same boundary data layers were used as the common boundary for all geospatial operations. The number of block groups included in the analyses for each city are listed in Table 3. The social variables used in this study were calculated at the CBG level using data from the 2000 US Census. Social variables include indicators of race, ethnicity, income, and educational attainment. For race variables we used percent White, Black, and Asian and, for ethnicity, percent Hispanic. For income, we used median household income. For educational attainment (age 25+) we used percent with no high school diploma and percent with a bachelor's degree or higher. Percent of houses occupied by renters, median housing age, the percent of the CBG classified as residential, and population density were also included in the analyses. Because our study focuses on environmental justice we highlight results for race, ethnicity, and income; the remaining variables act as controls in our regression analysis and are typical in empirical environmental justice research. In addition, the variables may further explain UTC cover density and distribution.

Table 3: Spearman's Correlation Results. http://dx.doi.org/10.1371/journal.pone.0122051.t003

	Baltimore		Los Angeles		New York		Philadelphia		Raleigh		Sacramento		Washington DC	
	CBG	Tract	CBG	Tract	CBG	Tract	CBG	Tract	CBG	Tract	CBG	Tract	CBG	Tract
Percent Asian	-0.01	-0.09	0.21**	0.17**	0.03**	-0.02	0.03	0.05	0.06	0.09	-0.21**	-0.35**	0.11*	0.06
Percent Black	-0.4	0.04	-0.32**	-0.33**	0.02	0.04	0.09**	0.08	-0.18*	-0.35**	-0.39**	-0.57**	-0.19**	-0.02
Percent Hispanic	-0.00	-0.09	-0.42**	-0.48**	-0.26**	-0.25**	-0.12**	-0.16**	-0.14	-0.17	-0.23**	-0.27*	-0.06	-0.08
Income	0.36**	0.38**	0.65**	0.67**	0.28**	0.23**	0.31**	0.45**	0.35**	0.38**	0.36**	0.31**	0.46**	0.32**
n	710	200	2449	839	5732	2216	1816	381	123	60	289	85	433	188

Note:
** $p < .01$.
* $p < .05$.

doi:10.1371/journal.pone.0122051.t003

Biogeophysical Data—High Resolution Tree Canopy

The percent of tree canopy cover for each CBG was calculated based on tree canopy data derived from high spatial resolution imagery. The imagery used included 0.6 m resolution pan-sharpened QuickBird satellite imagery, 1 m resolution NAIP (National Agriculture Imagery Program) near-infrared aerial imagery, and 0.15 m resolution natural color aerial imagery. LiDAR (light detection and ranging) data with 0.5 m or 1 m resolution were also used to aid in classification. Object-based classification approaches were used for tree canopy classification for all cities, except Los Angeles, where a pixel-based approach was used. Overall accuracy for the New York City classification was 96%, Sacramento was 92%, Philadelphia was 95%, and Baltimore was 94%. The accuracy of the Los Angeles tree cover classification was assessed using a stratified random sample of 100 parcels across the city, with an overall classification accuracy of 88.6% based on pixel-by-pixel comparison [48]. Although overall accuracy was not calculated for the remaining cities, similar data and procedures were used and we assume high accuracy for those classifications as well.

Statistical Analyses

Bivariate and multivariate techniques were used in the analyses. Spearman's correlation, ordinary least squares (OLS) regressions, and spatial autoregressive (SAR) techniques were used to investigate the relationship between key environmental justice variables (race, ethnicity, and income) and UTC cover. The CBG served as the unit of analysis. The percent of UTC cover was used as the response variable in subsequent statistical analyses.

Bivariate analysis and correlations between major variables has been a primary analytical tool in EJ studies. It is used to establish a baseline incidence of disparities in the distribution of environmental bads (or goods) and particular race, income and demographic characteristics. In our analysis, we use the Spearman correlation measure as a simple indicator of association between two variables (e.g. high values in one matching high values in another). We employ this type of analysis in our paper in order to formulate a baseline "picture" that is comparable across all cities in our analysis.

We next employ OLS in order to correct some of the obvious disadvantages that emerge from bivariate analysis. Regression allows us to examine the effect of a change in one variable on the value of another variable, controlling for other factors in the system under study. Empirically, the correlation coefficient between two variables can be different from a multiple regression coefficient when one of the variables is the dependent variable and the other, the independent variable. There is a theoretical possibility that a correlation coefficient between two variables is the same with the one you would get in regression analysis, but that happens only in the case of simple regression (one dependent and one independent variable) and only if the standard deviations of the two variables coincide. In multiple regression, the researcher captures the effects of other covariates (beyond the two that are included in the bivariate analysis) with the result of different correlation and regression coefficients. Thus, results from bivariate and multivariate (regression) analysis are expected to be different and not directly comparable. Moving away from simple co-variation, the latter is a methodology that tries to get closer to causation (but still is not causation).

The OLS method assumes that the error terms are independent. From a spatially-explicit perspective, this is an unrealistic assumption since according to Tobler's Law [49], "everything is related to everything else, but near things are more related than distant things." Many studies employing spatial data are typically hindered by the problem of spatial autocorrelation, which biases coefficient estimates of the variables employed in OLS regressions—typically described as providing best linear unbiased estimators [45].

Identification of spatial autocorrelation patterns requires the use of spatial autoregression modeling and the comparison of the results from the two regression approaches. Anselin [45] classifies spatial econometric models in two broad categories: models of spatial dependence (capturing auto-correlative effects of distance in spatial processes) and spatial heterogeneity (capturing the lack of stability or relationships across space). Spatial autoregression methods bring the spatial interrelationship structure of the units of analysis into the standard multiple regression in order to correct for problems of spatial autocorrelation. They quantify the spatial relationship of a (dependent) variable by a $n \times n$ matrix of spatial weights with each element of the matrix representing the strength of the interaction of two locations based on a proximity measure such as contiguity or distance. Spatial regression models have become increasingly popular in a variety of fields such as geography, economics, and political science [50], demography [51], ecology [52], and other fields that rely on spatially explicit data.

There are two spatial autoregressive techniques that have foundations in theoretical and practical considerations: the spatial lag (SLAG) model and spatial error (SEM) model [45]. The SLAG assumes that the imposed spatial structure affects the dependent variable, introducing a spatially lagged dependent variable. Typical theoretical explanations of the importance of the SLAG include issues of externalities and spill-over effects. In our study, socioeconomic conditions, local climatology, and ecological processes could account for such effects. A first order mixed regressive-spatial autoregressive SLAG model takes the matrix algebra model form of:

$$y = \rho Wy + X\beta + \varepsilon \qquad \textbf{(1)}$$

where Wy is a n×1 vector of the spatially lagged response variable, ρ is the spatial autoregressive coefficient, X is a n×k vector of explanatory variables, β is the k×1 vector of regression coefficients and ε is a n×1 vector of independently and identically distributed (iid) errors.

The SEM model incorporates the spatial effects into the error terms capturing the spatial structure of unobserved variation e.g. from missing variables that are spatially autocorrelated. This model is essentially capturing a spatial form of the problem of heteroscedasticity. The SEM model takes the mixed regressive—autoregressive form of

$$y = X\beta + (I - \lambda W)^{-1} \varepsilon \qquad \textbf{(2)}$$

where λ is the spatial autoregressive coefficient, W is the n×n spatial weight matrix, I is the n×n identity matrix and ε is a n×1 vector of iid errors. This paper reports the first type of spatial modelling (spatial dependence) using spatial autoregression methods.

We use a queen contiguity-based spatial weight matrix *W* (in the order of one) for all cities in our study. This approach in creating spatial weight matrices imposes a spatial structure in the units of observation that takes into account the topology of shared borders or vertices of the irregular polygon data. All regression models and spatial statistics used for measuring spatial autocorrelation are estimated using maximum likelihood estimation and are run on GeoDa 0.9.5-i software platform [53].

The spatial models are compared through the Akaike Information Criterion (AIC) and log-likelihood statistic [45]. A higher value of log-likelihood and a lower value of AIC point to a model with a better fit. Only the model with the best fit (using spatial lag or spatial error) is reported in the results. The model with the best fit is indicated in table 4.

Table 4: Ordinary Least Square (OLS) and Spatial Autoregressive (SAR) Results.

	Baltimore		Los Angeles		New York		Philadelphia		Raleigh		Sacramento		Washington DC	
	OLS	SAR (SLAG)	OLS	SAR (SEM)	OLS	SAR (SEM)	OLS	SAR (SLAG)	OLS	SAR (SEM)	OLS	SAR (SEM)	OLS	SAR (SLAG)
Percent Asian	-0.10	-0.17	-0.03*	-0.03*	-0.05**	-0.08**	-0.00	-0.01	-0.03	-0.04	-0.12**	0.01	0.10	0.08
Percent Black	0.07**	0.02*	-0.10**	-0.01	0.01*	0.02*	0.05**	0.03**	0.13*	0.11	-0.11*	-0.02	0.04	0.05
Percent Hispanic	-0.10	0.03	0.03*	-0.00	-0.00	-0.04**	0.04**	0.03*	0.12	0.11	0.03	0.03	0.24**	0.09
Income (in thousands)	0.09*	0.05	0.07**	0.07	-0.05**	-0.04**	0.17**	0.09**	-0.18*	-0.17*	0.03	0.10**	0.20**	0.10**
R squared	0.56	0.75	0.54	0.82	0.18	0.42	0.40	0.60	0.55	0.56	0.56	0.79	0.46	0.70
n	710	710	2449	2449	5732	5732	1816	1816	123	123	289	289	433	433

Note: Results are shown for Census Block Group.
** p <. 01.
* p <. 05.

doi:10.1371/journal.pone.0122051.t004

RESULTS

Bivariate Analysis

In some cities, the bivariate analyses reveal negative and significant relationships among race, ethnicity, and UTC cover. However, the signs of the estimated coefficients are not consistent across all cities. Specifically, the relationship between UTC cover and percent Asian is positive and significant for Los Angeles and negative and significant for Sacramento (Table 3). The relationship between UTC cover and percent Black is negative and significant for Los Angeles, Raleigh, Sacramento, and Washington D.C. (Table 3, Figs. 1 and 2). The relationship between UTC cover and percent Hispanic is negative and significant for Los Angeles, New York City, Philadelphia, and Sacramento (Table 3). Consistent across all cities is a positive and significant relationship between UTC cover and median household income (Table 3).

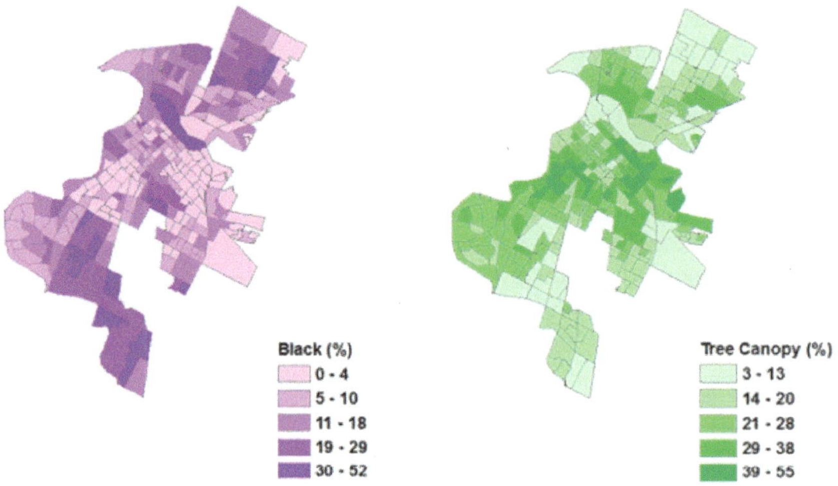

Figure 1: A spatially-explicit map depicting the percent of the population that self-identifies as black (left panel) and the percent of UTC cover for Sacramento City, CA (right panel). http://dx.doi.org/10.1371/journal.pone.0122051.g001

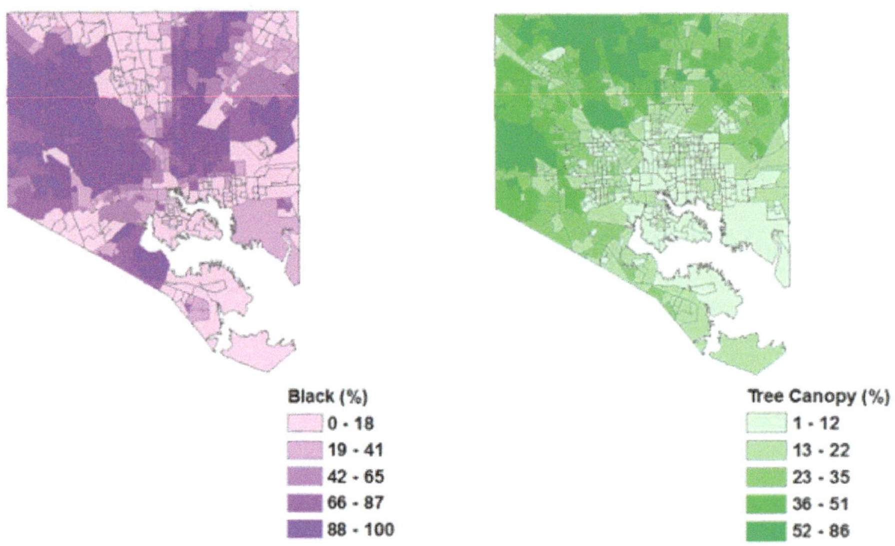

Figure 2: A spatially-explicit map depicting the percent of the population that self-identifies as black (left panel) and the percent of UTC cover for Baltimore, MD (right panel). http://dx.doi.org/10.1371/journal.pone.0122051.g002

Multivariate Analyses

In general, the multivariate regressions reveal that with the addition of other control variables the relationships between race or ethnicity and UTC are not as strong likely due to collinearity among the variables included in the regression. Negative and statistically significant associations between percent Asian and UTC cover are observed in Los Angeles and New York City although the regression coefficients are very small (Table 4). In Baltimore, Philadelphia, and New York City, there is a significant and positive association between percent Black and UTC cover but regression coefficients are very small (Table 4). Relationships between percent Black and UTC are significant and weakly positive in Raleigh and negative in Los Angeles but only using OLS models; they are not significant using SAR models. Finally, there is a significant and positive association between percent Hispanic and UTC cover in Los Angeles (OLS), Philadelphia (OLS & SAR), and Washington D.C. (OLS) although the regression coefficients are very small (Table 4).

Income, which was positively correlated with UTC and highly significant across all cities in the bivariate analyses, shows positive and statistically significant in most cities for the multivariate regressions but the magnitude of the coefficients is not large (Table 4). We estimate that an increase of median household income by $1,000 (in 2000 dollars) brings about a range of 0.05–0.20 point increase in the percent of UTC for the majority of our cities. Philadelphia and Washington D.C. present the highest positive income coefficients. New York and Raleigh are the only two cities with a negative effect of income; the magnitude of that effect is very close to zero for New York—even if statistically significant—while it is more substantial in the case of Raleigh.

We also examine the effects of income through the use of SAR models. The SAR models that account for the spatial structure of the data consistently provide more robust results compared to the OLS models, as evidenced by higher r-square values across all cities (Table 4). Overall, while the SAR models do not substantially alter our findings about the signs of income coefficients, differences occur in terms of magnitude and statistical significance. In the case of Los Angeles, the statistical significance of the estimated positive income coefficient is reduced. The SAR models also reduce the size of the positive income coefficient for Philadelphia and Washington, D.C. In the case of Raleigh, the difference between the estimated income coefficients for OLS models and the SAR models are minimal. Our findings suggest collectively that in addition to our variables exhibiting collinearity they are also spatially clustered.

DISCUSSION

Correlations between the distribution of benefits and socio-economic variables can vary across cities. Baltimore is not the same as Philadelphia, but the cities may share policies and regulations that have the potential to create environmental injustice. However, factors such as climate, demographics, and city size may filter the outcome that shared processes have on the distribution of environmental amenities and burdens. Demographics or, more specifically, racial and ethnic diversity differs among the represented cities and may represent an alternative explanation to climate. Along with an arid climate, Sacramento and Los Angeles have greater racial and ethnic diversity compared to the other cities with a larger portion of the population self-identifying as Asian or Hispanic. The only city in the Northeast with comparable racial and ethnic diversity is New York City, which demonstrates negative correlations between UTC cover and ethnicity, but not race. We have approached our analyses using various statistical methods, each one revealing something new about our data. The most significant finding from each method is listed in Table 5.

Table 5: Statistical Methods Used. http://dx.doi.org/10.1371/journal.pone.0122051.t005

Question	Statistical Method	Finding	Implication
Is UTC cover distributed equally in the cities examined?	Spearman's Correlation	Positive correlation with income across all cities. Strongest correlations among UTC and race occur in arid cities.	Regardless of what drives the pattern, the pattern exists—UTC cover is not equally distributed in regards to income, and in some cases, race.
What other variables drive the distribution of UTC cover?	Ordinary Least Squares Regression	There are no strong consistent drivers across all cities.	This is likely due to collinearity among the variables included in the regression providing less explanatory power.
Do the data have significant spatial structure?	Spatial Autoregressive Model	SAR models result in a better fit, evidenced by higher r-square values.	This suggests that in addition to our variables exhibiting collinearity, they are also spatially clustered. Accounting for the spatial structure improves fit.

doi:10.1371/journal.pone.0122051.t005

Cities are patchy landscapes, and this unevenness can lead to inequitable distribution of environmental benefits and burdens [54,55]. We expected all of the seven cities to show that neighborhoods with lower-income, less education, and high percentages of people of color would have low UTC cover. Overall, our results confirm that UTC cover is related to social characteristics of neighborhoods but not consistently across cities that vary in climate, size, and racial and ethnic composition. We also found that the ability to detect patterns in the data is affected by statistical method (OLS vs. SAR). Accounting for spatial autocorrelation produces more robust models and in certain cases corrects the aspatial specification bias of regular OLS estimation (Table 4).

The most striking pattern in the data comes from the bivariate analyses which reveal a strong positive relationship between UTC cover and income across all cities. Our multivariate regression analysis, controlling for other factors, reveals a positive relationship between UTC cover and income for most cities in our study (with the exception of New York and Raleigh). As our methodology section explains, it is theoretically possible to have two variables with positive correlations through bivariate analysis that become negatively related in a multiple regression context; controlling for other variables adjusts for the potential bias in the coefficient of the income variable, with the resulting sign reversal in the case of two cities in our study. The relationship between UTC cover and income may be the result of a feedback loop where high amounts of UTC cover increase property values and further attract households with high incomes. This positive feedback loop may support the continued maintenance of UTC cover in neighborhoods with high-income households and high levels of homeownership. Similarly, areas with low UTC cover have low property values, and residents may have less access to resources or incentive to increase property values because they are renters or on fixed incomes. Residents in low income neighborhoods might reasonably resist increases in UTC cover to avoid gentrification and rising rents [56]. The cost of tree maintenance, such as leaf clean-up, watering, and pruning, is another disincentive that can be particularly acute in low-income neighborhoods [33].

Income is so strongly associated with UTC cover that it must be controlled for in the analyses in order to detect additional drivers. Race/ethnicity was not consistently a significant factor in the bivariate analyses (Table 3) and few factors in addition to income were significant in the multivariate analyses (Table 4). However, in the bivariate analyses of our California cities, we did see evidence of a link between race/ethnicity and UTC (Table 3). For Sacramento and Los Angeles, percent black and percent Hispanic are strongly and negatively correlated with UTC cover, but the relationships are weak or insignificant for the other cities. A distinguishing characteristic of the California cities is their arid climate. Trees planted in more arid climates require irrigation in order to survive, while in temperate regions of the United States, trees can grow without water subsidies on unmanaged lands. Given the high water resource requirements in arid cities, the dominant role of income in driving UTC cover likely creates a greater potential for environmental injustice in cities receiving little precipitation. Natural growth and regeneration of trees in cities with greater precipitation could dampen any effects of different levels of resources allocated towards growing and maintaining UTC cover.

In arid climates, trees place a direct demand on limited resources, such as water, and an indirect demand on additional resources, such as energy used to transport that water, making trees more costly from an environmental, social, and economic perspective. In arid climates, treeless areas may represent less investment in UTC cover and/or management, while in temperate areas, treeless areas may not be a reliable indicator of low investment. In contrast, treeless areas may represent areas of greater economic and/or community investment, such as the costs of maintaining grassy areas. Nevertheless, the results of this study cannot be interpreted without consideration of some important caveats, which we outline below.

Other variables may be better predictors of tree canopy cover

This study was framed by environmental justice, which focuses on race and ethnicity as well as other social status characteristics (e.g., income, education, homeownership) as the variables of concern. While the models included population density and housing density as proxies for building density, other variables such as urban morphology [57], impervious surface, land rents, past tree canopy cover, water budgets, tree planting policies, and dominant tree species might generate more robust models.

Examining patterns between race and amenities does not capture intent

Identifying patterns of inequity are important and bring attention to the inequitable distribution of environmental goods or bads but it is equally important to understand the processes driving the inequitable distribution. Consideration of both distributional and procedural equity can help explain why the expected patterns predicted by environmental justice theory are not always present [15,58,59]. For instance, the high amount of green vegetation in predominantly African American neighborhoods in Baltimore may be reflective of the increased number of vacant lots in those neighborhoods, which resulted in part from decades of deliberate disinvestment and discrimination [6,60].

Vegetation structure and social structure may be mismatched

Our study is a comparison of current vegetation structure and how it relates to current social structure. However, trees are long-lived organisms that can take a very long time to establish and grow. In contrast, the social structure of cities can change more rapidly. Recent studies have highlighted this phenomenon describing the current landscapes we observe as legacies of past consumption patterns [11]. Specifically Boone et al. [11] found that vegetation operating at short time scales (i.e. lawns) reflect current lifestyle characteristics, while vegetation that takes longer to establish (i.e. trees) are reflections of the characteristics of past residents. Research that does not include a historical analysis is still valuable as it represents current patterns and potential inequities associated with environmental amenities, but it is also important to keep in mind that present day patterns could be the result of inherited landscapes. This may be especially true for systems where trees are not part of the native landscape and are typically planted at the time of development. A more complete picture of the correlations between tree canopy cover and social characteristics could be obtained from analyzing multiple time periods that correspond to the life expectancy of dominant species of trees in the study area. However, incorporating the importance of legacies and inherited landscapes into studies such as these requires multiple and compatible datasets.

Tree cover canopy is treated as homogeneous across the unit of analysis

In reality, tree canopy cover may be clustered with some residents in census block groups experiencing greater coverage and others experiencing less. In addition, our analysis does not distinguish between tree canopy cover on public versus private lands or residential versus non-residential lands. This may be important as management regimes likely differ on public versus private and residential versus non-residential lands [43]. This distinction may also speak to the total plantable surfaces that are available for increasing UTC cover. In addition, by not distinguishing between residential and non-residential we may be inflating or decreasing tree canopy cover. For a CBG that contains a small residential area and a large park, the sum of all the trees within the tract will be correlated with the social characteristics that correspond to only the residential area of the CBG.

Trees are not always an environmental amenity

While trees are generally accepted as an environmental amenity, evidenced by the many public and private investments supporting increased tree canopy cover; trees are sometimes considered a disamenity, especially in places of disinvestment [9]. Trees may place a direct demand on limited resources, such as water, and an indirect demand on additional resources, such as energy used to transport water. In addition, there is the *perception* that increased vegetation facilitates crime [36,61] although recent research in Baltimore has shown a decrease in crime with increasing UTC cover [62]. Residents may also incur maintenance costs [36]. Trees can also be a source of allergens and pollution precursors (VOCs). While some of the benefits attributed to increased tree canopy cover are well supported in the scientific literature, others, such as the removal of atmospheric pollutants, are more tenuous [34]. For many of the purported ecosystem services attributed to increased tree cover the data simply do not exist in order to adequately evaluate and quantify the service provided [34]. More research is needed in order to accurately evaluate the ecosystem services provided by increased tree canopy cover and evaluate the tradeoffs.

SUPPORTING INFORMATION

S1 Table: Full Model Results. The full model results of the OLS and spatial regressions at the CBG level omitting the intercept. Median household income is reported in dollars.

Baltimore, MD OLS

R square: 0.56

Number of Observations: 710

Variable	Coefficient	P value
Percent Asian	-0.09865593	0.4708787
Percent Black	0.06635435	0.0000023
Percent Hispanic	-0.09514766	0.4156902
Percent Bachelor's or Higher	0.2316971	0.0000000
Percent No High School	-0.05814618	0.1346190
Median Household Income	8.554848e-005	0.0267933
Percent Renter Occupied	0.1323241	0.0000000
Housing Density	-0.003158398	0.0000000
Median Housing Age	-0.01110426	0.0000043
Percent Residential	0.319631	0.0000000
Population Density	-0.0007175005	0.0000013

Baltimore, MD SAR (SLAG)

R square: 0.75

Number of Observations: 710

Variable	Coefficient	P value
Percent Asian	-0.1694023	0.0996490
Percent Black	0.02301605	0.0285573
Percent Hispanic	0.02626179	0.7653314
Percent Bachelor's or Higher	0.1322238	0.0000062
Percent No High School	0.006274743	0.8312878
Median Household Income	5.453843e-005	0.0603328
Percent Renter Occupied	0.07997306	0.0000016
Housing Density	-0.00126021	0.0000075
Median Housing Age	-0.008362998	0.0000036
Percent Residential	0.1221853	0.0000000
Population Density	-0.0004504699	0.0000492

Los Angeles, CA OLS

R square: 0.54

Number of Observations: 2449

Variable	Coefficient	P value
Percent Asian	-0.03447574	0.0256903
Percent Black	-0.100818	0.0000000
Percent Hispanic	0.03121855	0.0416715
Percent Bachelor's or Higher	0.1819017	0.0000000
Percent No High School	-0.05586807	0.0052837

Median Household Income	7.002071e-005	0.0000000
Percent Renter Occupied	-0.03134991	0.0023170
Housing Density	-0.001692299	0.0000000
Median Housing Age	-0.001255932	0.2331398
Percent Residential	1.425553	0.0000000
Population Density	0.0001703862	0.1312023

Los Angeles, CA SAR (SEM)

R square: 0.82

Number of Observations: 2449

Variable	Coefficient	P value
Percent Asian	-0.02908873	0.0429848
Percent Black	-0.009505632	0.5179749
Percent Hispanic	-0.00409466	0.7372137
Percent Bachelor's or Higher	0.1173555	0.0000000
Percent No High School	-0.02333432	0.1086863
Median Household Income	7.420392e-006	0.3870335
Percent Renter Occupied	-0.03290984	0.0000130
Housing Density	-0.001308047	0.0000000
Median Housing Age	-0.0005290653	0.4598202
Percent Residential	1.267915	0.0000000
Population Density	7.981155e-005	0.3246201

New York City, NY OLS

R square: 0.18

Number of Observations: 5732

Variable	Coefficient	P value
Percent Asian	-0.04702724	0.0000032
Percent Black	0.01095483	0.0208121
Percent Hispanic	-0.002188744	0.7637434
Percent Bachelor's or Higher	0.1183658	0.0000000
Percent No High School	-0.0541594	0.0000183
Median Household Income	-4.983553e-005	0.0000001
Percent Renter Occupied	-0.04463426	0.0000000
Housing Density	-0.0002112892	0.0000004
Median Housing Age	-0.001323661	0.0276709
Percent Residential	0.1535623	0.0000000
Population Density	-8.834976e-006	0.6585055

New York City, NY SAR (SEM)

R square: 0.42

Number of Observations: 5732

Variable	Coefficient	P value
Percent Asian	-0.0779728	0.0000000

Percent Black	0.01797417	0.0130568
Percent Hispanic	-0.03748614	0.0000308
Percent Bachelor's or Higher	0.078313	0.0000000
Percent No High School	-0.004395142	0.6975087
Median Household Income	-3.882175e-005	0.0000108
Percent Renter Occupied	-0.009087931	0.1971653
Housing Density	-8.598879e-005	0.0541989
Median Housing Age	-0.001792347	0.0010110
Percent Residential	0.131136	0.0000000
Population Density	-8.37889e-005	0.0000466

Philadelphia, PA OLS

R square: 0.40

Number of Observations: 1816

Variable	Coefficient	P value
Percent Asian	-0.002619466	0.9130164
Percent Black	0.05285299	0.0000000
Percent Hispanic	0.0388327	0.0045986
Percent Bachelor's or Higher	0.1391117	0.0000000
Percent No High School	-0.04905669	0.0049420
Median Household Income	0.0001675523	0.0000000
Percent Renter Occupied	0.07002264	0.0000000
Housing Density	-0.0007096402	0.0001674
Median Housing Age	-0.003258964	0.0001590
Percent Residential	0.1180562	0.0000000
Population Density	-0.0004932415	0.0000006

Philadelphia, PA SAR (SLAG)

R square: 0.60

Number of Observations: 1816

Variable	Coefficient	P value
Percent Asian	-0.01338241	0.4927785
Percent Black	0.02670228	0.0000000
Percent Hispanic	0.02816893	0.0115499
Percent Bachelor's or Higher	0.0518488	0.0001805
Percent No High School	-0.03852974	0.0067634
Median Household Income	9.323095e-005	0.0000000
Percent Renter Occupied	0.02667037	0.0029457
Housing Density	-0.0002589358	0.0915474
Median Housing Age	-0.0007899644	0.2621500
Percent Residential	0.009511328	0.4237571
Population Density	-0.0002074168	0.0104884

Raleigh, NC OLS

R square: 0.55

Number of Observations: 123

Variable	Coefficient	P value
Percent Asian	-0.02515213	0.9383915
Percent Black	0.1268416	0.0304209
Percent Hispanic	0.1223582	0.3174488
Percent Bachelor's or Higher	0.08951561	0.3628973
Percent No High School	-0.1610129	0.2077138
Median Household Income	-0.0001786685	0.0289689
Percent Renter Occupied	-0.1539209	0.0093991
Housing Density	-0.006663367	0.1160905
Median Housing Age	0.007863337	0.1747404
Percent Residential	0.4096791	0.0000000
Population Density	0.002256158	0.2529118

Raleigh, NC SAR (SEM)

R square: 0.56

Number of Observations: 123

Variable	Coefficient	P value
Percent Asian	-0.03859736	0.9013044
Percent Black	0.1051943	0.0732561
Percent Hispanic	0.1113269	0.3274092
Percent Bachelor's or Higher	0.06047299	0.5284501
Percent No High School	-0.1587924	0.2077453
Median Household Income	-0.0001725719	0.0255022
Percent Renter Occupied	-0.1539177	0.0054694
Housing Density	-0.006580671	0.1053713
Median Housing Age	0.008441119	0.1302369
Percent Residential	0.4111514	0.0000000
Population Density	0.002274611	0.2181045

Sacramento, CA OLS

R square: 0.56

Number of Observations: 289

Variable	Coefficient	P value
Percent Asian	-0.1245937	0.00083
Percent Black	-0.1092469	0.01126
Percent Hispanic	0.02616373	0.55740
Percent Bachelor's or Higher	0.2993236	0.00000
Percent No High School	0.1203542	0.01446
Median Household Income	3.052681e-005	0.56198
Percent Renter Occupied	0.02355939	0.45280
Housing Density	0.003322599	0.00067

Median Housing Age	-0.006061257	0.04170
Percent Residential	0.1970188	0.00000
Population Density	-0.0005213956	0.07004

Sacramento, CA SAR (SEM)

R square: 0.79

Number of Observations: 289

Variable	Coefficient	P value
Percent Asian	0.01273681	0.68077
Percent Black	-0.02369551	0.54529
Percent Hispanic	0.03276404	0.31286
Percent Bachelor's or Higher	0.1579269	0.00015
Percent No High School	-0.03080654	0.40487
Median Household Income	0.0001039211	0.00296
Percent Renter Occupied	0.01373081	0.51988
Housing Density	0.002616896	0.00084
Median Housing Age	-0.003613357	0.06247
Percent Residential	0.1828874	0.00000
Population Density	-0.0002772249	0.14260

Washington, DC OLS

R square: 0.46

Number of Observations: 433

Variable	Coefficient	P value
Percent Asian	0.1015783	0.6215174
Percent Black	0.03952639	0.4409423
Percent Hispanic	0.2388635	0.0097067
Percent Bachelor's or Higher	0.07101428	0.3453675
Percent No High School	-0.166635	0.0512760
Median Household Income	0.0002001232	0.0000005
Percent Renter Occupied	0.1154585	0.0009326
Housing Density	-0.0007224304	0.3126912
Median Housing Age	-0.006354714	0.1403559
Percent Residential	0.288331	0.0000000
Population Density	-0.001309582	0.0022574

Washington, DC SAR (SLAG)

R square: 0.70

Number of Observations: 433

Variable	Coefficient	P value
Percent Asian	0.07901413	0.6021071
Percent Black	0.0459065	0.2246929
Percent Hispanic	0.09276362	0.1711534
Percent Bachelor's or Higher	0.0318401	0.5660820

Percent No High School	-0.05936688	0.3451256
Median Household Income	9.604378e-005	0.0009649
Percent Renter Occupied	0.02772408	0.2782260
Housing Density	7.992388e-005	0.8797373
Median Housing Age	-0.0036835	0.2454111
Percent Residential	0.03903467	0.2150817
Population Density	-0.0005928335	0.0597203

CONCLUSION

Money may not grow on trees, but this study suggests that in a way, trees grow on money. Our findings show that high-income neighborhoods in our selected cities are more likely than low-income neighborhoods to have high tree canopy cover. We did not find, however, the expected pattern of low tree canopy cover in neighborhoods with a high percentage of racial and ethnic minorities for all cities; only in the California cities of Sacramento and Los Angeles did we see such a result. Since these cities depend on irrigation to maintain tree canopy cover, this distinguishing factor of aridity may amplify socioeconomic differences in neighborhood tree cover more than in the humid temperate cities we analyzed. In-depth analyses of the costs and benefits of maintaining tree cover in arid and humid environments along with qualitative analyses of residential land management practices could provide some explanation for these differences.

Increasing UTC has become a widespread goal, often incorporated into municipal sustainability plans. It has been proposed as a way to mitigate impacts from human-dominated systems on the immediate (e.g. shade and cooling) and global (e.g. carbon capture) environment. Sacramento Tree Foundation has pledged to plant five million trees by the year 2025, an effort that would double the region's tree canopy cover. Philadelphia has established a goal of increasing tree canopy cover to 30% by the year 2025 (www.phila.gov/green/trees). New York City, Baltimore, and Los Angeles have also announced extensive tree planting initiatives (www.milliontreesnyc.org, www.baltimorecity.gov, www.milliontreesla.org). In addition to regional efforts, there are national and global efforts to bring more awareness to the benefits of UTC cover (Urban Environmental Accord 2005,www.sfenvironment.org/downloads/library/accords.pdf; http://www.plant-for-the-planet-billiontreecampaign.org/Partners/VariousPartners/TreePlanting.aspx). One of the implications of embedding tree canopy goals in sustainability plans is that environmental justice is frequently included as an objective of the plans, and sometimes explicitly linked to UTC. For example, Philadelphia's 2009 GreenWorks Plan includes goals of increasing tree canopy cover*in all neighborhoods* highlighting the desire for the equitable distribution of UTC cover (www.phila.gov/green/greenworks/2009-greenworks-report.html).

If the equity dimensions of sustainability are put into practice, UTC goals can help to redress environmental injustices. However, it is important to note that tree planting schemes are not a panacea for environmental justice. Often overlooked is that the presence of trees can generate disservices, and in some cases costs may exceed local benefits or local desire and capacity to care for trees and other green infrastructure [33,34,63]. Benefits from tree canopy cover must therefore be assessed in spatially explicit ways, since in some circumstances tree cover will provide strong positive services while in others tree cover may provide negligible benefits or disservices.

The association between income and tree canopy cover has important implications for urban sustainability plans, many of which include increased UTC cover as a goal. If UTC and income are positively reinforced (through property values, investments, or other mechanisms) any public or private interventions to increase tree canopies might first consider the needs of low income communities, provided that the economic and ecological benefits do not outweigh the costs and that the initiatives are supported by local residents. In addition to considering distributional equity, urban sustainability goals need to be supported by stronger science on perceived ecosystem services provided by UTC cover so that justice can be considered in terms of the distribution of ecosystem services, not just the equitable distribution of trees.

AUTHOR CONTRIBUTIONS

Conceived and designed the experiments: KS MF CGB WZ MRM JMG JOD JPM GLB DC LO SP DP AW MLC. Analyzed the data: KS MF CGB WZ JOD. Wrote the paper: KS MF CGB WZ MRM JMG JPM MLC.

REFERENCES

1. Bullard RD. Dumping in Dixie: Race, Class, and Environmental Quality. Boulder, CO: Westview Press; 2000.
2. Adamson J, Evans MM, Stein R. The Environmental Justice Reader: Politics, Poetics, and Pedagogy. Tucson, AZ: University of Arizona Press; 2002.
3. Boone CG. Environmental justice as process and new avenues for research. Environ Justice. 2008;1: 149–154. doi: 10.1089/env.2008.0530
4. Low S, Taplin D, Scheld S. Rethinking Urban Parks: Public Spaces and Cultural Diversity. Austin, TX: University of Texas Press; 2005.
5. Wolch JR, Wilson JP, Fehrenbach J. Parks and park funding in Los Angeles: an equity-mapping analysis. Urban Geogr. 2005;26: 4–35. doi: 10.2747/0272-3638.26.1.4
6. Boone CG, Buckley GL, Grove JM, Sister C. Parks and people: an environmental justice inquiry in Baltimore, Maryland. Ann Assoc Am Geogr. 2009;99: 767–787. doi: 10.1080/00045600903102949
7. Cutts BB, Darby KJ, Boone CG, Brewis A. City structure, obesity, and environmental justice: an integrated analysis

of physical and social barriers to walkable streets and park access. Soc Sci Med. 2009;69: 1314–1322. 10.1016/j.socscimed.2009.08.020. pmid:19751959

8. Pedlowski MA, Da Silva VAC, Adell JJC, Heynen NC. Urban forest and environmental inequality in Campos dos Goytacazes, Rio de Janeiro, Brazil. Urban Ecosyst. 2002;6: 9–20.

9. Heynen NC, Perkins HA, Roy P. The political ecology of uneven urban green space. Urban Aff Rev. 2006;42: 3–25. doi: 10.1177/1078087406290729

10. Landry SM, Chakraborty J. Street trees and equity: evaluating the spatial distribution of an urban amenity. Environ Plan A. 2009;41: 2651–2670. doi:10.1068/a41236.

11. Boone CG, Cadenasso ML, Grove JM, Schwarz K, Buckley GL. Landscape, vegetation characteristics, and group identity in an urban and suburban watershed: why the 60s matter. Urban Ecosyst. 2010;13: 255–271. doi: 10.1007/s11252-009-0118-7

12. Downey L. Environmental racial inequality in Detroit. Soc Forces. 2006;85: 771–796. pmid:21874071 doi: 10.1353/sof.2007.0003

13. Mohai P, Saha R. Racial inequality in the distribution of hazardous waste: a national level assessment. Soc Probl. 2007;54: 343–370. doi: 10.1525/sp.2007.54.3.343

14. Su JG, Larson T, Gould T, Cohen M, Buzzelli M. Transboundary air pollution and environmental justice: Vancouver and Seattle compared. GeoJournal. 2010;75: 595–608. doi: 10.1007/s10708-009-9269-6

15. Downey L. US metropolitan-area variation in environmental inequality outcomes. Urban Stud. 2007;44: 953–977. pmid:21909171 doi: 10.1080/00420980701256013

16. Tooke TR, Klinkenber B, Coops NC. A geographical approach to identifying vegetation-related environmental equity in Canadian cities. Environ Plan B Plan Des. 2010;37: 1040–1056. doi: 10.1068/b36044

17. Clark LP, Millet DB, Marshall JD. National patterns in environmental injustice and inequality: outdoor NO_2 air pollution in the United States. PLoS One. 2014;9: e94431.10.1371/journal.pone.0094431. pmid:24736569

18. Saha R, Mohai P. Historical context and hazardous waste facility siting: understanding temporal patterns in Michigan. Soc Probl. 2005;52: 618–648. doi: 10.1525/sp.2005.52.4.618

19. Dwyer JF, McPherson GE, Schroeder HW, Rowntree RA. Assessing the benefits and costs of the urban forest. J Arboric. 1992;18: 227–234.

20. McPherson GE, Rowntree RA. Energy conservation potential of urban tree planting. J Arboric. 1993;19: 321–331.

21. McPherson GE. Net benefits of healthy and productive urban forests. In: Bradley GA, editor. Urban Forest Landscapes: Integrating Multidisciplinary Perspectives. Seattle, WA: University of Washington Press; 1995. pp. 180–194.

22. McPherson GE. Urban forest landscapes: how greenery saves greenbacks. In: Wagner C, editor. Proceedings of the Annual Meeting of the American Society of Landscape Architects. Washington DC; 1996. pp. 27–29.

23. Xiao Q, McPherson GE. Rainfall interception by Santa Monica's municipal urban forest. Urban Ecosyst. 2002;6: 291–302. doi: 10.1023/b:ueco.0000004828.05143.67

24. Xiao Q, McPherson GE, Simpson JR, Ustin SL. Rainfall interception by Sacramento's urban forest. J Arboric. 1998;24: 235–244. pmid:10410677

25. Nowak DJ, Crane DE. Carbon storage and sequestration by urban trees in the USA. Environ Pollut. 2002;116: 381–389. pmid:11822716 doi: 10.1016/s0269-7491(01)00214-7

26. Nowak DJ, Crane DE, Stevens JC. Air pollution removal by urban trees and shrubs in the United States. Urban For Urban Green. 2006;4: 115–123. doi: 10.1016/j.ufug.2006.01.007

27. Bell JF, Wilson JS, Liu GC. Neighborhood greenness and 2-year changes in body mass index of children and youth. Am J Prev Med. 2008;35: 547–553.10.1016/j.amepre.2008.07.006. pmid:19000844

28. Kaplan R, Kaplan S. The Experience of Nature: A Psychological Perspective. New York: Cambridge University Press; 1989.

29. Wolf KL. Urban ecosystem services: metro nature and human well-being. In: Laband DN, editor. Emerging Issues along Urban/Rural Interfaces III: Linking Science and Society. Atlanta, GA; 2010. p. 45.

30. Kuo FE. The role of arboriculture in a healthy social ecology. J Arboric. 2003;29: 148–155.

31. Westphal LM. Urban greening and social benefits: a study of empowerment outcomes. J Arboric. 2003;29: 137–147.

32. Holtan MT, Dieterlen SL, Sullivan WC. Social life under cover: tree canopy and social capital in Baltimore, Maryland. Environ Behav. 2014; 0013916513518064. doi: 10.1177/0013916513518064

33. Pincetl S, Bunje P, Holmes T. An expanded urban metabolism method: Toward a systems approach for assessing urban energy processes and causes. Landsc Urban Plan. Elsevier B.V.; 2012;107: 193–202. doi:10.1016/j.landurbplan.2012.06.006.

34. Pataki DE, Carreiro MM, Cherrier J, Grulke NE, Jennings V, Pincetl S, et al. Coupling biogeochemical cycles in urban environments: ecosystem services, green solutions, and misconceptions. Front Ecol Environ. 2011;9: 27–36. doi:10.1890/090220.

35. Pincetl S, Gillespie T, Pataki DE, Saatchi S, Saphores JD. Urban tree planting programs, function or fashion? Los Angeles and urban tree planting campaigns. GeoJournal. 2013;78: 475–493. doi: 10.1007/s10708-012-9446-x

36. Pincetl S. Implementing municipal tree planting: Los Angeles Million-Tree Initiative. Environ Manage. 2010;45: 227–238. 10.1007/s00267-009-9412-7. pmid:20016982

37. Harlan SL, Brazel AJ, Prashad L, Stefanov WL, Larsen L. Neighborhood microclimates and vulnerability to heat stress. Soc Sci Med. 2006;63: 2847–2863. pmid:16996668 doi: 10.1016/j.socscimed.2006.07.030

38. Cutter SL, Holm D, Clark L. The role of geographic scale in monitoring environmental justice. Risk Anal. 1996;16: 517–526. doi: 10.1111/j.1539-6924.1996.tb01097.x
39. Openshaw S, Taylor PJ. A million or so correlation coefficients: three experiments on the modifiable areal unit problem. In: Wrigley N, editor. Statistical Applications in the Spatial Sciences. London: Pion; 1979. pp. 127–144.
40. Moskal LM, Styers DM, Richardson J, Halabisky M. Seattle hyperspectral land use/land cover (LULC) from LiDAR and near infrared aerial imagery using object-based image analysis (OBIA). Seattle, WA; 2011.
41. Homer C, Huang C, Yang L, Wylie B, Coan M. Development of a 2001 national land-cover database for the United States. Photogramm Eng Remote Sens. 2004;70: 829–840. doi: 10.14358/pers.70.7.829
42. Cadenasso ML, Pickett STA, Schwarz K. Spatial heterogeneity in urban ecosystems: reconceptualizing land cover and a framework for classification. Front Ecol Environ. 2007;5: 80–88. doi:10.1890/1540–9295(2007)5[80:SHIUER]2.0.CO;2.
43. Grove JM, Troy AR, O'Neil-Dunne JPM, Burch WR Jr., Cadenasso ML, Pickett STA. Characterization of households and its implications for the vegetation of urban ecosystems. Ecosystems. 2006;9: 578–597. doi:10.1007/s10021–006–0116-z. pmid:17354937
44. Gillespie TW, Pincetl S, Brossard S, Smith J, Saatchi S, Pataki DE, et al. A time series of urban forestry in Los Angeles. Urban Ecosyst. 2012;15: 233–246. doi: 10.1007/s11252-011-0183-6
45. Anselin L. Spatial Econometrics: Methods and Models. Dudrecht, The Netherlands: Kluwer Academic Publishers; 1988.
46. Anselin L. Spatial effects in econometric practice in environmental and resource economics. Am J Agric Econ. 2001;83: 705–710. doi: 10.1111/0002-9092.00194
47. Wagner HH, Fortin M- J. Spatial analysis of landscapes: concepts and statistics. Ecology. 2005;86: 1975–1987. doi: 10.1890/04-0914
48. McPherson GE, Simpson JR, Xiao Q, Wu C. Million Trees Los Angeles canopy cover and benefit assessment. Landsc Urban Plan. 2011;99: 40–50. doi: 10.1016/j.landurbplan.2010.08.011
49. Tobler WR. A computer movie simulating urban growth in the Detroit region. Econ Geogr. 1970;46: 234–240. doi: 10.2307/143141
50. Anselin L. The future of spatial analysis in the social sciences. Geogr Inf Sci. 1999;5: 67–76. doi: 10.1080/10824009909480516
51. .Chi G, Zhu J. Spatial regression models for demographic analysis. Popul Res Policy Rev. 2008;27: 17–42. doi: 10.1007/s11113-007-9051-8
52. Lichstein JW, Simons TR, Shriner SA, Franzreb KE. Spatial autocorrelation and autoregressive models in ecology. Ecol Monogr. 2002;72: 445–463. doi: 10.2307/3100099
53. Anselin L. Exploring spatial data with GeoDa: a workbook. Urbana, IL; 2005 p. 61801.
54. Pickett STA, Cadenasso ML, Grove JM. Resilient cities: meaning, models, and metaphor for integrating the ecological, socio-economic, and planning realms. Landsc Urban Plan. 2004;69: 369–384. 10.1016/j.landurbplan.2003.10.035.
55. Pickett STA, White PS. The Ecology of Natural Disturbance and Patch Dynamics. New York: Academy Press; 1985. pmid:17816073
56. Battaglia MJ. A multi-methods approach to determining appropriate locations for tree plantings in two of Baltimore's tree-poor neighborhoods. Ohio University. 2010.
57. .Bigsby KM, McHale MR, Hess GR. Urban morphology drives the homogenization of tree cover in Baltimore, MD and Raleigh, NC. Ecosystems. 2014;17: 212–227. doi: 10.1007/s10021-013-9718-4
58. Boone CG. An assessment and explanation of environmental inequity in Baltimore. Urban Geogr. 2002;23: 581–595. doi: 10.2747/0272-3638.23.6.581
59. Mennis JL, Jordan L. The distribution of environmental equity: exploring spatial nonstationarity in multivariate models of air toxic releases. Ann Assoc Am Geogr. 2005;95: 249–268. doi: 10.1111/j.1467-8306.2005.00459.x
60. Troy AR, Grove JM, O'Neill-Dunne JP, Pickett STA, Cadenasso ML. Predicting opportunities for greening and patterns of vegetation on private urban lands. Environ Manage. 2007;40: 394–412. pmid:17602257 doi: 10.1007/s00267-006-0112-2
61. .Kuo FE, Sullivan WC. Environment and crime in the inner city: does vegetation reduce crime? Environ Behav. 2001;33: 343–367. doi: 10.1177/00139160121973025
62. Troy AR, Grove JM, O'Neill-Dunne JP. The relationship between tree canopy and crime rates across an urban-rural gradient in the greater Baltimore region. Landsc Urban Plan. 2012;106: 262–270. doi: 10.1016/j.landurbplan.2012.03.010
63. Lyytimäki J, Petersen LK, Normander B, Bezák P. Nature as a nuisance? Ecosystem services and disservices to urban lifestyle. Environ Sci. 2008;5: 161–172. doi: 10.1080/15693430802055524

CHAPTER 11 ENHANCEMENT OF TREE PRODUCTIVITY FROM FOREST TO URBAN LAND COVERS

Brittain M. Briber, Lucy R. Hutyra, Andrew B. Reinmann, Steve M. Raciti, Victoria K. Dearborn, Christopher E. Holden, Allison L. Dunn

INTRODUCTION

Terrestrial ecosystems are an important and dynamic component of the global carbon cycle. In recent decades, terrestrial ecosystems have sequestered approximately 25% of the carbon emitted to the atmosphere by human activities [1]. Patterns of human development, including deforestation and urbanization, have changed the spatial distribution, structure, and extent of terrestrial ecosystems. Understanding the consequences of urban land use and land cover changes on terrestrial productivity and carbon stocks is critical for modeling the local, regional and global carbon cycle [2,3].

Population growth, migration into cities, and sprawling forms of land development are increasing urban populations and the spatial extent of cities. Between 1950 and 2010, the global population grew from 2.6 to 6.9 billion people, with the percentage of people living in cities increasing from 29% to 52% [4]. Urban areas comprise nearly 3% of global land area [5], an extent comparable to that of temperate forests [6], and the urban extent is projected to double by 2050 [4]. In the U.S., approximately 80% of the population now lives in urban areas [7] and urban land cover expanded by 50–380% between 1974 and 2002, depending on how urban land cover is defined [8–10].

Urban land cover is inherently heterogeneous, with roads, buildings, and vegetation co-occurring within small patches of land. Urban areas can include significant quantities of vegetation [11], and often have a complex land use history [12]. These areas represent a patchwork of vegetated and developed land covers across which ecosystem function, productivity, and structure vary significantly [13,14].

The global environmental impacts of cities are growing [15–17]. Cities are estimated to consume 67% of global energy and emit 71% of energy-related carbon emissions [18]. Further, urban development results in additional carbon emissions associated with the clearing and paving of land [17,19,20]. Studies also suggest that urban areas have unique meteorological and atmospheric properties [21], distinct flora and faunal diversity [22], and higher levels of air and water pollution due to increased human activity and waste production [16]. These ecological impacts of urbanization extend far beyond individual city limits through environmental teleconnections and demand for agricultural and energy related goods and services [23,24].

Vegetation in urban areas experiences modified growing conditions [14,25]. Urban areas typically exhibit 3–12°C higher mean annual temperatures than adjacent rural locations

Citation: BM, Hutyra LR, Reinmann AB, Raciti SM, Dearborn VK, Holden CE, et al, Enhancement of Tree Productivity from Forest to Urban Land Covers, PLoS ONE 10(8): e0136237. doi:10.1371/journal.pone.0136237

[26–28] and this "urban heat island" effect has been shown to increase growing season length by 15–31 days in New England cities [29,30]. Fertilizer applications and enhanced urban atmospheric N deposition have been associated with increased soil N concentrations [31]. Atmospheric N inputs have been positively correlated with on-road CO_2 emissions [32] and proximity to major roadways [33], highlighting the influence of human activity on plant nutrient inputs. Furthermore, urban areas often exhibit atmospheric CO_2 concentrations that are 10–50 ppm above ambient, which can stimulate plant productivity [29,34,35]. Elevated concentrations of urban atmospheric aerosols increase diffuse beam radiation, thereby potentially enhancing photosynthetic efficiency of vegetation [36,37]. While increases in N inputs, atmospheric CO_2, and light availability coupled with a longer growing season may enhance plant productivity in urban areas, carbon deficient and compacted soils [38–40], foliar ozone damage [41], and reduced water availability due to runoff and the presence of hydrophobic soils [13] may limit plant productivity. Our understanding of the net effect of urbanization on plant growth remains uncertain. The few existing *in situ* studies focusing on urban vegetation productivity found higher growth rates for urban trees than for trees in adjacent rural areas [42,43]. These studies focused on seedlings and saplings and did not consider land use history in their sampling.

In this study we assessed the effect of land cover conversion from forest to a variety of urban land covers on vegetation dynamics. Specifically, we used a combination of remote sensing, field biometric measurements, dendrochronology, and tax assessor records to quantify how vegetation structure and growth rates changed as a function of urban development type and development intensity. We hypothesized that urbanization would simultaneously decrease terrestrial carbon pools and increase the relative ecosystem productivity of the remaining vegetation.

MATERIALS AND METHODS

Study Area

This study was conducted across a range of land covers in eastern Massachusetts (Fig 1). Boston is the largest urban area in New England, both in terms of population and land extent. The population within the Boston Metropolitan Statistical Area (MSA) is approximately 4.6 million, making it the 10th largest MSA in the United States. Since 1950, the Boston MSA population has increased by an average of nearly 1% yr^{-1} [44]. The region has a humid continental climate (Köppen classification Dfa), which is characterized by cold, snowy winters and hot, humid summers: Boston's mean January and July temperatures are -1.5° and 23.3°C, respectively. There is a 0.25–1.5°C mean annual temperature gradient between our coastal and inland field sites. Mean annual precipitation is 1056 mm and is evenly distributed throughout the year [45]. Vegetation in eastern Massachusetts is dominated by temperate mixed-hardwood forests that include red oak (*Quercus rubra*), white pine (*Pinus strobus*), red maple (*Acer rubrum*), and paper birch (*Betula papyrifera*) [46]. Soils in eastern Massachusetts are glacial in origin and are dominated by sand and silt loams and, to a lesser extent, rocky outcroppings [47].

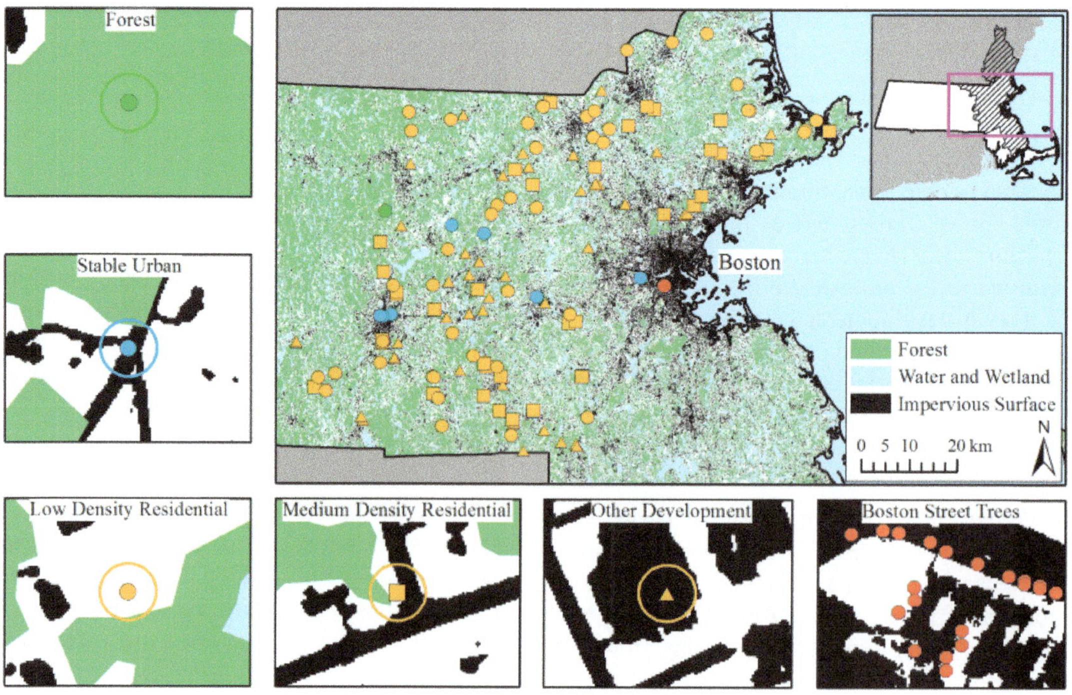

Figure 1: The study area. The maps show the distribution of forested land use/cover and impervious surfaces across eastern Massachusetts. The hatched area on the inset denotes the extent of the Boston MSA. Each point on the map reflects field plot locations and the area for street tree sampling. All of the orange points were converted from forest to urban land uses between 1971 and 2012; the shapes of the points denote the intensity of urban development, which include low density residential (LDR; circle), medium density residential (MDR; square), and other development (OTH; dominated by commercial, industrial, high density residential, and parking lot areas; triangle). The green, blue, and red points indicate plot locations for stable forests, stable urban land covers, and City of Boston street trees, respectively. The panels surrounding the central map illustrate the land cover characteristics of the different plot types, with white areas representing all other land covers (e.g. lawns, isolated trees, etc.). Example field plots are shown to scale (30 m diameter) and point locations for street trees correspond to sample trees. http://dx.doi.org/10.1371/journal.pone.0136237.g001

Land use and land cover in eastern Massachusetts vary considerably and reflect evolving land use practices and development patterns over the last several hundred years. Most of the forested land in eastern Massachusetts was cleared for agriculture or human settlement in the 19th century. Agricultural abandonment in the early 20th century resulted in widespread reforestation across the region [48]. While eastern Massachusetts has a large and dense population, it also includes patches of extensive vegetation cover, with 41.7% of the Boston MSA forested [49].

To quantify changes in vegetation structure and productivity that occur after urban development, we established 135 field plots (706.9 m^2 each) across eastern Massachusetts. All plots were located on land with a coarse-loamy, mixed, mesic typic dystrochrepts soil, the most common soil family in eastern Massachusetts [47]. All of these plots were converted from forest to an urban land cover between 1971 and 2012. We also revisited existing plots from Raciti et al. [49] that were on the same soil family and had red oak present. Additionally, and in collaboration with the Morton Arboretum, we analyzed street tree growth rates in the City of Boston (Scharenbroch, unpublished data). For the purposes of data comparison and extrapolation, our final study area included five eastern and central Massachusetts counties: Essex, Suffolk, Middlesex, Norfolk and Worcester.

Plot Selection and Stratification

Definitions of "forest" and "urban" areas vary geographically and with the context of analysis (e.g. Raciti et al. [49]). For the purposes of this analysis, we used two different remote sensing-based land use and land cover change datasets to identify candidate locations for field sampling and categorize land cover. The MacConnell dataset was based on manual photo interpretation and included land use maps for 1971, 1985, and 1999 [50]. In addition, we used the Continuous Change Detection and Classification (CCDC) [51] algorithm applied to a timeseries of Landsat observations in order to identify changes between 1984 and 2012. Combining the two datasets provided a forty-year time window for land cover change analysis.

The MacConnell land use maps were created by classifying 21 distinct land covers/uses using 1:25,000 aerial photographs across Massachusetts. The resulting data layers were developed explicitly for land cover change analysis and contain nested polygons at a 0.4 ha minimum mapping unit, with each polygon containing classification values for all three time periods. Polygons containing change after 1971 were subset and relabeled in a nested, hierarchical fashion to allow direct comparison across time periods. Using the CCDC algorithm, we analyzed the spectral reflectance from all available Landsat observations between 1984 and 2012 at each 30m pixel to identify the timing of spectral disturbances. We categorized change in land cover by classifying the land cover of stable time periods before and after a disturbance and comparing pixel-level reflectances to their stable model statistics [51]. These land cover change products were combined to stratify field plots based on the intensity of urban development and time since development.

The MacConnell and CCDC products had different mapping units (0.4 ha and 0.09 ha, respectively) and different output data given their focus on land use and land cover, respectively. To facilitate comparison, the CCDC maps were filtered to include only land cover change pixels that were neighbored by at least two additional change pixels, changing from forest to any type of developed land cover. Filtering by such pixel groups helped to reduce high frequency errors and spectral changes that did not represent actual land cover change while increasing the effective minimum mapping unit to a level comparable with the MacConnell maps. In addition, we spatially dilated the area of land cover change when there were three or more change pixels adjacent to one another. Dilating areas of land cover change identified by CCDC reduced change pixel omission along the edges of these patches and created continuous polygons that more closely reflect the land use change polygons mapped by MacConnell. For example, while forest edges adjacent to new suburban developments might not have changed their spectral reflectance—and thus would not have been picked up by CCDC—the land use may change from a forest to a developed, residential use.

Using land use and land cover change locations identified by the MacConnell and CCDC maps, we stratified field plots into three urban intensity categories; low density residential (LDR), medium density residential (MDR), and other development (OTH). These classifications, which were generated before fieldwork began, were also used to analyze and parse processed data. These classes were defined by combining a collapsed version of a Landsat based Massachusetts training dataset [51] with a 2005 1m impervious Massachusetts GIS data layer [52]. Areas around and including the LDR plots exhibited 0–50% impervious surface area (ISA) which contained single family residences; areas around and including the MDR plots exhibited 0–80% ISA and typically contained multi-story and condominium residences; areas around and including the OTH plots exhibited greater than 80% ISA and were dominated by parking lots and commercial and industrial buildings. There was overlap in the ISA range for the LDR and MDR classes, but the land use composition of those classes differed in housing density. The urban categorization was confirmed with field observation. All field plots converted from forest to urban development and were further stratified into three time windows of conversion: 1971–1985, 1985–1999, and 1999–2012. This sample design resulted in nine different land use and time since land cover change categories.

With a goal of sampling at least 15 plots in each of these nine change categories, we generated a stratified random sample of 1,500 candidate plot locations from both public and private lands sharing the same soil type. Following the methods outlined in Raciti et al. [49], we used digital parcel data to derive ownership information and mailed letters requesting permission to sample. We visited only those sites for which permission to conduct the study on that site was granted, including sites on public and private property. Access limitations prevented us from using a truly random sample. In order to prevent a potential bias towards plots where access was easier (e.g., commercial parking with no vegetation), we sampled public and private properties in the same proportion as found among the first 15 randomly chosen plots in each category. The final number of plots in this "land cover change" group included a minimum of 14 plots in each of the nine categories for a total of 135 plots (Fig 1). For most of the analysis, the plots were aggregated based on the land use conversion category: forest-to-LDR, forest-to-MDR, and forest-to-OTH. The minimum number of plots in each of these three categories was 43.

Field Sampling

Field sampling took place from June through August 2013. Each sample plot was circular with a 15 m radius (706.9 m^2). Plot centers were located using satellite basemaps in ArcMap 10.0 (ESRI, Redlands, CA) and handheld GPS units with accuracies of ~3 m. A TruPulse 200 Professional Range Finder and Hypsometer (Laser Technology, Centennial, CO, U.S.) and field measuring tapes were used to determine the plot boundaries. The tree canopy cover was visually estimated from the ground at each plot and ground cover fraction was visually estimated for lawn, garden, forest, degraded forest (defined as small, unmanaged tracts that were often surrounded by development), weedy, bare ground, and paved surfaces.

All trees with a diameter at breast height (DBH; measured at 1.37 m aboveground) ≥ 5 cm were measured and identified to species, whenever possible. Coarse woody debris (CWD) with a minimum diameter of 10cm and 1m length were measured within the plot area using either a Haglöfs (Langsele, Sweden) 40 cm analogue caliper or a DBH tape measure. See Raciti et al. [49] for a complete methodological description of the aboveground live and dead biomass estimations.

Tree Cores

We extracted one tree core in each plot at breast height from a red oak ≥ 20 cm in DBH with a 200 mm long Haglöfs 5 mm increment borer. If no red oaks were present within a sample plot, cores were extracted from a similarly sized red oak within ~50m of the plot center. Only one core was extracted from each plot due to property owner concerns about tree damage. In an attempt to control for differences in light conditions across plots, we preferentially extracted cores from canopy trees that were located along forest edges. These trees frequently had crowns that were directly abutted on one side by other trees' crowns. Cores were mounted and glued onto grooved wood blocks in the field.

A total of 75 tree cores were extracted. After drying, each core was sanded on a belt sander to a broad, flat surface using coarse, 220 grit sand paper and then sanded by hand with fine, 400 grit paper to create a smooth surface. Cores were scanned using a high-resolution color scanner (Epson Perfection V700 Photo) and ring increments were measured using WinDENDRO 2012 (Regent Instruments, Inc., Sainte-Foy, Quebec, Canada) image analysis software. All increment measurements were independently verified for their ring count and position by at least two observers. When chronologies were long enough, initial cross-dating was performed by identifying statistically significant reductions in growth during the gypsy moth outbreak of 1980 [53]. All measurements, regardless of chronology length, were cross-dated using the program COFECHA [54,55].

DBH (cm) for each year in a core's chronosequence was calculated as

$$DBH_i = DBH_{2013} - 2 * RD_i \tag{1}$$

where DBH_{2013} is the DBH measured in 2013 DBH and RD is the radial distance from the year i to the bark. Biomass was calculated using a forest-based red oak allometric equation [56] with biomass in kg and DBH in cm:

$$Biomass = 0.113 * (DBH^{2.4572}) \tag{2}$$

Basal area was estimated as follows with basal area in cm^2 and DBH in cm:

$$Basal\ area = \pi * (DBH/2)^2 \tag{3}$$

Basal Area Increment (BAI) and biomass increment (BI) were calculated for each year (Y) as:

$$BAI_Y = basal\ area_Y - basal\ area_{Y-1} \tag{4}$$

$$BI_Y = biomass_Y - biomass_{Y-1} \tag{5}$$

where the units of BAI and BI are cm^2 yr^{-1} and Kg yr^{-1}, respectively.

Land cover change dates for each core were determined using a combination of Massachusetts parcel data from tax assessor records [57], remote sensing CCDC results [51], and aerial imagery from Google Earth [58]. In cases where the tax assessor record was incomplete, we used alternative commercial real estate databases to supplement the parcel records (www.thewarrengroup.com, www.zillow.com). In addition, time-lapse imagery in Google Earth was used to confirm and identify land cover changes for each plot between 1993 and 2012.

We required a minimum of five years of increment data before and after land cover change for each core to obtain robust growth estimates. We assessed the suitability of requiring 10 years of rings before and after development, but found that 1) this reduced the number of available cores from 59 to 46; 2) the 10 year window yielded aggregated before and after BAI means within 2% and 18%, respectively, of those using the 5 year window; and 3) the 5-year window acceptably captured typical inter-annual variability (S1 Fig). Of the 75 cores extracted in 2013 for analysis in the "land cover change" category, four cores were excluded from analysis as they did not have a minimum of five years of growth increment data both before and after the established land cover change date and nine additional cores were removed due to severe rot and/or unreadable rings. Five cores were reclassified from the land cover change category to the "stable urban" category (see below) as these plots likely had no vegetation pre-dating land cover change or we could not locate any physical records that the plot had been modified during the last 50 years. Two cores from the Raciti et al. [49] sampling effort were found to have converted from forest-to-LDR since 1971 and were reclassified accordingly. In total, we had 59 useable cores for our land cover change analysis, 57 of which came from our 2013 field campaign and two of which came from our 2010 field campaign.

Background, Stable Urban, and Street Tree Increment Data

In order to develop background regional tree growth rates for forest trees we used Forest Inventory and Analysis data [59] from 99 red oaks within our study area. The spatial distribution of these FIA data were not as even as the spatial distribution of our land cover change cores: 82 of the FIA data originated from trees in Worcester County, while the remaining 11 came from locations north and northwest of Boston. Annual increment data was not available from the FIA; we estimated an average increment by dividing the most recent DBH by the tree age.

Tree growth rates for street trees within the City of Boston were estimated using 17 red oak cores collected by the Morton Arboretum in 2012 (Scharenbroch, personal communication). The Boston street trees were all open grown, and most were in high exposed light environments. The street tree cores were prepared (mounted, sanded, etc.) and scanned by the Morton Arboretum; we processed the images with WinDENDRO using the methods described above.

We also obtained tree growth rates for our "stable urban" category by revisiting 13 plots from the work described in Raciti et al. [49]. Eleven of these plots did not undergo land cover or use change from 1971 to 2012 and were classified as stable urban. The two plots that underwent recent forest to urban land cover change were analyzed with our 2013, land cover change data. Including the four reclassified cores from our 2013 field campaign, we analyzed a total of 15 stable urban cores. One of the 15 plots was also identified as being completely forested (seeFig 1).

All 2010 and 2013 plots, FIA, and Boston street tree data originated from within 55 miles of Boston. What distinguished our urban, land cover change plots from our background, forest plots was that our urban plots were in locations that had experienced anthropogenic land cover change. A list of all plots from the land cover change, stable urban, Boston street tree, and FIA categories with location information is provided in the supplemental materials (S1 Table).

Statistical Analysis

Statistical analysis was conducted using R Statistical Software 3.1.3 [60]. Unless specifically noted otherwise, all reported errors in the text and figures reflect 95% confidence intervals. Confidence intervals were calculated by bootstrap analyses [61] due to heteroscedasticity within the data distributions. Bootstrap samples were drawn 1000 times with replacement to estimate 95% confidence intervals around mean. Error estimates do not include allometric or spatial scaling errors. An alpha-value of 0.05 was used to denote significance.

RESULTS

Plot Biomass, Ecosystem Structure, and Land Cover in Urban Plots

Mean plot-level DBH, total live aboveground biomass, stem density, and canopy cover all declined as development intensity increased such that LDR > MDR > OTH (Table 1 and Fig 2). Across all forest-to-urban land cover conversion pathways, we found a mean, area-weighted aboveground biomass of 42.7 ± 14.7 Mg C ha^{-1}, approximately 54% of the mean eastern MA forest biomass (78.6 Mg C ha^{-1}) reported by the FIA [59]. While the changes in ecosystem structural characteristics were broadly consistent across all land cover change categories, differences between the forest-to-LDR and forest-to-MDR categories were more pronounced than differences between the forest-to-MDR and forest-to-OTH categories (Table 1). Moreover, as development intensity increased from LDR to OTH, aboveground biomass decreased by about 65% from 52.3 ± 16.0 to 18.4 ± 9.5 Mg C ha^{-1}, respectively. Biomass differences were more pronounced than accompanying differences in canopy cover and stem density across the same change in development intensity.

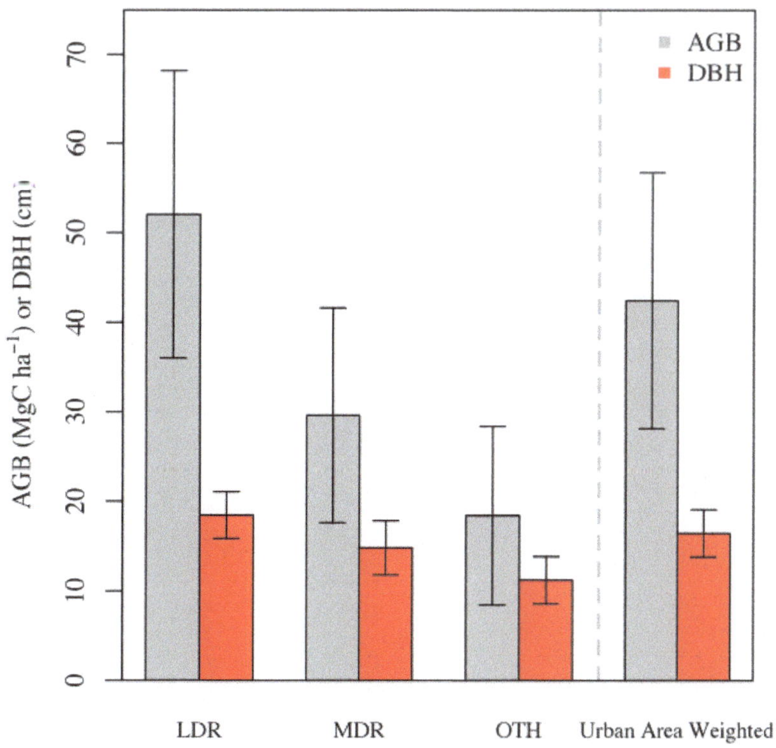

Figure 2: Ecosystem structure. Aboveground biomass (AGB) and diameter at breast height (DBH) for all low density residential (LDR), medium density residential (MDR), and other (OTH) 2013 land cover change field plots. The area weighted urban estimate was based on the areal extents from Table 1. Error bars are 95% confidence intervals. http://dx.doi.org/10.1371/journal.pone.0136237.g002

Table 1: Ecosystem characteristics and land conversion rates. Ecosystem structure characteristics of all plots, parsed by forest conversion pathway. Land area within the eastern Massachusetts study area includes all five counties in which we had study plots and is approximately 8,800 km^2. Stem density and canopy cover estimates across all development types are area weighted based on the current areal extent of each development category. Land conversion rates were based on the difference in areal extent within each category between 1971 and 1999. Current values were generated from 2013 field and FIA data. http://dx.doi.org/10.1371/journal.pone.0136237.t001

	Land Conversion Rate (km^2 yr^{-1})	Current Areal Extent (km^2)	Percent of Study Area	Number of Plots	Current Stem Density (stems ha^{-1})	Current Canopy Cover (%)
Forest-to-LDR (0–50% ISA)	20.3	2028	23.0	46	359 ± 109	45.2 ± 9.7
Forest-to-MDR (0–80% ISA)	1.3	110	1.3	43	241 ± 96	30.0 ± 9.8
Forest-to-OTH (>80% ISA)	0.9	75	8.5	46	232 ± 94	25.2 ± 10.4
All Development Types	22.5	2890	32.8	135	322 ± 99	39.4 ± 9.9

doi:10.1371/journal.pone.0136237.t001

We found that trees ≥ 20cm in DBH (minimum size for coring) had a stem density of 106.1 ± 18.3 stems ha^{-1}, represented an area-weighted mean biomass of 39.1 ± 7.7 Mg C ha^{-1}, and had a mean post-development growth rate of 16.8 ± 0.6 kg C $tree^{-1}$ yr^{-1}. Excluding stem recruitment, planting, and mortality and assuming all trees exhibit the same growth response to development as those cored for this study, our data suggest an aboveground biomass growth rate of 1.8 ± 0.4 Mg C ha^{-1} yr^{-1}, representing a 4.6% yr^{-1} increase in biomass (ranging from 3 to 7% yr^{-1}).

Visual estimates of ground cover were highly variable across all urban land cover types. As development intensity increased from LDR to OTH development types, plot-level lawn extent showed no statistically significant change. However, there was a significant increase in plot-level ISA from 20.2 ± 6.8% to 53.0 ± 10.2% [52] as development intensity increased from LDR to OTH development types. We found no significant relationship between the amount of ISA on the plot and stem density, live biomass, or tree-level BAI.

Land cover conversion rates between 1971 and 1999 indicate reductions in forest cover that are concomitant with increases in the coverage of urban development, especially in the form of forest-to-LDR land cover change (see Table 1). Between 1971 and 1999 forest cover within the study area decreased 13% from 5,023 to 4,451 km^2 [50]. Given that most of this forested land was converted to some kind of urban development, we estimated that the resulting forest area lost was roughly 0.23% yr^{-1} [50]. However, independent, county-level FIA estimates suggest that forest area loss may have been higher at 0.35% yr^{-1} for 1985–1998, decreasing to 0.19% yr^{-1} between 2005 and 2012 [62]. Based on the 2011 National Land Cover Database, forested and developed land covers (LDR, MDR, OTH) comprised approximately 54% and 33%, respectively, of total land cover within our study area [63].

Tree Growth Rates

We analyzed a total of 91 tree cores for annual growth increment. Cores from plots that underwent conversion from forest-to-urban land covers since 1971 (n = 59 cores) had a mean DBH at the time of sampling of 45.6 ± 3.2 cm, with a mean BAI of 25.2 ± 3.7 cm^2 yr^{-1}. Cores from open-grown Boston street trees (n = 17 cores) had a mean DBH of 39.6 ± 6.5 cm at the time of sampling, with a mean BAI of 36.9 ± 9.5 cm^2 yr^{-1}. Cores from the stable urban plots that did not undergo any land cover change since 1971 (n = 15 cores) had a mean DBH of 45.9 ± 9.2 cm at the time of sampling, with a mean BAI of 30.0 ± 7.1 cm^2 yr^{-1}. Data for the background growth category were based on 99 trees from FIA plots with a mean DBH of 31.4 ± 1.2 cm at the time of sampling and a mean BAI of 13.4 ± 1.1 cm^2 yr^{-1}.

Of the 59 cores analyzed from sites that underwent land cover change, there were 28, 18, and 13 cores in the LDR, MDR, and OTH subcategories, respectively. Differences in increment growth rates between development trajectories (i.e., urban subcategories) were not statistically significant and were aggregated for analysis. Following conversion of forest to urban development, the mean overall BAI increased from 17.1 ± 3.0 cm^2 yr^{-1} to 35.8 ± 4.7 cm^2 yr^{-1}(Fig 3). Fifty-three of the 59 tree cores showed a significant increase in BAI, while 18 cores showed an increase of over 200% after conversion (9.6 ± 3.9 to 39.7 ± 11.7 cm^2 yr^{-1}). Generally, the cores with a lower initial BAI showed a larger growth response following development (Fig 3). Three cores showed no significant change and three cores showed a decrease in growth post-development. Overall mean annual biomass increment increased by 138%, from 6.5 ± 1.6 to 16.9 ± 3.0 kg C $tree^{-1}$ yr^{-1} after the land conversion date.

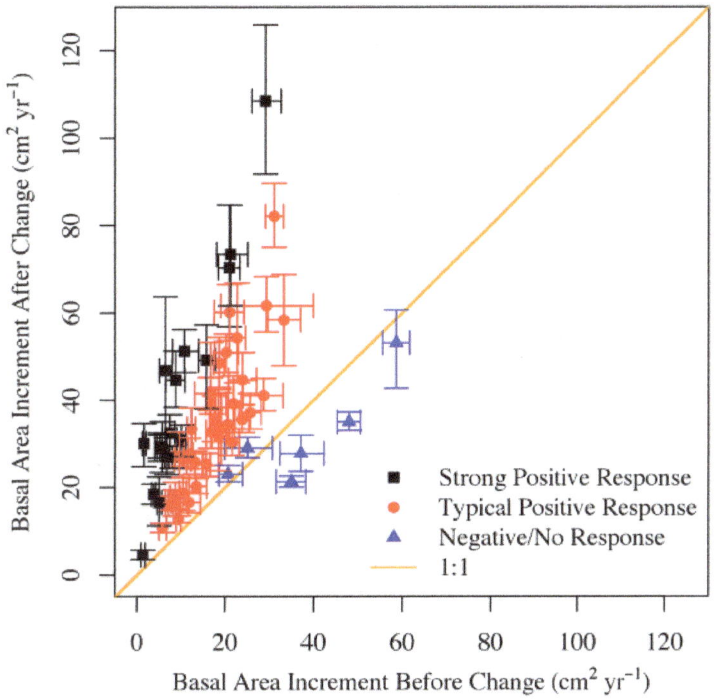

Figure 3: Basal area increment before and after land cover change. Mean basal area increment before and after conversion from forest to urban land cover for each tree. The 95% confidence intervals reflect both inherent variability in growth rates and changes in the number of rings before and after land conversion. Individual cores are color coded to denote the difference in before vs. after basal area increment and are categorized as a "Negative/No Response", "Typical Positive", and "Strong Positive." Trees which exhibited a "strong" response were those that had a ≥ 200% increase in growth rates following land use change. http://dx.doi.org/10.1371/journal.pone.0136237.g003

Across all cores, the average BAI indicated an increase in productivity after land cover change (Fig 4). The mean growth rates for the period after land conversion, street trees, and stable urban plots were similar at 35.8 ± 4.7, 36.9 ± 9.5, 30.0 ± 7.1 cm^2 yr^{-1}, respectively. Annual growth rates from these urban sites were approximately double the growth rates of the forest-grown red oaks sampled from the FIA plots and the pre-development BAI. Estimated growth rates in the FIA trees (13.4 ± 1.1 cm^2 $tree^{-1}$ yr^{-1}) were similar to the pre-conversion growth rates at our sites undergoing land cover change (17.1 ± 3.0 cm^2 $tree^{-1}$ yr^{-1}).

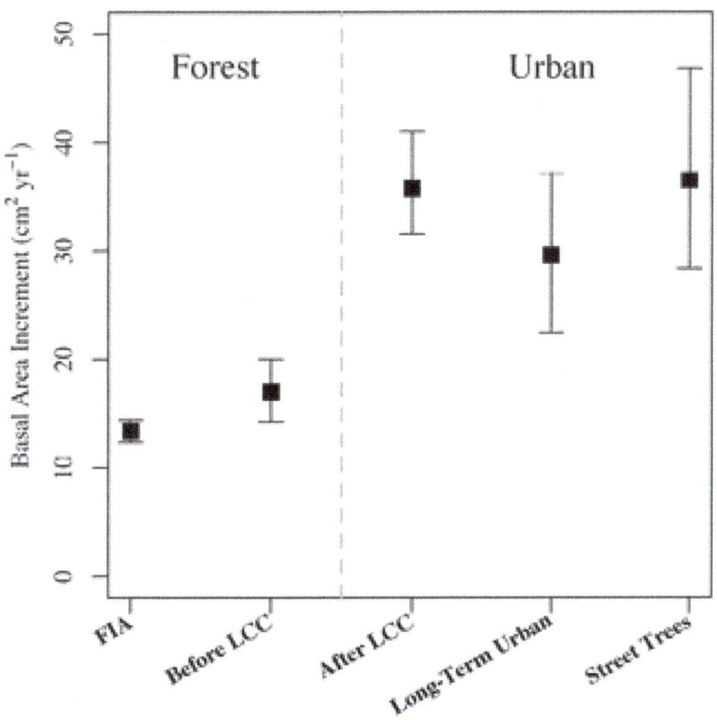

Figure 4: Basal area increment for forest and urban land covers. Mean Basal Area Increment (BAI) within forested areas was based on the FIA regional plots and growth rates before land cover change. Urban BAI includes growth rates after development, trees from stable urban plots, and open-grown Boston street trees. http://dx.doi.org/10.1371/journal.pone.0136237.g004

The mean DBH at the time of conversion was 28.5 ± 3.6 cm, while the mean DBH at the time of sampling (i.e., 2013) was 45.6 ± 3.2 cm. To test for tree size-related biases in the observed before-after land cover change BAI, we also calculated BAI as a function of size and found that, on average, trees within a given DBH class grew faster after the conversion than before the conversion ($p < 0.01$ for paired student t-test; Fig 5). While the variability within each size class is substantial, this result suggests that changes in size class did not drive the observed increases in productivity. Note, while the basal area of a tree scales non-linearly with DBH, BAI is a function of growth in any given year and is independent of tree DBH.

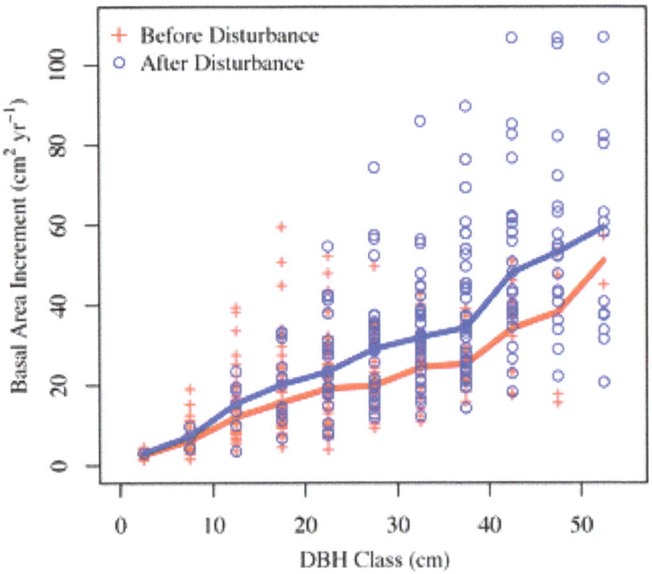

Figure 5: Basal area increment as a function of DBH class. Basal area increment as a function of diameter at breast height (DBH) class before and after conversion to urban land covers. DBH classes represent binned DBH values spaced at 5cm intervals. The solid lines represent the medians for each category. http://dx.doi.org/10.1371/journal.pone.0136237.g005

DISCUSSION

Plot-level Changes in Biomass with Land Cover Change

We found that aboveground biomass was negatively correlated with development intensity (e.g. OTH < MDR < LDR). Reductions in aboveground biomass with increasing development intensity were likely due to a combination of biomass removal associated with development and ensuing land management practices, including tree maintenance and removal. Hutyra et al. [11] and Raciti et al. [49] found similarly large decreases in aboveground biomass with increasing intensity of urban development. We also found that as development intensity increased, the percentage of paved and lawn land covers increased while the percentage forested cover decreased. Further, plots that underwent a forest-to-OTH land cover change typically had fewer and smaller trees, and consequently far less aboveground biomass than the other urban categories. While we observed a ~100% increase in tree growth rates following development, we found that there was little relationship between plot biomass and growth rates. The biomass present within a plot at any point in time is a product of growth, recruitment, planting, and mortality rates integrated over time. Further, tree growth rates reflect both inherent biological processes and active land management. Since these processes and management choices likely change through time, we were only able to capture the current aboveground biomass pools and the pre- and post- development growth rates for one tree species (red oak).

We found no significant relationship between plot-level ISA and biomass. The ISA-biomass relationship was influenced by plots that simultaneously contained high canopy cover (approaching 100%), high biomass, and high ISA due to a small number of large trees whose expansive crowns overtopped the paved areas around them. This result is in contrast to the significant decreases of biomass with increases in ISA found by Raciti et al. [49] at larger spatial scales.

Tree-level productivity and land cover change

Tree productivity is a function of many biotic and abiotic factors, including: species, age, resource availability, growing season length, and competition [64,65]. Searle et al. [42] found an eight-fold increase in tree growth rates in red oak seedlings in New York City compared to seedlings in rural New York that they attributed to local temperature differences. Gregg et al. [43] suggested that lower levels of atmospheric ozone in urban areas, compared to adjacent suburban and rural areas, were responsible for elevated urban tree growth rates in New York City *Populus deltoides* seedlings.

Despite a wide range of counterbalancing factors that might lead to increased or decreased vegetation productivity in urban areas, we found that the productivity (in BAI) of individual trees not removed during urbanization doubled with the conversion of forest to urban land cover. While we are not aware of other studies that explicitly quantify tree growth in response to urbanization, several forest gap studies have found that mature red oak trees growing in gaps created in a Connecticut forest during tree harvesting experienced a 25–47% increase in DBH 5–12 years following disturbance [66,67]. Ward [66,67] attributed increased growth rates to crown release, as the trees in those studies had 25–100% of their canopy released, which is likely comparable to the increased light conditions experienced by residual trees on recently developed land. Tree crown release could be responsible for the similar increases in growth observed in our study, as greater than 80% of the trees that we cored were within 10 meters of an edge or clearing, at the edge itself, or were completely isolated from other trees (e.g., in a lawn). Trees growing in these conditions would have been exposed to more light due to less competition and, depending on land management practices, could have also been irrigated, fertilized, and given treatment for pests and diseases.

In order to control for drivers of tree growth other than land cover change, we compared red oak BAI after land cover change with BAI from trees in stable urban conditions (i.e., no known land cover changes since at least 1971), from Boston street trees, and from stable forests (i.e., from FIA plots). Before land cover change, tree growth rates were comparable to FIA forest plots. In contrast, the average growth rates for post-land cover change trees, Boston street trees, and the stable urban trees were statistically indistinguishable from one another, and all were enhanced relative to FIA and pre-disturbance BAI. Elevated light availability is an environmental condition shared by urban trees generally, which may explain the higher growth rates in these locations compared to forest trees.

The suite of observational data from our study is consistent with the hypothesis that reduced competition for light and nutrients after land cover change stimulates tree growth. The timing of tree growth response was closely correlated with our carefully determined date of land cover change, over half of the plots experiencing land cover change showed a large increase in tree growth within five years of the land cover change. In several cases where a strong increase in growth did not fall within the 5-year window of the land cover change date, we suspect that an earlier land cover change may have occurred (see S1 Fig). Further, the increases in growth after land cover change exceeded inter-annual variation. Annual tree ring data can be noisy, but for over 3/4 of all cores analyzed we observed a flat trend (slope ~0) in increment prior to development, followed by a dramatic increase in growth rates after development. In several instances growth rates increased over 1000% after land cover change. We do not know of any other abrupt anthropogenic changes in growing conditions during the 1971–2012 time period, other than the identified land cover change. During this study interval, there were climatological and ecological disturbances, like the 1980 Gypsy Moth infestation. However, the resulting changes in growth associated with these natural disturbances were typically short lived and/or several orders of magnitude smaller than those seen after land cover change.

We observed the strongest growth response in the trees with the slowest growth prior to land cover change. The three trees that did not experience a significant enhancement in growth following development had high pre-development growth rates, suggesting that growing conditions for these trees may have been altered prior to the estimated land cover change date or that resources made available after land cover change were not previously limiting. Further, for two of the three cores that showed no response in growth after development, the estimated land cover change date fell more than three years outside of the identified CCDC/MacConnell land cover change time period (see S1 Fig). This suggests that we may have missed the timing of growth release in the cores that exhibit no change after our estimated date of land cover change.

Effect of elevated productivity on carbon stocks in urban areas.

While there remains considerable uncertainty regarding C fluxes and terrestrial C stocks in urban areas [68,69], recent studies have suggested that urban vegetation plays an important role in C cycling. Studies that used field and remote sensing data to map vegetation have found extensive canopy cover and large aboveground C stocks in urban areas [70,71]. Furthermore, the vegetative fluxes in urban areas may be increasing due to urbanization-induced changes to regional climate. For example, it has been estimated that gross primary productivity has increased by 13% in the heavily developed eastern U.S., primarily through increases in downwelling shortwave radiation and temperature [10]. Briber et al. [29] estimated that urban heat islands and resultant longer growing seasons may account for as much as a 50% increase in net biogenic exchange in the City of Boston, though there is still large uncertainty regarding the role of growing season length in C sequestration [72].

Despite the substantial extent of urban vegetation and potentially high urban C densities, many remote sensing products and C mapping initiatives discount or completely ignore vegetation in urban areas (e.g. [71,73–75]). It should be noted that many remote sensing products were never intended to map vegetation in urban areas, rather their focus is typically on forested land covers. In the case of the NASA Carbon Management System forest biomass maps for biosphere flux estimation [76], ~ 30% of Massachusetts is effectively assumed to have no vegetative fluxes due to its degree of urbanization. We found that aboveground biomass in urban land covers stored as much as 42.7 ± 14.7 Mg C ha^{-1}, which is nearly 1/3 the aboveground C stocks in mature New England forests such as Harvard Forest [77]. These urban C pools coupled with the rapid tree growth observed in this study highlight the importance of urban vegetation for C storage and large biological fluxes from highly developed regions such as the northeastern United States.

In order to explore the role of urban vegetation in net C sequestration, we must further our understanding of urban tree mortality and recruitment/planting rates [78–80]. For example, Nowak et al. [78] found that mortality rates for urban trees were 6.6% yr^{-1}, resulting in an overall net decrease in the number of live trees by 4.2% yr^{-1}. Urban tree mortality is often higher for smaller, more recently planted trees. Nowak et al. [79] found mortality rates for newly planted street trees can be as high as 19% yr^{-1}. However, over 50% of the red oaks that we cored during our 2013 campaign date back to at least 1950, suggesting that mortality may affect street trees more than residual trees left standing or trees planted after land cover change.

Considering only trees large enough for coring in this study (≥ 20 cm in DBH), we found 39.1 ± 7.7 Mg C ha^{-1} of biomass in developed areas and an aboveground biomass growth rate of 1.8 + 0.4 Mg C ha^{-1} yr^{-1}, representing a 4.6% yr^{-1} increase in biomass (ranging from 3.8 to 5.7% yr^{-1}). For comparison, we extracted data from the Harvard Forest data archives [81] for a mature mixed-hardwood forest in the Environmental Measurement Site at Harvard Forest in central Massachusetts. Using aboveground biomass accumulation of trees ≥ 20 cm in DBH over a 16 year period (1998 through 2013) we found a 1.4 ± 0.3% yr^{-1} increase in biomass (accounting for recruitment and mortality), with 1.5 ± 0.34 Mg C ha^{-1} yr^{-1} biomass accumulation and 117 ± 7.9 Mg C ha^{-1} in aboveground live biomass [81]. Despite having ~1/3 the aboveground biomass, the biomass growth rate in developed areas is comparable to that of Harvard Forest.

The high growth rates in developed areas likely reflect favorable growing conditions, but this analysis does not fully inform us of the annual vegetative C balance since it is likely that recruitment rates will be lower and mortality rates will be higher within developed areas. Moreover, urban tree biomass estimates in this paper were based on forest allometries which may not accurately reflect urban tree forms [82]. Nonetheless, the high observed tree growth rates highlight that vegetation within developed areas is very dynamic and has the potential to be a significant C sink at some points in time and space.

Our results clearly show that the conversion of forests to developed areas have the potential to considerably alter both ecosystem productivity and the regional C balance. It is likely that urban ecosystems could sequester a significant amount of C, depending on residential and municipal landscape management choices. Given global forecasts for continued population growth, patterns of urban expansion and the resulting land cover changes, urban areas will continue to be a critical component of the changing global carbon cycle [68]. Additional measurements of vegetation growth, mortality, planting, and recruitment rates within developed areas will be necessary for constraining our estimates of urban regional C balance and informing our land development choices. Moreover, more accurate urban tree allometries are needed to better estimate terrestrial C stocks in urban areas. The consequences of development will vary among pre-development ecosystems (e.g., forest vs. grassland) and land use conversion pathways (e.g., forest to development vs. agriculture to development), improved our global estimates of both land cover change and biomass are needed to assess the C impacts of different development typologies [83].

SUPPORTING INFORMATION

S1_Table.xlsx

	1	2	3	4
1	FIA Forest Tree Cores			
2	Location Name	Coordinates	Nu. Of samples	
3	Worcester County	NA	82	
4	Norfolk County	NA	6	
5	Middlesex County	NA	7	
6	Essex County	NA	4	
7	...			
8	Land Cover Change Plots			
9	Location Name	Longitude	Latitude	Nu. Of samples
10	Pepperell	-71.64	42.65	1
11	Lunenburg	-71.74	42.63	1
12	Tewksbury	-71.27	42.62	1
13	Hopkinton	-71.57	42.2	1
14	Boxborough	-71.51	42.49	1
15	Southborough	-71.49	42.32	1
16	Charlton	-71.93	42.17	1
17	Tewksbury	-71.24	42.61	1
18	Tewksbury	-71.23	42.63	1
19	Dracut	-71.29	42.66	1
20	Charlton	-71.95	42.13	1
21	Millbury	-71.81	42.19	1
22	Westborough	-71.63	42.24	1
23	Charlton	-71.97	42.16	1
24	Haverhill	-71.07	42.800000000000004	1
25	Sterling	-71.81	42.42	1
26	Danvers	-70.95	42.59	1
27	Franklin	-71.42	42.1	1
28	Andover	-71.18	42.64	1
29	Hopedale	-71.55	42.12	1
30	Stoneham	-71.09	42.47	1
31	North Andover	-71.11	42.67	1
32	Northbridge	-71.68	42.13	1
33	Franklin	-71.42	42.1	1
34	Natick	-71.33	42.26	1
35	Milford	-71.55	42.19	1

36	Gloucester	-70.67	42.63	1
37	Mendon	-71.51	42.1	1
38	West Boylston	-71.81	42.36	1
39	Leominster	-71.74	42.57	1
40	Upton	-71.61	42.17	1
41	Bellerica	-71.25	42.53	1
42	Auburn	-71.81	42.22	1
43	Essex	-70.82	42.62	1
44	Bellerica	-71.27	42.53	1
45	Marlborough	-71.59	42.35	1
46	Saugus	-71.04	42.47	1
47	Southborough	-71.53	42.31	1
48	Sterling	-71.76	42.45	1
49	Danvers	-70.97	42.59	1
50	Westford	-71.45	42.59	1

A table of all plots used in this analysis, with Massachusetts town name, longitude, latitude, and the number of samples per plot. Note that FIA data locations were confidential, and were not released. For those plots, data were grouped by county.

S1 Fig. Raw increment data.

Time series of basal area increment for each individual core in our analysis. The header at the top of each page indicates the type of core. The cores under the "Strong Response" (≥ 200% increase in BAI), "Typical Response" (statically significant increase in BAI), and "Negative/No Response" (statistically significant decrease in BAI or no significant change in BAI) headers comprise the cores from trees where known land cover changed occurred. For these cores, the orange vertical line indicates the estimated disturbance date and the grey area indicates the time period over which the land cover change should have taken place, according to a combination of the CCDC and MacConnell datalayers. Before and after land cover change mean basal area increment is displayed above the plots along with the p-value for the difference between these means (95% confidence). Also included are the cores from the "Stable Urban" and Boston "Street Tree" locations for comparison. Mean basal area increment for the entire time series is displayed above the plots.

doi:10.1371/journal.pone.0136237.s001

S1 Table. Plot coordinates.

A table of all plots used in this analysis, with Massachusetts town name, longitude, latitude, and the number of samples per plot. Note that FIA data locations were confidential, and were not released. For those plots, data were grouped by county.

doi:10.1371/journal.pone.0136237.s002

ACKNOWLEDGMENTS

We would like to thank Pontus Olofsson and Curtis Woodcock for help with site selection for our 2013 field campaign. We also thank Mike Dietze for his support in tree ring processing, and Mark Friedl for analytical support. We would like to thank Jackie Getson for all of her help with tree core analysis, GIS analysis, and figures. We would like to thank Bryant Scharenbroch for graciously providing tree core data from the Boston area. Lastly, we would like to thank Ronald Salmonson, Nicole Volk, and Cleo Woodcock for help collecting field data during the summer of 2013.

AUTHOR CONTRIBUTIONS

Conceived and designed the experiments: LRH BMB ABR SMR CEH ALD. Performed the experiments: BMB LRH ABR VKD SMR CEH ALD. Analyzed the data: BMB LRH ABR SMR ALD CEH VKD. Contributed reagents/materials/analysis tools: BMB LRH ABR SMR CEH ALD. Wrote the paper: BMB LRH ABR SMR VKD CEH ALD.

REFERENCES

1. Stocker TF, Qin D, Plattner G-K, Tignor M, Allen SK, Boschung J, et al. (2013) Climate change 2013: The physical science basis. Intergovernmental Panel on Climate Change, Working Group I Contribution to the IPCC Fifth Assessment Report (AR5)(Cambridge Univ Press, New York).
2. Christen A (2014) Atmospheric measurement techniques to quantify greenhouse gas emissions from cities (in press). Urban Climate. doi: 10.1016/j.uclim.2014.04.006
3. Churkina G (2008) Modeling the carbon cycle of urban systems. Ecological Modelling 216: 107–113. doi: 10.1016/j.ecolmodel.2008.03.006
4. UNDESA (2012) United Nations: Department of Economic and Social Affairs: World urbanization prospects: the 2011 revision: United Nations Publications. Available:http://esa.un.org/wpp/ppt/CSIS/WUP_2011_CSIS_4.pdf. Accessed 1 May 2013.
5. Schneider A, Friedl MA, Potere D (2009) A new map of global urban extent from MODIS satellite data. Environmental Research Letters 4: 168–182. doi: 10.1088/1748-9326/4/4/044003
6. Food Agriculture Organization (2001) State of the World's Forests 2001: FAO.

7. United States Census (2010) Unites States Census Bureau: 2010 Census Data. Available: https://www.census.gov/compendia/statab/2010/. Accessed May 1 2013.

8. Fulton WB, Pendall R, Nguyen M, Harrison A (2001) Who sprawls most?: How growth patterns differ across the US: Brookings Institution, Center on Urban and Metropolitan Policy Washington, DC.

9. Jenerette GD, Wu JG (2001) Analysis and simulation of land-use change in the central Arizona-Phoenix region, USA. Landscape Ecology 16: 611–626.

10. Zhao TT, Brown DG, Fang HL, Theobald DM, Liu T, Zhang T (2012) Vegetation productivity consequences of human settlement growth in the eastern United States. Landscape Ecology 27: 1149–1165. doi: 10.1007/s10980-012-9766-8

11. Hutyra LR, Yoon B, Alberti M (2011) Terrestrial carbon stocks across a gradient of urbanization: a study of the Seattle, WA region. Global Change Biology 17: 783–797. doi: 10.1111/j.1365-2486.2010.02238.x

12. Foster CHW (2005) Forests in time: The environmental consequences of 1,000 years of change in New England. Journal of Interdisciplinary History 36: 270–271. doi: 10.1162/0022195054741442

13. McDonnell MJ, Pickett STA, Groffman PM, Bohlen P, Pouyat RV, Zipperer WC, et al. (1997) Ecosystem processes along an urban-to-rural gradient. Urban Ecosystems: 21–36. doi: 10.1007/978-0-387-73412-5_18

14. Pickett STA, Cadenasso ML, Grove JM, Boone CG, Groffman PM, Irwin E, et al. (2011) Urban ecological systems: Scientific foundations and a decade of progress. Journal of Environmental Management 92: 331–362. doi: 10.1016/j.jenvman.2010.08.022. pmid:20965643

15. Foley JA, DeFries R, Asner GP, Barford C, Bonan G, Carpenter SR, et al. (2005) Global consequences of land use. Science 309: 570–574. pmid:16040698 doi: 10.1126/science.1111772

16. Grimm NB, Faeth SH, Golubiewski NE, Redman CL, Wu JG, Bai XM, et al. (2008) Global change and the ecology of cities. Science 319: 756–760. doi: 10.1126/science.1150195. pmid:18258902

17. Seto KC, Guneralp B, Hutyra LR (2012) Global forecasts of urban expansion to 2030 and direct impacts on biodiversity and carbon pools. Proceedings of the National Academy of Sciences of the United States of America 109: 16083–16088. doi: 10.1073/pnas.1211658109. pmid:22988086

18. International Energy Agency (2008) World Energy Outlook: 2008. Paris: International Energy Agency.

19. Hutyra LR, Yoon B, Hepinstall-Cymerman J, Alberti M (2011) Carbon consequences of land cover change and expansion of urban lands: A case study in the Seattle metropolitan region. Landscape and Urban Planning 103: 83–93. doi: 10.1016/j.landurbplan.2011.06.004

20. Raciti SM, Fahey TJ, Thomas RQ, Woodbury PB, Driscoll CT, Carranti FJ, et al. (2012) Local-Scale Carbon Budgets and Mitigation Opportunities for the Northeastern United States. Bioscience 62: 23–38. doi: 10.1525/bio.2012.62.1.7

21. Coutts AM, Beringer J, Tapper NJ (2007) Characteristics influencing the variability of urban CO2 fluxes in Melbourne, Australia. Atmospheric Environment 41: 51–62. doi: 10.1016/j.atmosenv.2006.08.030

22. Hope D, Gries C, Zhu WX, Fagan WF, Redman CL, Grimm NB, et al. (2003) Socioeconomics drive urban plant diversity. Proceedings of the National Academy of Sciences of the United States of America 100: 8788–8792. pmid:12847293 doi: 10.1073/pnas.1537557100

23. DeFries RS, Rudel T, Uriarte M, Hansen M (2010) Deforestation driven by urban population growth and agricultural trade in the twenty-first century. Nature Geoscience 3: 178–181. doi: 10.1038/ngeo756

24. Seto KC, Reenberg A, Boone CG, Fragkias M, Haase D, Langanke T, et al. (2012) Urban land teleconnections and sustainability. Proceedings of the National Academy of Sciences of the United States of America 109: 7687–7692. doi: 10.1073/pnas.1117622109. pmid:22550174

25. Kaye JP, Groffman PM, Grimm NB, Baker LA, Pouyat RV (2006) A distinct urban biogeochemistry? Trends in Ecology & Evolution 21: 192–199. doi: 10.1016/j.tree.2005.12.006

26. Arnfield AJ (2003) Two decades of urban climate research: A review of turbulence, exchanges of energy and water, and the urban heat island. International Journal of Climatology 23: 1–26. doi: 10.1002/joc.859

27. Bonan GB (2008) Ecological climatology: concepts and applications. Cambridge; New York: Cambridge University Press. xvi, 550 p. p.

28. Oke TR (1982) The Energetic Basis of the Urban Heat-Island. Quarterly Journal of the Royal Meteorological Society 108: 1–24. doi: 10.1002/qj.49710845502

29. Briber BM, Hutyra LR, Dunn AL, Raciti SM, Munger JW (2013) Variations in Atmospheric CO2 Mixing Ratios across a Boston, MA Urban to Rural Gradient. Land 2: 304–327. doi: 10.3390/land2030304

30. Zhang XY, Friedl MA, Schaaf CB, Strahler AH, Schneider A (2004) The footprint of urban climates on vegetation phenology. Geophysical Research Letters 31. doi: 10.1029/2004gl020137

31. Raciti SM, Groffman PM, Jenkins JC, Pouyat RV, Fahey TJ, Pickett STA, et al. (2011) Accumulation of Carbon and Nitrogen in Residential Soils with Different Land-Use Histories. Ecosystems 14: 287–297. doi: 10.1007/s10021-010-9409-3

32. Rao P, Hutyra LR, Raciti SM, Finzi AC (2013) Field and remotely sensed measures of soil and vegetation carbon and nitrogen across an urbanization gradient in the Boston metropolitan area. Urban Ecosystems 16: 593–616. doi: 10.1007/s11252-013-0291-6

33. .Bettez ND, Marino R, Howarth RW, Davidson EA (2013) Roads as nitrogen deposition hot spots. Biogeochemistry 114: 149–163. doi: 10.1007/s10533-013-9847-z

34. George K, Ziska LH, Bunce JA, Quebedeaux B (2007) Elevated atmospheric CO2 concentration and temperature across an urban-rural transect. Atmospheric Environment

41: 7654–7665. doi: 10.1016/j.atmosenv.2007.08.018

35. Strong C, Stwertka C, Bowling DR, Stephens BB, Ehleringer JR (2011) Urban carbon dioxide cycles within the Salt Lake Valley: A multiple-box model validated by observations. Journal of Geophysical Research-Atmospheres 116: 1–12. doi: 10.1029/2011jd015693
36. Jacovides CP, Timbios F, Asimakopouolos DN, Steven MD (1997) Urban aerosol and clear skies spectra for global and diffuse photosynthetically active radiation. Agricultural and Forest Meteorology 87: 91–104. doi: 10.1016/s0168-1923(97)00031-2
37. Takagi M, Gyokusen K (2004) Light and atmospheric pollution affect photosynthesis of street trees in urban environments. Urban forestry & urban greening 2: 167–171. doi: 10.1078/1618-8667-00033
38. Golubiewski NE (2006) Urbanization increases grassland carbon pools: Effects of landscaping in Colorado's front range. Ecological Applications 16: 555–571. pmid:16711044 doi: 10.1890/1051-0761(2006)016[0555:uigcpe]2.0.co;2
39. Lorenz K, Lal R (2009) Biogeochemical C and N cycles in urban soils. Environment International 35: 1–8. doi: 10.1016/j.envint.2008.05.006. pmid:18597848
40. Pongratz J, Reick CH, Raddatz T, Claussen M (2009) Effects of anthropogenic land cover change on the carbon cycle of the last millennium. Global Biogeochemical Cycles 23. doi: 10.1029/2009gb003488
41. Novak K, Skelly JM, Schaub M, Krauchi N, Hug C, Landolt W, et al. (2003) Ozone air pollution and foliar injury development on native plants of Switzerland. Environmental Pollution 125: 41–52. pmid:12804826 doi: 10.1016/s0269-7491(03)00085-x
42. Searle SY, Turnbull MH, Boelman NT, Schuster WSF, Yakir D, Griffin KL (2012) Urban environment of New York City promotes growth in northern red oak seedlings. Tree Physiology 32: 389–400. doi: 10.1093/treephys/tps027. pmid:22491523
43. Gregg JW, Jones CG, Dawson TE (2003) Urbanization effects on tree growth in the vicinity of New York City. Nature 424: 183–187. pmid:12853954 doi: 10.1038/nature01728
44. U.S. Census 2012 (2012) United States Census Bureau: 2012 Census of Population and Housing. Available: http://www.census.gov/compendia/statab/. Accessed May 1 2013.
45. National Climatic Data Center (2014): National Oceanic and Atmospheric Administration. Available: http://www.ncdc.noaa.gov. Accessed 1 May 2013.
46. Forest Inventory Analysis: Tree Count Standard Report (2014): USDA Forest Service. Available: http://apps.fs.fed.us/fia/fido/index.html. Accessed 15 October 2014.
47. USDA NRCS (2009) Soil survey of Middlesex County, Massachusetts. Washington, D.C., USA. Available:http://www.nrcs.usda.gov/Internet/FSE_MANUSCRIPTS/massachusetts/MA017/0/middlesex.pdf. Accessed 20 April 2013.
48. Foster DR, Aber JD (2004) Forests in time: the environmental consequences of 1,000 years of change in New England. New Haven: Yale University Press. xiv, 477 p. p.
49. Raciti SM, Hutyra LR, Rao P, Finzi AC (2012) Inconsistent definitions of "urban" result in different conclusions about the size of urban carbon and nitrogen stocks. Ecological Applications 22: 1015–1035. pmid:22645829 doi: 10.1890/11-1250.1
50. MassGIS Datalayers: Land Use Available: http://www.mass.gov/anf/research-and-tech/it-serv-and-support/application-serv/office-of-geographic-information-massgis/datalayers/lus.html. Accessed 5 June 2013.
51. Zhu Z, Woodcock CE (2014) Continuous change detection and classification of land cover using all available Landsat data. Remote Sensing of Environment 144: 152–171. doi: 10.1016/j.rse.2014.01.011
52. MassGIS Datalayers: Impervious Surface. Available:http://www.mass.gov/anf/research-and-tech/it-serv-and-support/application-serv/office-of-geographic-information-massgis/datalayers/impervioussurface.html. Accesed 15 December 2012.
53. Naidoo R, Lechowicz MJ (2001) Effects of gypsy moth on radial growth of deciduous trees. Forest Science 47: 338–348.
54. Grissino-Mayer HD (2001) Evaluating crossdating accuracy: a manual and tutorial for the computer program COFECHA.
55. Holmes RL (1983) Computer-assisted quality control in tree-ring dating and measurement. Tree-ring bulletin 43: 69–78.
56. TerMikaelian MT, Korzukhin MD (1997) Biomass equations for sixty-five North American tree species. Forest Ecology and Management 97: 1–24. doi: 10.1016/s0378-1127(97)00019-4
57. MassGIS Datalayers: Level 3 Assessors' Parcels. Available:http://www.mass.gov/anf/research-and-tech/it-serv-and-support/application-serv/office-of-geographic-information-massgis/datalayers/l3parcels.html. Accessed 1 June 2013.
58. Google Inc. (2005) Google Earth.
59. Forest Inventory Data Online: Aboveground Biomass Standard Report: USDA Forest Service. Available: http://apps.fs.fed.us/fia/fido/index.html. Accessed 15 October 2014.
60. R Core Team (2012) R: A Language and Environment for Statistical Computing. Vienna, Austria: R Foundation for Statistical Computing.
61. Efron B, Tibshirani RJ (1994) An introduction to the bootstrap: CRC press.
62. Forest Inventory Data Online: Area Standard Report: USDA Forest Service. Available:http://apps.fs.fed.us/fia/fido/index.html. Accessed 15 October 2014.
63. .Jin SM, Yang LM, Danielson P, Homer C, Fry J, Xian G (2013) A comprehensive change detection method for updating the National Land Cover Database to circa 2011. Remote Sensing of Environment 132: 159–175. doi: 10.1016/j.rse.2013.01.012
64. Iakovoglou V, Thompson J, Burras L, Kipper R (2001) Factors related to tree growth across urban-rural

gradients in the Midwest, USA. Urban Ecosystems 5: 71–85. doi: 10.1023/a:1021829702654

65. McDonald RI, Urban DL (2004) Forest edges and tree growth rates in the North Carolina Piedmont. Ecology 85: 2258–2266. doi: 10.1890/03-0313

66. Ward JS. CROP TREE RELEASE INCREASES GROWTH OF RED OAK SAWTIMBER IN SOUTHERN NEW ENGLAND: 12-YEAR RESULTS; 2008. pp. 457–465.

67. Ward JS. Stand and individual tree growth of mature red oak after crop tree management in southern New England: 5-year results; 2011. pp. 502–513.

68. Hutyra LR, Duren R, Gurney KR, Grimm N, Kort EA, Larson E, et al. (2014) Urbanization and the carbon cycle: Current capabilities and research outlook from the natural sciences perspective. Earth's Future. doi: 10.1002/2014ef000255

69. Pataki DE, Carreiro MM, Cherrier J, Grulke NE, Jennings V, Pincetl S, et al. (2011) Coupling biogeochemical cycles in urban environments: ecosystem services, green solutions, and misconceptions. Frontiers in Ecology and the Environment 9: 27–36. doi: 10.1890/090220

70. Davies ZG, Edmondson JL, Heinemeyer A, Leake JR, Gaston KJ (2011) Mapping an urban ecosystem service: quantifying above-ground carbon storage at a city-wide scale. Journal of Applied Ecology 48: 1125–1134. doi: 10.1111/j.1365-2664.2011.02021.x

71. Raciti SM, Hutyra LR, Newell JD (2014) Mapping carbon storage in urban trees with multi-source remote sensing data: Relationships between biomass, land use, and demographics in Boston neighborhoods. Science of The Total Environment 500: 72–83. doi: 10.1016/j.scitotenv.2014.08.070. pmid:25217746

72. Wu CY, Gonsamo A, Chen JM, Kurz WA, Price DT, Lafleur W, et al. (2012) Interannual and spatial impacts of phenological transitions, growing season length, and spring and autumn temperatures on carbon sequestration: A North America flux data synthesis. Global and Planetary Change 92–93: 179–190. doi: 10.1016/j.gloplacha.2012.05.021

73. Bondeau A, Smith PC, Zaehle S, Schaphoff S, Lucht W, Cramer W, et al. (2007) Modelling the role of agriculture for the 20th century global terrestrial carbon balance. Global Change Biology 13: 679–706. doi: 10.1111/j.1365-2486.2006.01305.x

74. Sitch S, Smith B, Prentice IC, Arneth A, Bondeau A, Cramer W, et al. (2003) Evaluation of ecosystem dynamics, plant geography and terrestrial carbon cycling in the LPJ dynamic global vegetation model. Global Change Biology 9: 161–185. doi: 10.1046/j.1365-2486.2003.00569.x

75. Zhang XY, Friedl MA, Schaaf CB, Strahler AH, Hodges JCF, Gao F, et al. (2003) Monitoring vegetation phenology using MODIS. Remote Sensing of Environment 84: 471–475. doi: 10.1016/s0034-4257(02)00135-9

76. NASA Carbon Monitoring System (2014) CMS Biomass Pilot. Available:http://carbon.nasa.gov/biomass.html. Accessed 1 December 2014.

77. Urbanski S, Barford C, Wofsy S, Kucharik C, Pyle E, Budney J, et al. (2007) Factors controlling CO2 exchange on timescales from hourly to decadal at Harvard Forest. Journal of Geophysical Research-Biogeosciences 112: G02020. doi: 10.1029/2006jg000293

78. Nowak DJ, Kuroda M, Crane DE (2004) Tree mortality rates and tree population projections in Baltimore, Maryland, USA. Urban Forestry & Urban Greening 2: 139–147. doi: 10.1078/1618-8667-00030

79. Nowak DJ, McBride JR, Beatty RA (1990) Newly planted street tree growth and mortality. Journal of Arboriculture 16: 124–129.

80. Roman LA, Battles JJ, McBride JR (2014) The balance of planting and mortality in a street tree population. Urban Ecosystems 17: 387–404. doi: 10.1007/s11252-013-0320-5

81. Munger W, Wofsy S (1999) EMS—Biomass inventories. Harvard Forest Data Archive, HF069 [Database]. Available: http://harvardforest fas harvard edu 8080.

82. Troxel B, Piana M, Ashton MS, Murphy-Dunning C (2013) Relationships between bole and crown size for young urban trees in the northeastern USA. Urban Forestry & Urban Greening 12: 144–153. doi: 10.1016/j.ufug.2013.02.006

83. Romero-Lankao P, Gurney KR, Seto KC, Chester M, Duren RM, Hughes S, et al. (2014) A critical knowledge pathway to low-carbon, sustainable futures: Integrated understanding of urbanization, urban areas, and carbon. Earth›s Future. doi: 10.1002/2014ef000258

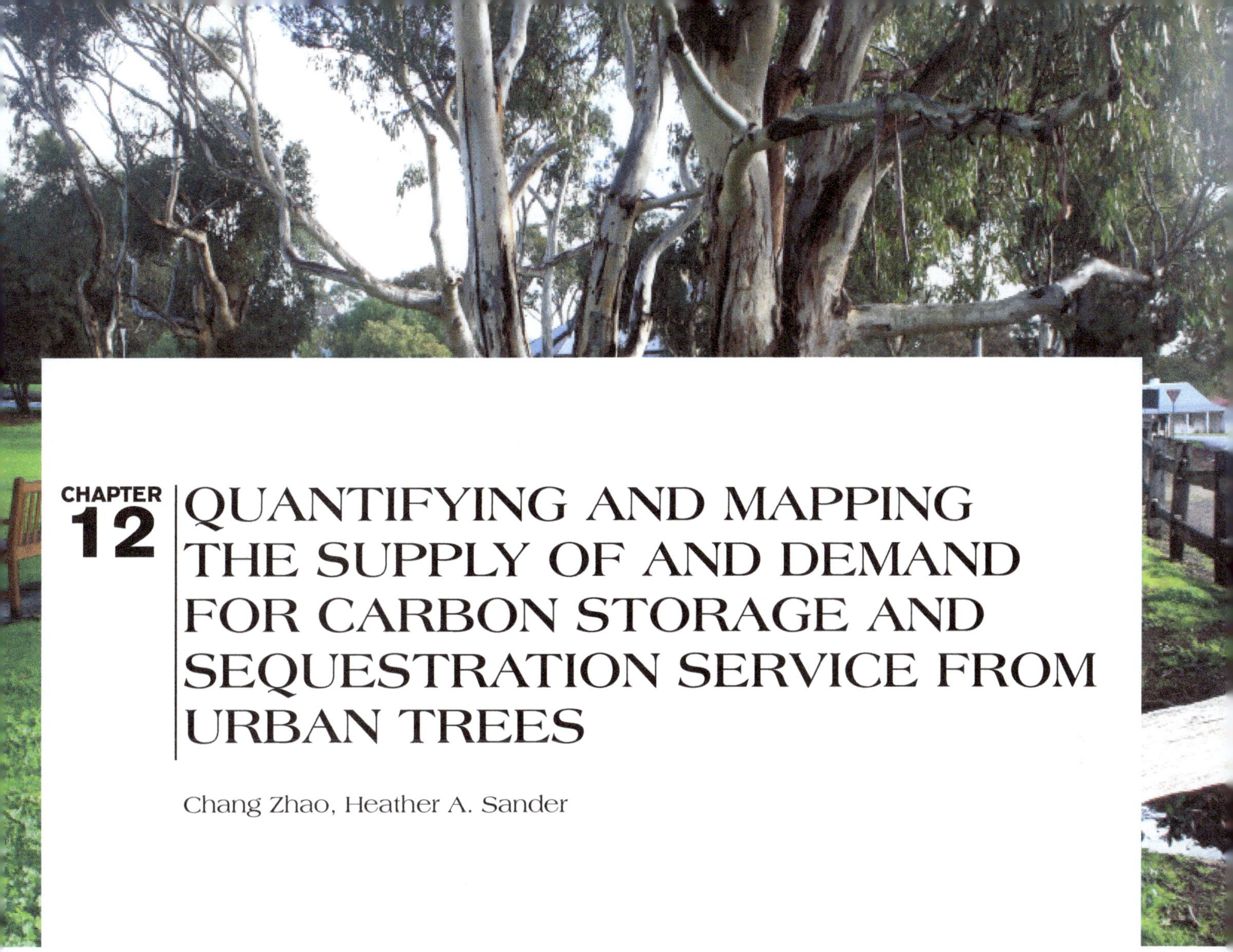

CHAPTER 12

QUANTIFYING AND MAPPING THE SUPPLY OF AND DEMAND FOR CARBON STORAGE AND SEQUESTRATION SERVICE FROM URBAN TREES

Chang Zhao, Heather A. Sander

INTRODUCTION

Ecosystem services are the benefits people obtain from ecosystems upon which human well-being largely depends [1]. A wide variety of services exist, falling generally into four categories: regulating services (e.g., pollination, climate regulation, air quality regulation, water purification), provisioning services (e.g., food production, biomass, freshwater, mineral resources), supporting services (e.g., nutrient cycling, primary production, soil formation) and cultural services (e.g., recreation, aesthetic views, cultural heritage) [1]. Despite their wide recognition in biological conservation and natural resource management, ecosystem services are not well-integrated in land-use planning and management [2]. One of the key hurdles to considering ecosystem services in decision making is the present incomplete understanding of how services are delivered from ecosystems to social systems (i.e., the supply side of ecosystems services), and the extent to which human well-being relies on ecosystem services (i.e., the demand for ecosystem services), as well as the relationship between these two sides.

Improving this understanding requires us to distinguish among the aspects of ecosystems that influence service delivery (e.g., ecosystem structures, ecosystem functions) and ecosystem service potential, flow, and demand [3]. Ecosystem structures are the set of components that make up an ecosystem, while ecosystem functions are the processes and cycles that occur in an ecosystem [4]. For example, with respect to carbon storage and sequestration, an urban forest ecosystem may be characterized in terms of its autotrophs, heterotrophs, water and soils through which ecological functions such as photosynthesis and carbon cycling and nutrient cycling occur. While ecosystem structures and functions are essential to support overall ecosystem integrity and resilience, they may not directly contribute to human well-being. For example, water uptake by plants, infiltration of soil, soil stabilization and nutrient cycling maintain the integrity and resilience of riparian systems, but, depending on the system considered, may not supply services (e.g., flood mitigation, provision of clean water for drinking) unless these products are actually used by humans. Ecosystems services thus are the products of ecosystem structures and functions that actually do contribute to human well-being [5,6].

The notion of human dependence on nature implies that ecosystem services do not exist without beneficiaries, as goods and benefits must be either directly or indirectly consumed, used, or desired by humans [7]. Some researchers have distinguished between ecosystem service potential and flow.

Citation: Zhao C, Sander HA (2015) Quantifying and Mapping the Supply of and Demand for Carbon Storage and Sequestration Service from Urban Trees. PLoS ONE 10(8): e0136392. doi:10.1371/journal.pone.0136392

While ecosystem service potential is commonly referred to as the total capacity of ecosystems to generate services [3,8], the term ecosystem service flow is much more ambiguous, referring either to the spatial connections between service provisioning areas (SPA) and service benefiting areas (SBA) [9–11], or the actual production or use of ecosystem services at a given location [3,12,13]. Note that potential and flow as defined above may also be referred to as supply and demand, although with slightly different meanings. While supply refers to the type and quantity of ecosystem services associated with the geophysical and ecological characteristics of ecosystems [14], demand for ecosystem services reflects those services actually consumed or desired by beneficiaries [11]. The distinction between these terms is crucial for selecting appropriate methods and indicators for quantifying and mapping the different dimensions of ecosystem services and provides clarity in the use of associated products (e.g., statistics, maps) that support decision making. Therefore, to improve our understanding of ecosystem services and to better incorporate this concept in policy making, it is necessary to capture the full ecosystem service dynamic moving from service supply to demand by assessing various aspects of ecosystem service delivery.

Following the completion of the MEA in 2005, several international initiatives called for attention to the economic benefits of natural capital and biodiversity (e.g. The Economics of Ecosystems and Biodiversity, Intergovernmental Platform on Biodiversity and Ecosystem Services) [1]. In response the number of studies measuring the supply of ecosystem services grew exponentially. These studies use various methods to assess ecosystem service supply, including empirical models, process-based models (e.g. hydrological models, climate models), GIS-based approaches, expert knowledge, and decision-support tools (e.g. Integrated Valuation of Ecosystem Services and Tradeoffs, InVEST; Artificial Intelligence for Ecosystem Services, ARIES; Global Unified Model of the BiOsphere, GUMBO) [15–20]. While some of these tools implicitly consider demand via their economic valuation components (e.g., InVEST) or explicitly account for ecosystem service flows from SPA to SBA via simulations (e.g. ARIES), modeling approaches that assess the demand side of ecosystem services remain rare [5]. The majority of ecosystem service studies focus solely on service supply and lack a systematic, integrated consideration of demand, often as a result of data limitations with respect to human needs and well-being. For example, potential recreational services may be estimated for a forest based on landscape characteristics, but the actual ability of populations to access and enjoy these services may not be considered. Such a lack of consideration for ecosystem service demand creates a knowledge gap that impacts the quality of management and policy decisions made, in this example, regarding forest management.

Given the importance of considering both the supply and demand sides of ecosystem services in decision making, several conceptual frameworks for analyzing ecosystem services in an integrated manner have been produced in recent years [12–14]. Although the literature that examines both the supply and demand of ecosystem services has recently grown [9,21–25], studies which specifically distinguish among ecosystem service supply and demand remain rare. Demand in existing studies is commonly conceptualized from three perspectives. Firstly, demand for provisioning services may be quantified as the use or consumption of those services. In such cases, population size and average consumption rates are often combined to produce indicators of demand. This, for example, was the case in a recent study that calculated demand for the energy provisioning service based upon statistical data on the energy consumption per unit area associated with different land cover types [26]. Secondly, for most cultural services, demand has been assessed based largely on social preference or on ecosystem service valuation [27–29]. For example, participatory mapping of ecosystem service demand via interviews has been used to assess preferences for and perceptions of various cultural services at the community level, recognizing that ecosystem service demand is driven by lifestyle choices, demographic characteristics and cultural beliefs [28]. Lastly, demand for supporting and regulating services is most often expressed in terms of the social desire or need to reduce risks or increase service benefits [22,24,25,30]. For example, demand for flood regulating services may be quantified using the potential vulnerability of assets and monetary risk associated with loss as indicators [22]. However, estimating demand for maintaining desirable environmental conditions is often challenging because doing so requires information about the need for risk reduction and protection. In most cases this entails quantitative modeling of many factors, including potential risks, levels of exposure and degrees of social vulnerability [11].

Quantification of the demand side of regulating services is particularly challenging because these services are neither directly used nor consumed. In addition, demand for regulating services is often underestimated by society due to the complexity of the natural and social processes associated with their delivery. Process-based models of vulnerability to risk and accessibility to service supply are often used to determine the degree of demand for regulating services from a risk-reduction perspective. For example, previous studies have quantified flood regulation service demand based on the presence of a flood hazard, possibility of exposure and vulnerability of assets [22,25]. Similarly, exposure to natural hazards, people and assets has been estimated to assess demand for coastal protection services [31]. This risk reduction perspective is less desirable for quantifying demand for carbon storage and sequestration. This is because climate change, the major risk that can be reduced or avoided through carbon storage and sequestration, is difficult to explicitly model and may further be perceived as a multiplier of other risks (e.g., reduced water availability, intensified urban heat islands). Given that this risk is entangled with negative consequences that are linked to other ecosystem services (e.g., water production, microclimate regulation), directly modeling demand for carbon storage and sequestration based on risk could lead to double-counting issues.

Proxy-based methods offer an alternative to the use of process-based methods to model desire for reducing the risk of climate change. These methods use a proxy, or indicator, to identify ecosystem service delivery or demand levels. Proxies are useful when the state and trend of targeted ecosystem services are not directly monitored or measurable. Previous studies have used proxies for mapping the distribution of ecosystem service supply (e.g., land use types, net primary productivity, ecosystem components and functions) [32], and they may also be used in estimating demand in a way that avoids many of the issues related to double-counting discussed above.

We present an approach to assessing ecosystem service supply and demand in a spatially-explicit manner focusing on carbon storage and sequestration provided by urban trees. In so doing, we seek to improve our understanding of the balance between supply and demand for this service and our ability to consider this service in local land-use decision making. Urban carbon storage and sequestration provides a highly policy-relevant arena for exploring the balance between service supply and demand. Although climate change mitigation via carbon storage and sequestration is delivered globally, understanding the degree of spatial mismatch between supply and demand for this service at the local level remains critical to its management because current global policies fail to effectively regulate local carbon emissions. Thus, setting policies which seek to mitigate climate change, for example those aimed at producing carbon-neutral cities, requires us to move to regional and local spatial scales. These scales are more relevant to local policy makers as they provide spatially-explicit estimates of the supply of and demand for carbon storage and sequestration within their jurisdictions. Such estimates could thus facilitate carbon balance assessments for cities and improved policy making.

For the purposes of this study, we define an ecosystem service flow as the actual services delivered from SPA to SBA. Given that carbon storage and sequestration is "used" to mitigate climate change, the supply of this service constitutes a flow from SPA to SBA. We thus refer to this flow as a "supply". We also distinguish between ecosystem service supply and demand by defining demand for carbon sequestration as the benefits desired or expected by beneficiaries which, in return, is driven by factors including biophysical supply, population size, consumption patterns, cultural perceptions and valuation of ecosystem services [3,11].

Our approach begins with the assessment and mapping of the supply side of this service, conceptualizing supply of carbon storage as the total amount of carbon stored in urban trees and supply of carbon sequestration as the amount of carbon sequestered by trees annually. We then assess and map demand using a proxy for the social desire for this service, quantifying benefits expected by society as a function of local annual anthropogenic CO_2emissions. To facilitate comparison of supply and demand for carbon storage and sequestration and to identify critical locations where imbalances in supply and demand occur, we lastly conduct a spatially-explicit supply-to-demand ratio analysis. In this way, our approach facilitates not only identification of the general carbon storage and sequestration service supplied by urban trees and the demand for this service across the study area, but also the specific locations to which management and policy might be targeted to improve the provision of this service. As such, our approach could enable policy makers to clearly target land use planning and management to locations where it would most increase service supply or where critical actions to reduce demand are needed.

OVERVIEW OF CARBON STORAGE AND SEQUESTRATION MODELING

Carbon storage and sequestration provided by trees in urban areas has been addressed in a number of studies. Field-collected data on urban forest structure (e.g., species; diameter at breast height, dbh; height; crown width) are most commonly used to estimate the supply of carbon storage and sequestration provided by urban trees [33–37]. The frequently-used i-Tree Eco model, which uses field-generated tree inventory data to estimate ecosystem services from urban trees, has been applied in numerous US cities to measure carbon storage and sequestration [38]. For instance, urban trees in Brooklyn, New York were estimated to store approximately 172,000 metric tons of carbon with a value of $3.5 million [39] while urban trees on public lands in Corvallis, Oregon were estimated to store about 71,000 metric tons valued at $1.45 million [40].

Remote sensing-based methods are also used to estimate carbon storage and sequestration by trees and can be helpful where forestry applications require quickly-updated data over larger extents. A number of remote-sensing technologies are now available to expedite the estimation of urban forest structure and corresponding carbon storage and sequestration [41]. Compared to passive sensors which excel at detecting the two-dimensions of ground features, terrestrial and airborne Light Detection and Ranging (LiDAR) are particularly promising, allowing for examination of the vertical structures of natural and built environments with high precision to facilitate assessments of aboveground forest biomass and bioenergy [42,43].

While studies that use these methods quantify the supply side of carbon storage and sequestration by urban trees, they do not compare this with demand for this service, making it difficult to understand ecosystem service dynamics and the relationship between service delivery and human well-being. Some studies, however, have compared estimated supply to demand [44,45]. For example, Baró et al. (2014) applied the i-Tree Eco model to quantify the biophysical supply of carbon storage and sequestration within the Barcelona metropolitan area in Spain and compared it to carbon emissions to assess the relative contribution of this service to climate change mitigation under targeted environmental policies [45]. These authors, however, did not examine the degree of balance in the supply and demand of carbon storage and sequestration in a spatially-explicit manner, but focused on the metropolitan area as a whole and did not identify key locations of provision and demand for this service.

As noted above, forest inventory data together with field sampling are generally used for biomass estimation. These data are often available at a coarse grain over large extents and very few fine-grained datasets exist for local and regional extents [43]. Urban tree inventory data are particularly lacking because urban trees are not systematically monitored by government agencies and field data collection is expensive, labor-intensive and time consuming [46]. Where these data exist, they are often limited to selected sites and may not be representative of urban forest characteristics across the larger region. Interpolation techniques can be used to overcome this problem, but, because of the heterogeneous nature of urban forests, are likely to introduce error to the modeling processes and increase uncertainty in carbon storage and sequestration estimation. As described above, the i-Tree Eco model is widely used for quantifying ecosystem services associated in urban trees. Since detailed urban forest data, such as species, dbh, total height, crown width and tree health condition are needed to implement carbon calculations using this model [45], it suffers from the same disadvantages as general field-data based methods discussed above.

Remotely-sensed LiDAR data may be analyzed to provide a faster, efficient and reliable way to identify urban forest characteristics for the assessment carbon storage and sequestration by urban trees. Airborne laser scanning technology enables LiDAR to detect numerous ground features over a large spatial extent. LiDAR emits an intense and focused laser pulse and measures the elevation of ground features by computing the time it takes for the sensor to detect laser returns, the angle at which the laser pulse is emitted, and the sensor location. Multiple laser returns are typically used to characterize both the horizontal and vertical structure of trees [43]. A first return is typically generated from the uppermost portion of the tree canopy, followed by multiple less-intense returns down through the canopy and a last return of the underlying terrain. Since LiDAR data sources are publicly available in many municipalities, using LiDAR datasets to assess urban forest structure can facilitate analyses of the supply of carbon storage and sequestration service provided by urban trees. This is particularly useful in dense urban environments because it facilitates efficient biomass and carbon calculations without the time, economic, and labor expenditures associated with intensive field sampling. LiDAR technology also has great potential to be used to generate reliable estimates of the supply of carbon storage and sequestration by trees because of its high accuracy in extracting three dimensional attributes of trees which might be obscured by cloud or shadows of other structures in aerial photography (e.g., height, crown width, dbh).

METHODS

We develop a method for quantifying, mapping and comparing the supply of and demand for carbon storage and sequestration by urban trees. We implement this method in our study area, a portion of the Twin Cities Metropolitan Area (TCMA) of Minnesota, USA (Fig 1). Our basic approach involves four steps: (1) estimating and mapping the supply of carbon storage using LiDAR technology and allometric models, (2) estimating and mapping the annual supply of carbon sequestration based on carbon gains from biomass growth and carbon loss from tree mortality and decay, (3) estimating and mapping demand for carbon sequestration based on the intensity of CO_2 emissions by a set of human activities under the assumption that demand for carbon sequestration service increases with anthropogenic carbon emissions, and (4) analyzing the relationship between the supply of this service (carbon sequestration) and the demand for it (Fig 2).

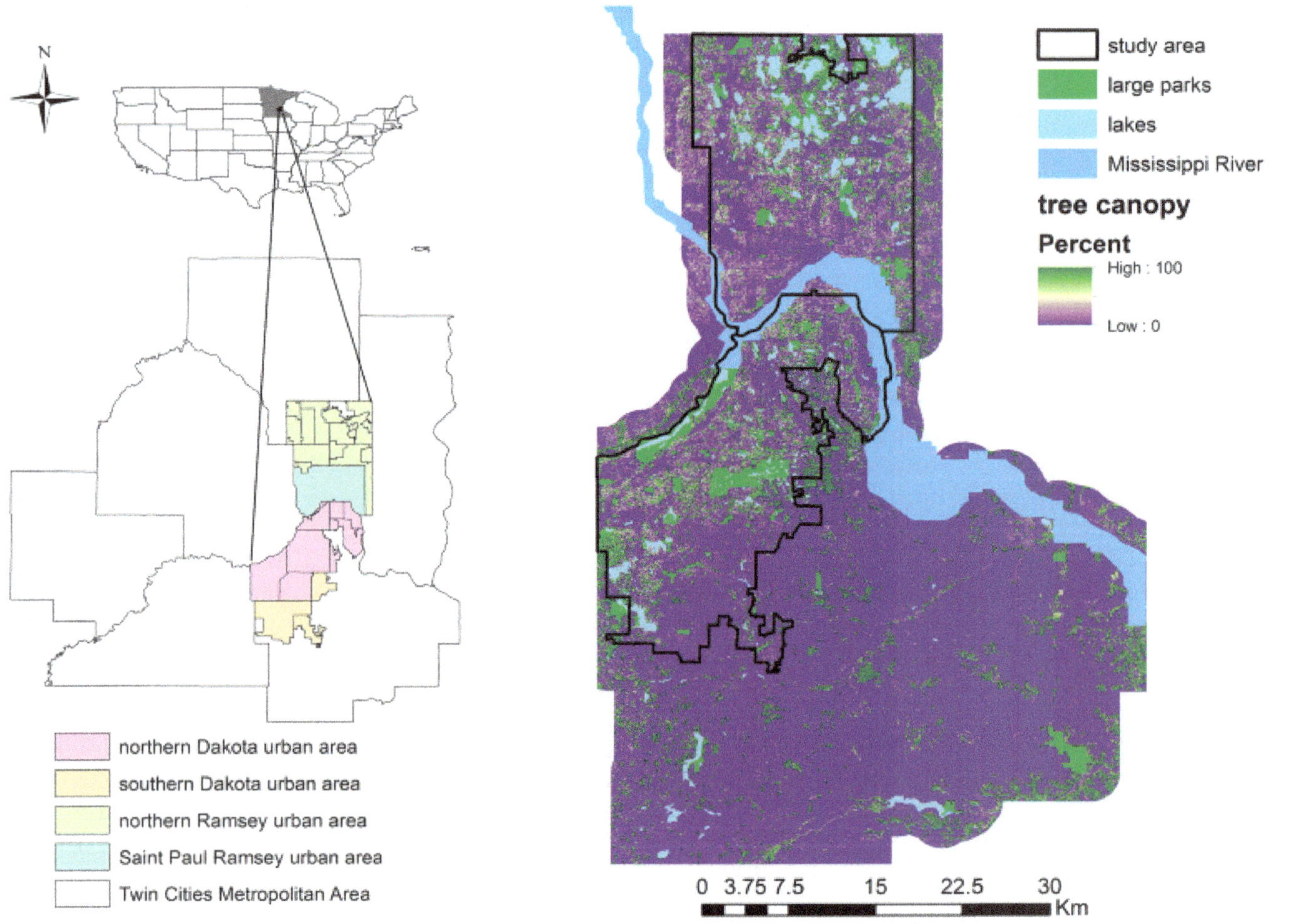

Figure 1: Location of the study area, urbanized areas of Dakota and Ramsey County, MN, including tree canopy coverage, parks and water body locations. This study area is divided into four regions (i.e., northern Dakota, southern Dakota, northern Ramsey, St. Paul Ramsey) to coincide with the regions for which tree abundance data used in the study were collected [47]. http://dx.doi.org/10.1371/journal.pone.0136392.g001

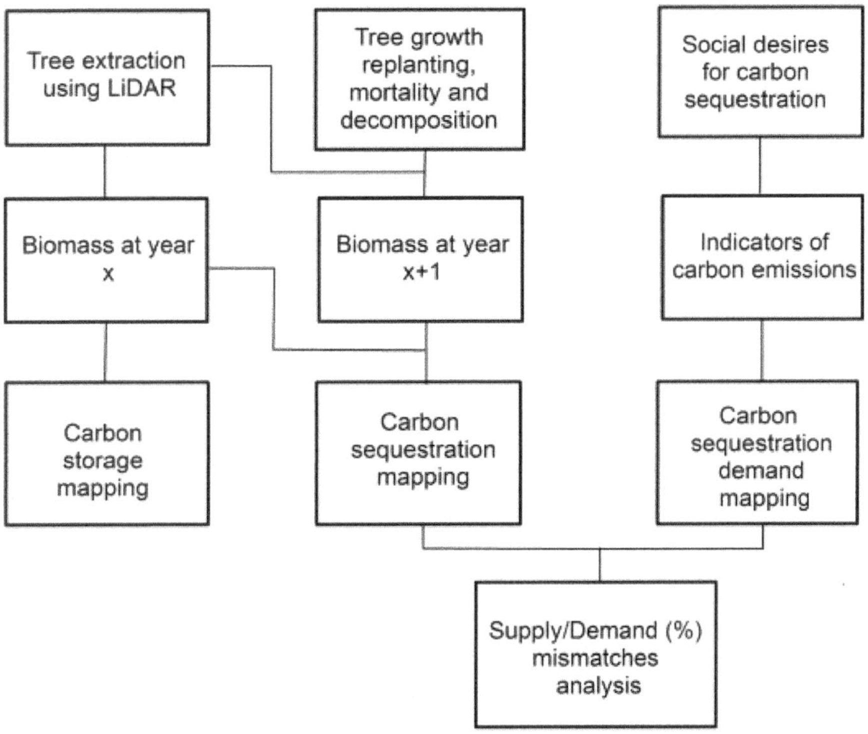

Figure 2: Methodological approach to the assessment and mapping of the supply and demand for carbon storage and sequestration. http://dx.doi.org/10.1371/journal.pone.0136392.g002

A. Study area

We illustrate this integrated approach using the urbanized areas of Dakota and Ramsey County in the TCMA (Fig 1). This study area includes an urban land area of 856 km^2 with a population of 359,365 and is comprised of 35 cities and towns, among which the state capital, St. Paul, is the second-most populous city in Minnesota. This urban land area was defined using the Urbanized Extents defined by the US Census and includes areas that are predominately, but not entirely, urban in their character. For the 2010 Census, an urban area comprises densely settled census tracts that have at least 2,500 people, along with adjacent land containing non-residential urban land uses as well as other land use types with low population density (e.g. forests, agriculture, non-developed land) [48]. In this study area, Dakota County is characterized by a mixture of urbanized land and forest reserves with dense tree coverage in the north, while southern portions of the county are dominated by agriculture with sparse trees. Similarly, northern Ramsey is mainly characterized by natural landscapes with dense tree coverage, while southern Ramsey is dominated by dense developed areas with sparser, street trees.

Because urbanization patterns in this study area are typical of those in most of the US, this location provides a representative location for quantifying and mapping supply and demand for urban carbon storage and sequestration from urban trees, given both the high number of urban trees and beneficiaries who receive their services in this location.

B. Tree extraction with LiDAR data

We retrieved 1175 tiles containing LiDAR point-cloud data collected in November, 2011, with a vertical accuracy of 5 cm for Ramsey County and 10.8 cm for Dakota County from the Minnesota Geospatial Information Office (MnGeo) [49]. This high-density dataset contains 8 points/m^2 for Ramsey and northern Dakota County and 2 points/m^2 for southern Dakota County. These data contain information on return number and number of returns, intensity of return signal, coordinates, scan direction and point cloud classifications, which are useful for tree feature extraction. This dataset uses oversampling techniques and has a vertical accuracy of up to 50 cm and a horizontal accuracy of up to 30 cm.

We used the LiDAR Analyst 5.0 extension [50] of ArcGIS 10.1 [51] to identify biophysical attributes of trees in the study area that are relevant to the estimation of biomass. This proprietary automated feature extraction-based tool (AFE) uses polygon decimation techniques, including shape-matching and machine-learning approaches, for LiDAR analysis [52]. LiDAR processing can provide two distinct surfaces: a Digital Elevation Model (DEM) which contains bare earth elevations, and a Digital Surface Model (DSM) which contains elevation information for tree canopy and buildings as well as other features (e.g., shrubs, bridges, elevated roadways). Extraction processes for trees require both an accurate DEM and DSM. Typically, a DSM is modelled by removing any point clouds belonging to the ground. Buildings also need to be correctly identified because they are critical for effectively modeling and displaying trees on an accurate DSM. The tree-extraction process thus begins with the extraction of bare earth and buildings from a sample tile, followed by adjustment of the extraction parameters controlling the detection process to ensure accurate feature extraction. In our analysis, default parameter settings were sufficient for extracting bare earth elevations as no significant distortion or raised features were observed. We classified LiDAR point clouds to ground and non-ground through bare earth extraction. A point-cloud-based method was selected for building extraction. We isolated and filtered trees based upon the bare earth DEM and corrected building features using a fixed-window search often applied to the analysis of dense urban forests. We set a minimum tree height of 3 m and typical tree height of 20 m, since this identified most tree and forest features. This presents a limitation as a tree under 3 m tall can contain a large amount of biomass. Nevertheless, considering the tendency for LiDAR to omit understory vegetation, particularly in dense urban areas, our 3 m threshold for tree extraction ensures high parameter prediction accuracy. This process produced a point shapefile, representing individual trees with their attributes, including tree height, crown width and dbh. Once we set workflow and extraction parameters with the sample tile in this way, we used batch processing to extract tree features from other tiles in the dataset based on the AFE algorithm.

Since tree extraction is based on a shape-matching approach, ground features similar to trees may be misclassified as trees. Infrared beams emitted by LiDAR sensors also tend to be absorbed by water and return very weak to no signals, resulting in poorer extraction performance. To validate LiDAR results, we visually compared the extracted results with a 2012 orthophoto of the TCMA seven-county metropolitan area from the Minnesota Department of Natural Resources (MNDNR). We deleted misclassified trees on water and roof tops, and trees that were actually water towers or other round features. Tree records from the MNDNR Minnesota Native Big Tree registry [53] were used to help eliminate trees of abnormal height (>40 m), dbh (>3.187 m) or crown width (> 42.672 m).

To assess tree-extraction accuracy, we collected tree attribute measurements from the orthophotograph mentioned above, then selected 50 trees distributed evenly across the study area, and measured their crown width using ArcGIS 10.1. These trees had clear crown shapes to minimize measurement error. We fitted a linear regression model using the resulting dataset and calculated root mean squared error (RMSE) using our measured values and the LiDAR Analyst results to determine the accuracy of the LiDAR metric in predicting tree attributes.

C. Estimation of carbon storage

Estimation of tree dry-weight biomass is of great interest since such information indicates carbon storage and could enable policy makers to track carbon dynamics in biomass [54]. We used taxon-specific biomass equations to estimate above-ground biomass following the work of Jenkins et al. (2004) which used allometric equations and conversion factors to transfer above-ground biomass to dry-weight biomass based on dbh [37,54]. We based our calculations on existing field datasets that identified the relative abundances of common species by land use intensity and location in the study area (Table 1) [47,54]. Based upon these sources, we classified common tree species in the study area into seven classes: soft maple (Acer negundo, Acer saccharinum, Acer pensylvanicum and Acer rubrum) /birch (Betula spp.), aspen/alder/cottonwood/willow (Populus, Alnus and Salix spp.), hard maple (Acer saccharum,Acer plantanoides and Acer nigrum)/oak (Quercus spp.)/hickory (Carya spp.)/beech (Fagus spp.), other hardwood, cedar (Chamaecyparis spp.)/larch (Larix spp.), spruce (Picea spp.) and pine (Pinus spp.). The tree species groupings for this analysis present a possible limitation as differences in biomass carbon storage may exist among species in these groups and as tree composition in urban areas is normally diverse and may contain species that are not well-represented by this classification system. Nevertheless, this is the most specific grouping system possible given data and model limitations and is sufficient to illustrate the application of our methodology. Additionally, although calibrating allometric models for specific species is of interest, our current understanding of the relationship between tree biophysical attributes and biomass at a single-species level is insufficient to support such calibrations [46]. Moreover, allometric models for different species have very similar shapes, thus causing minimal differences in biomass predictions among species.

Table 1: Relative abundance of tree species groups by region and land use and parameter values for biomass estimation in Dakota and Ramsey urban areas, MN. http://dx.doi.org/10.1371/journal.pone.0136392.t001

Species	Relative abundance (%)[1]					β_0	β_1	Ave. dbh growth rate (cm/yr)
	Developed land				Undeveloped land			
	N. Dakota	S. Dakota	N. Ramsey	St. Paul Ramsey				
mb[2]	12.82	18.23	17.47	18.17	11	-1.9123	2.3651	0.152
aa[3]	6.52	6.99	3.72	0	13	-2.2094	2.3867	0.406
mo[4]	5.19	2.95	0	0	14	-2.0127	2.4342	0.328
oh[5]	63.09	54.91	68	76.64	64	-2.4800	2.4835	0.282
cl[6]	8.29	2.73	5.35	5.19	0	-2.0336	2.2592	0.185
sp[7]	4.09	14.19	5.46	0	0	-2.0773	2.3323	0.348
pi[8]	0	0	0	0	1	-2.5356	2.4349	0.345

[1] percent of total trees represented by each species group
[2] soft maple/birch
[3] aspen/alder/cottonwood/willow
[4] hard maple/oak/hickory/beech
[5] other hardwood
[6] cedar/larch
[7] spruce
[8] pine, β_0 and β_1 are parameters used in biomass equations for estimating total aboveground biomass for hardwood and softwood species in the United States.
Table was produced based upon previous studies [47,54,71].

Since LiDAR processing does not provide species information, we identified likely tree species for the trees characterized by LiDAR processing based upon the relative abundances of common species estimated for the study area in previous studies (Table 1) [47]. In so doing, we first classified trees as occurring on developed or undeveloped land and by location. Then, for each group and location, we randomly sampled a percentage of trees corresponding to the relative abundance estimated by the field studies for that group and location and assigned them to the specified species group. We repeated this for each species group and location until all trees were assigned a species group. Once tree species were assigned, we applied the following equation to calculate above-ground biomass for each tree:

$$B = Exp(\beta_0 + \beta_1 * \ln(dbh)) \quad \textbf{(1)}$$

Where

1. β_0, β_1 is based on species type (Table 1)
2. B = total aboveground biomass (kg) for trees 2.5 cm and larger in dbh

Although the root-to-shoot ratio varies somewhat from species to species, we assumed a fixed root-to-shoot ratio of 0.26 for all species groups and converted above-ground biomass to whole-tree, dry-weight biomass following past studies [35,55]. Total tree dry-weight biomass was multiplied by 0.5 to calculate the total stored carbon in trees which approximates the proportional mass of carbon in trees [37,56–58].

D. Estimation of carbon sequestration

Our estimates of carbon sequestration firstly consider the probability of tree mortality. To generate a subset of trees surviving to the next year, we used tree mortality rates by dbh from a study of Chicago's urban forest ecosystem [34]. Although this study dates from 1994 and occurred outside of our study area, it is appropriate for mortality estimation in this analysis since our study area shares the same ecoregion (Eastern broadleaf forest) and has similar land use and climate conditions and tree composition [59]. Additionally, no other recent study of which we are aware exists for this region. This study estimated annual tree mortality rates of 2.1% for trees between 16 and 46 cm dbh, 2.9% for trees between 47 to 61 cm dbh, 3.0% for trees between 62 and 76 cm dbh, and 5.4% for trees greater than 77 dbh. Based upon this information, we randomly sampled trees in each dbh class to identify an appropriate percentage of trees surviving to the next year for use in estimating sequestration rates.

Our consideration of decomposition is based on best evidence and is conservative. Tree decomposition rates vary with their physical characteristics, location and environment. Within the city of St. Paul only, because almost all dead trees are removed annually and ground for use in power generation (Zach Jorgensen, City of St. Paul Forestry Division, personal communication), we assumed that 100% of carbon stored in trees would be emitted to the atmosphere within a year of death. On the other hand, trees on forest, agricultural or undeveloped land outside the city typically remain intact and experience natural decomposition as little forestry management exists there. For these sites, we estimated carbon releases from decomposition in two ways: (1) delayed release from tree roots; (2) delayed release from aboveground biomass. For the former, based on existing studies, we assumed and estimated belowground biomass to decompose over 20 years [35] with 20% of carbon released in the first year [60]. For aboveground biomass, we modified an exponential model of annual biomass loss following Olson et al. (1963) [61]:

$$C_t = 0.5 * M * e^{-kt} \quad \textbf{(2)}$$

Where

1. C_t = weight of carbon left at time t (kgC)
2. M = initial aboveground biomass before decomposition (kgC)
3. k = decomposition rate constant
4. 0.5 = carbon concentration

Determining the value of k is a key step in this process. Decomposition rates vary by vegetation component and species which influences the timing and magnitude of carbon released [62]. It is generally assumed that the decay rate of logs is much lower than the decay rate of foliage and branches [63]. However, small twigs and branches do not always decay more rapidly than larger materials, depending mainly on site moisture content [63]. In addition to temperature and moisture, other factors that influence decay rates include soil nutrient content and microbial community [64,65]. Since regional climate plays a major role in determining the decay rate of dead wood [66], we searched for past studies that assessed the value of k in Minnesota. Although decay rate constants for many tree species were available in different geographic locations, little research existed that estimated species-specific decay rates for Minnesota specifically [66–68]. Therefore, to simplify calculations, an average decomposition rate constant was used for all species, without differentiating decay rates between logs, foliage and twigs. Specifically, we followed a large wood decomposition study from a temperate mixed forest ecosystem in north-central Minnesota and set k at 0.062 [67]. To compute the weight of carbon remaining, we multiplied the final biomass at time t by the percent carbon content, which is reported to remain at 50% throughout the decomposition process, and to be relatively constant among sites and species [69]. Lastly, for trees outside the city of St. Paul located on other land uses (e.g., developed land), which are typically chipped and applied as landscape mulch, we conservatively assumed that 80% of carbon was released in the first year following mortality [70].

Tree growth rate should also be considered in estimating carbon sequestration rates as this is critical to carbon sequestration. Using values from a northeastern US study which estimated growth rates based on individual tree species group [71], we identified the following average dbh growth rates: aspen/alder/cottonwood/willow, 0.41 cm/year; spruce and pine, 0.35 cm/year; maple/oak/hickory/beech, 0.33 cm/year; other hardwood, 0.29 cm/year; cedar/larch, 0.19 cm/year; and soft maple/birch, 0.15 cm/year. For trees surviving into the next year, the average observed annual growth by species was added to the tree dbh from the LiDAR-generated tree data to estimate tree dbh in year two. We then recalculated the biomass equation using the new dbh to identify carbon storage in trees in year two. We calculated annual carbon sequestration due to tree growth as the difference between carbon storage estimates for year 1 and 2.

Tree planting also warrants consideration in our estimates as planted trees sequester carbon. The City of St. Paul's Forestry

Unit oversees the majority of tree planting within city limits. We assumed that 15% of dead trees were replaced with trees of the same size and species in St. Paul based on the Forestry Unit's current management practices (Zach Jorgensen, City of St. Paul Forestry Division, personal communication). We used a conservative 10% replanting rate for developed land in other jurisdictions, assuming that some of them may not replant to the extent that St. Paul does, and a 0% replanting rate for undeveloped land, given that little forestry management occurs on such land.

Lastly, estimates of the supply of carbon storage and sequestration were aggregated from the individual plant level to the census tract level in order to facilitate comparisons with subsequent demand assessments. The final equations for calculating annual net carbon sequestration are as follows:

$$C_s = C_g - C_l + C_p \tag{3}$$

$$C_g = 0.5*\sum(B_{i,x+1} - B_{i,x}) = 0.5*\sum e^{\beta_0+\beta_1*\ln(dbh_i+\Delta dbh_i)} - e^{\beta_0+\beta_1*\ln(dbh_i)} \tag{4}$$

$$C_l = \left\{ \begin{array}{c} 0.5*\sum B_u*100\% \\ 0.5*\left\{\sum[B_{above,v}*(1-e^{-k})] + \sum B_{below,v}*20\%\right\} \\ 0.5*\sum B_w*80\% \end{array} \right\} \tag{5}$$

$$C_p = 0.5*\sum B_j \tag{6}$$

Where

1. C_s = net annual carbon sequestration per census tract (kgC)
2. C_g = total carbon gained from tree growth in year x+1 (kgC)
3. C_l = total carbon lost from tree decomposition in year x+1 (kgC)
4. C_p = total carbon gain from tree replanting in year x+1 (kgC)
5. i = identifier for a single living tree for year x+1
6. $B_{i,x}$ = total biomass for tree i in year x (kgC)
7. $B_{i,x+1}$ = total biomass for tree i in year x+1 (kgC)
8. u = identifier for a single tree that died in a year in St. Paul
9. v = identifier for a single tree that died in a year outside St. Paul on forest, agricultural or undeveloped land
10. w = identifier for a single tree that died in a year on undeveloped land uses
11. B_u = total biomass for tree u (kgC)
12. $B_{above,v}$ = aboveground biomass for tree v (kgC)
13. $B_{below,v}$ = belowground biomass for tree v (kgC)
14. B_w = total biomass for tree w (kgC)
15. j = identifier for a single tree planted in year x+1
16. B_j = total biomass for tree j (kgC)

E. Estimating demand for carbon sequestration

Demand for carbon storage and sequestration service is expressed based on local anthropogenic CO_2 emissions. Our fundamental assumption in measuring demand is that increasing CO_2 emissions are associated with a greater need for carbon storage and sequestration service to mitigate climate change and that this need is an indicator of demand for this service. We quantify this need by estimating total carbon emissions by census tract, a level at which sufficient data exist and that can support policy formation. We compare demand estimated in this way to carbon sequestration supply estimated above to identify the ability of local trees to offset emissions. This thus indicates the degree of balance in supply and demand for this service.

We estimated CO_2 emissions by sector. The Minnesota Pollution Control Agency (MPCA) groups emissions of carbon dioxide in Minnesota into seven sectors: agricultural, commercial, electric utility, industrial, residential, transportation and waste [72]. Average emissions intensity and energy consumption values for Minnesota were used in this estimation due to a lack of local-level data [72]. Estimated values were combined with information on population, number of vehicles, number of employees in industrial and commercial occupations and areas of agricultural land to calculate final demand by census tract (Table 2).

Table 2: Symbols and parameters used in carbon sequestration demand calculation (after MPCA, 2012 [72]). http://dx.doi.org/10.1371/journal.pone.0136392.t002

Source	Time	Scale	Symbol	Parameter
MPCA	2008	state	I_e	lbsCO_2/kWh consumed
	2008	state	I_t	Short Tons CO_2/vehicle
	2008	state	I_a	lbsCO_2/acres harvested
	2008	state	I_i	lbsCO_2/ industrial employee
	2008	state	I_r	lbsCO_2/capita
	2008	state	I_c	lbsCO_2/commercial employee
	2008	state	I_w	lbsCO_2 emitted from wastewater treatment/ capita
	2008	state	E	kWh consumed per year/capita
Generalized Land Use 2010 for the TCMA	2010	1:3000–1:1500	A_a	Agricultural land in m^2
U.S. Census 2007–2011 American Community Survey 5-Year Estimates	2007–2011	census tract	P	Number of people/census tract
	2007–2011	census tract	W_i	Number of employees in industrial sectors
	2007–2011	census tract	W_c	Number of employees in commercial sectors
2008–2012 American Community Survey 5-Year Estimates	2008–2012	census tract	V	Number of vehicles/census tract

doi:10.1371/journal.pone.0136392.t002

Our estimates of carbon emissions from the electric utility sector exclusively account for the combustion of nonrenewable energy sources, given that renewable energy sources, such as geothermal, hydro, wind and solar, emit nearly zero carbon. Based on Xcel Energy, the utility company that serves most of the study area, 70% of the electricity in the study area comes from natural gas and coal combustion [73]. Accordingly, we calculated carbon sequestration demand from this sector as follows:

$$D_e = I_e * E * P * 0.7 * 0.453592 \quad (7)$$

Where

1. D_e = Carbon sequestration demand from electric utility sector ($kgCO_2$)
2. I_e = Electric intensity measured as $lbsCO_2$/kWh consumed based on MPCA estimates
3. E = Energy consumed as kWh/year/capita
4. P = Number of people/census tract
5. 0.70 adjusts total emissions to account for the percentage of electricity from nonrenewable energy sources (e.g. natural gas, coal-fired plants)
6. 0.453592 is a conversion factor that coverts lbs. to kg

Our estimate of carbon emissions from transportation is based on the number of vehicles (i.e., cars, trucks and vans) in a census tract used by workers 16 years of age and over from the 2008–2012 American Community Survey 5-Year estimates. We multiplied this by the amount of CO_2 emitted per vehicle:

$$D_t = I_t * V * 907.185 \quad (8)$$

Where

1. D_t = Carbon sequestration demand from transportation sector ($kgCO_2$)
2. I_t = Transportation intensity measured as short ton CO_2/ vehicle based on MPCA estimates
3. V = Number of vehicles /census tract
4. 907.185 is a conversion factor that coverts short ton to kg

To estimate the demand for carbon sequestration from industrial and commercial sectors, we used total employment and carbon dioxide emission rates for each sector. Based on the 2011 U.S. American Community Survey, employment in industry included occupations in mining, construction, manufacturing, and transportation and warehousing. Likewise, wholesale trade, retail trade, information, finance, insurance, real estate rental and leasing, management, educational services, as well as entertainment, accommodation and food services jobs were grouped into commercial employment. We obtained the number of employees from the above occupations from 2011 US Census American Community Survey and used it to estimate carbon emissions from these sectors as follows:

$$D_i = I_i * W_i * 0.453592 \quad (9)$$

Where

1. D_i = Carbon sequestration demand from industrial sector ($kgCO_2$)
2. I_i = Industrial intensity measured as $lbsCO_2$/ industrial employee based on MPCA estimates
3. W_i = Number of employees in industrial sector
4. 0.453592 is a conversion factor that coverts lbs. to kg

$$D_c = I_c * W_c * 0.453592 \quad (10)$$

Where

1. D_c = Carbon sequestration demand from commercial sector ($kgCO_2$)
2. I_c = Commercial intensity measured as $lbsCO_2$/ commercial employee based on MPCA estimates
3. W_c = Number of commercial sector employees
4. 0.453592 is a conversion factor that coverts lbs. to kg

We used a 2010 land-use map for the TCMA available from the Twin Cities Metropolitan Council to calculate areas of agricultural land in each tract in m^2 [74]. We then computed agricultural carbon emissions as:

$$D_a = I_a * A_a * 0.000247105 * 0.453592 \quad (11)$$

Where

1. D_a = Carbon sequestration demand from agricultural sector ($kgCO_2$)
2. I_a = Agricultural intensity measured as $lbsCO_2$/acre harvested based on MPCA estimates
3. A_a = Area of agricultural land in m^2
4. 0.000247105 is a conversion factor used to convert m^2 to acre
5. 0.453592 is a conversion factor that coverts lbs. to kg

Similarly, we estimated carbon demand from residential and waste sectors using the population of each census tract from the 2011 U.S. Census American Community Survey as follows:

$$D_r = I_r * P * 0.453592 \quad (12)$$

Where

1. D_r = Residential sector carbon sequestration demand ($kgCO_2$)
2. I_r = Residential intensity measured as $lbsCO_2$/capita from MPCA estimates
3. P = Census tract population
4. 0.453592 is a conversion factor that coverts lbs. to kg

$$D_w = I_w * P * 0.453592 \quad (13)$$

Where

1. D_w = Waste sector carbon sequestration demand ($kgCO_2$)
2. I_w = Waste intensity measured as $lbsCO_2$ emitted from wastewater treatment/capita from MPCA estimates
3. P = Census tract population
4. 0.453592 is a conversion factor that coverts lbs. to kg

Lastly, CO_2 emissions from all sectors were summed to estimate the net demand, D_{net} (kgC), for the carbon sequestration service in each tract. Based on the composition of CO_2, we used a factor of 0.2729 to determine the mass of emitted carbon:

$$D_{net} = (D_e + D_t + D_i + D_c + D_a + D_r + D_w) * 0.2729 \quad (14)$$

F. Supply–to-demand ratio analysis

We followed a two-step process to compare estimated carbon sequestration supply to demand. Firstly, we mapped supply and demand for each tract to visualize spatial correspondence between supply and demand. Secondly, we estimated a supply-to-demand ratio for each tract and the full study area to indicate the ability of supply to meet demand. Although the supply of carbon sequestration was assessed at the individual tree level, census tracts were selected as the minimum mapping unit for the supply-to-demand ratio analysis because demand for carbon sequestration was estimated at this level. We used the following formula to calculate the supply-to-demand ratio for each tract:

$$R = S/D_{net} * 100\% \quad (15)$$

Where

1. R = ratio of supply to demand for carbon sequestration
2. S = supply of carbon sequestration service in kgC/census tract/year
3. D_{net} = demand for carbon sequestration in kgC/census tract/year

RESULTS

A. Urban trees characteristics

In all, LiDAR data processing identifies 7,291,140 trees in Dakota and Ramsey County urban areas. These trees are not evenly distributed by land-use type or region. Generally, northern Dakota County and northern Ramsey County have much denser tree canopy cover than the rest of the study area (Fig 3). The characteristics of trees also vary (Table 3). Overall, trees have an average height of 11.51 m (range: 3 m-440 m, SD = 4.62 m) and average crown width of 3.90 m (range: 2.6 m-16.93 m, SD = 1.07 m). Tree dbh ranges from 0.24 m-0.83 m (mean = 0.29 m, SD = 0.04 m). The LiDAR metric variable explains 76% of the variability in the crown width of trees (Adjusted R^2: 0.7559; p <0.001) (S1 Fig). The average bias between predicted and measured crown width is about 0.19 m, and the RMSE of the regression between them is 0.43 m (11% of the mean crown width).

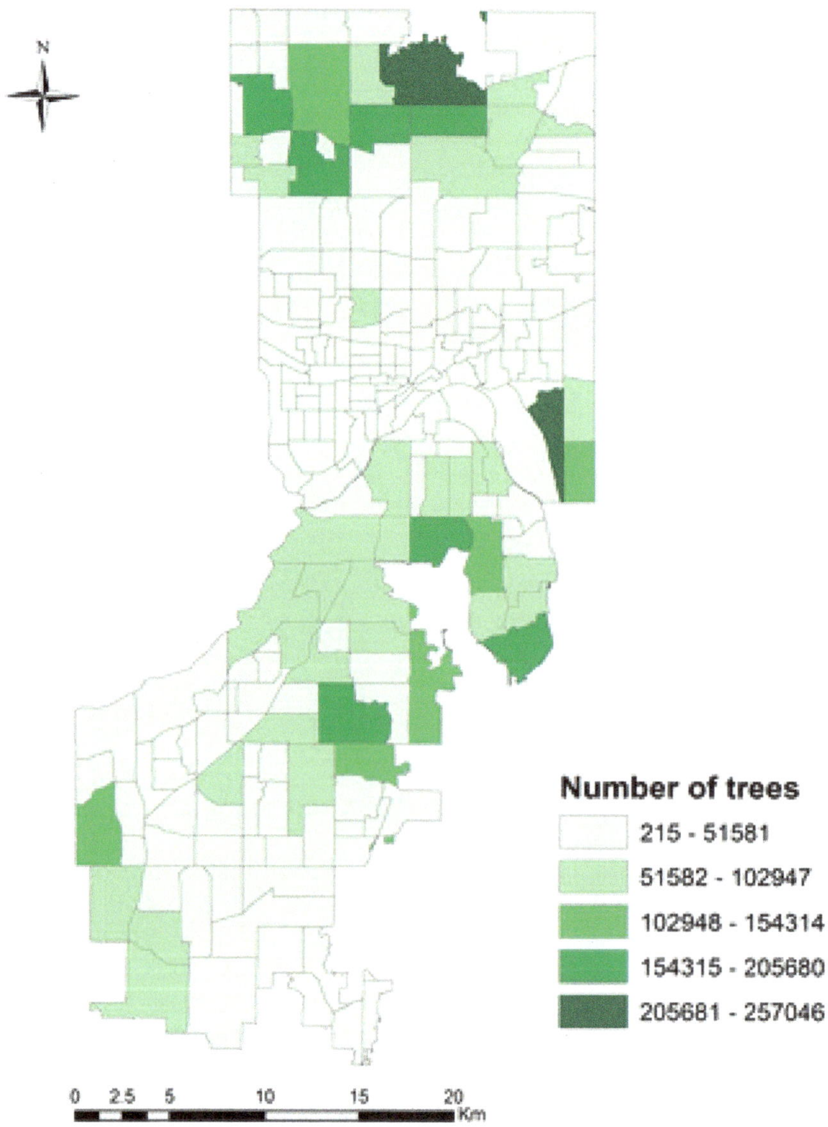

Figure 3: Number of urban trees by census tract for Dakota and Ramsey County, MN urban areas identified through LiDAR processing. A natural breaks (Jenks) classification system was used to more clearly represent trends in the data due to uneven distributions of values. http://dx.doi.org/10.1371/journal.pone.0136392.g003

Table 3: Statistics for individual trees identified via LiDAR processing in Dakota and Ramsey County, Minnesota. http://dx.doi.org/10.1371/journal.pone.0136392.t003

	Tree height (m)	Crown width (m)	dbh (m)
Mean	11.51	3.90	0.29
Minimum	3.00	2.60	0.24
Maximum	440.00	16.93	0.83
Standard Deviation	4.62	1.07	0.04

doi:10.1371/journal.pone.0136392.t003

B. Carbon storage

We estimate that urban tree resources in the study area store 1,735.69 million kgC of which 889.17 million kgC are stored by Dakota County trees and 846.52 million kgC are stored by Ramsey County trees. The carbon storage of individual trees ranges from 103.34 kg–3,402.61 kg. The average carbon storage per tree is 238.13 kg (range: 103.34 kg–3402.61 kg, SD = 106.81 kg, Table 4). Average carbon storage is 7.78 million kgC/census tract (range: 45,536 kg–58.71 million kg, SD = 10.64 million kg). This service is most abundant in northern Dakota County and in northern Ramsey County (Fig 4). In Dakota County, the highest carbon storage values occur in the county's northeastern corner, where 43 million kgC/census tract are stored by urban trees. This service is also relatively high along the northern and western borders of Dakota County and gradually decreases to 14 million kgC/census tract when it passes through more central urban areas. The lowest service levels occur in agricultural southern townships.

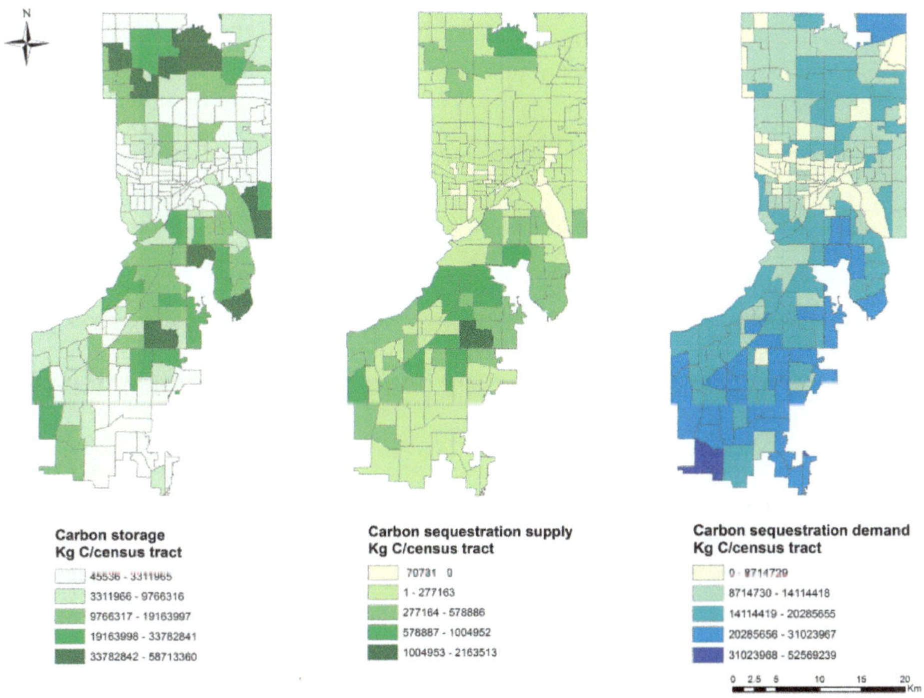

Figure 4: Carbon storage and carbon sequestration service supply and demand maps for Dakota and Ramsey County urban areas. A natural breaks (Jenks) classification system was used to more clearly represent trends in the data due to uneven distributions of values. http://dx.doi.org/10.1371/journal.pone.0136392.g004

Table 4: Statistics for carbon storage and sequestration service in urban areas in Dakota and Ramsey County, MN. http://dx.doi.org/10.1371/journal.pone.0136392.t004

	Carbon sequestration (kgC /census tract/ year)		Carbon storage (kgC /census tract)	Carbon storage (kgC/tree)
	Supply (flow)	Demand	Supply	
Mean	149,897	13,845,718	7,783,356	238.13
Min.	-70,731	0	45,536	103.34
Max.	2,163,513	52,569,239	58,713,360	3,402.61
SD	254,434	6,711,763	10,637,338	106.81
Total	33,427,076	3087,595,137	1,735,688,447	1,735,688,447

doi:10.1371/journal.pone.0136392.t004

In Ramsey County, this service decreases moving from north to south. The highest carbon storage levels occur in the northern region, where 58.7 million kgC/census tract are stored. Higher levels of carbon storage (mean = 22 million kgC) also occur in other areas of northern Ramsey County. This decreases to 7 million kgC/census tract moving towards central Ramsey County. The lowest levels occur in southern Ramsey County, where trees are more sparsely-distributed and store only 0.6 million kgC/census tract.

C. Supply of carbon sequestration

Net annual carbon sequestration for the study area is estimated to be 33.43 million kgC/year. On average, 149,897 kgC/census tract are sequestrated by trees annually (Table 4). Annual carbon sequestered due to tree growth is 49.3 million kgC, and carbon storage gained from planting is 3.4 million kgC. Loss from tree mortality is 19.3 million kgC. The highest carbon sequestration service occurs in eastern Dakota County, where large forested parks exist (Fig 4) and trees capture 2.1 million kgC/year. Higher carbon sequestration levels occur in northern Ramsey County and in Dakota County where carbon captured by tree growth offsets carbon lost to tree mortality (mean = 0.7 million kgC/census tract/year, SD = 0.36 million kgC). Conversely, the carbon sequestration service is negative on some land in the St. Paul where trees are sparsely distributed and tree mortality and decay rates are higher.

D. Demand for carbon sequestration

In total, 3087.60million kgC or 7.3% of total 2010 GHG emissions in the state of Minnesota (155.6 million CO2-equivalent tons [72]) are emitted annually through economic development and human consumption activities in the study area. The electric utility sector is the primary source of carbon emissions (1564.55 million kgC), followed by agriculture (618.69 million kgC), transportation (467.26 million kgC), residential (411.52 million kgC), waste (23 million kgC), industrial (2.23 million kgC) and commercial activities (0.33 million kgC). Moreover, commercial carbon emissions are concentrated in Ramsey County, while agricultural and industrial carbon emissions are distributed unevenly in Dakota County.

The spatial pattern exhibited by demand for carbon sequestration differs from that of supply (Fig 4). The highest demand occurs in southernmost Dakota County, where 55.2052.57 million kgC are emitted, but only 2596.83 kgC are sequestered. This emissions level is more than three times the average level for the overall study area (13.85 million kgC emissions/census tract). In general, greater demand occurs in southern Dakota County and northern Ramsey County, where census tracts have an average emission of 21.74 million kgC (SD = 4.20 million kgC). Most areas of lower demand occur in St. Paul (mean = 8.23 million kgC emissions/census tract, SD = 3.40 million kgC).

E. Comparing supply with demand

A considerable mismatch exists between supply and demand for the carbon sequestration service in the study area (Fig 4). Firstly, a substantial carbon sequestration deficit exists as demand far exceeds the carbon sequestration supply in the area as a whole (3087.60million kgC/year versus 33.43 million kgC/year, Table 4). Secondly, with the exception of some census tracts in northern Dakota County where supply and demand are more coincident, a very high spatial mismatch occurs between demand and carbon sequestration by trees. A substantial undersupply exists in southern Dakota County in particular, where high demand is observed.

The magnitude of the difference between carbon sequestration supply and demand makes comparing them in absolute quantities challenging. The ratio map however, facilitates the comparison of supply with demand by demonstrating the relationship between them in a quantitative and easily-visualized manner (Fig 5). For example, the highest ratio occurs in north-central Dakota County where 13% of carbon emissions are sequestered by trees, approximately equaling the US average rate of GHG reduction provided by forest ecosystems (10%) [75]. Along the Mississippi River Corridor Critical Area (MMRRA) region of Dakota County, a heavily-forested area where 116 km of the Mississippi River and 21,853 hectares of adjacent corridor lands are designated and managed as natural areas, approximately 4% of demand is offset by trees. A similarly high ratio is observed in northern Ramsey County, where forested recreational areas cluster. Most of central Ramsey County and southern Dakota County, however, have a nearly zero carbon offset, with a ratio of 0.3%, less than half of the average ratio for the study area (0.9%, SD = 1.6%). Critically, supply-to-demand ratios below zero (-0.5%–0%) occur in some areas in St. Paul, indicating negative supply of carbon sequestration (i.e., that trees are releasing CO_2).

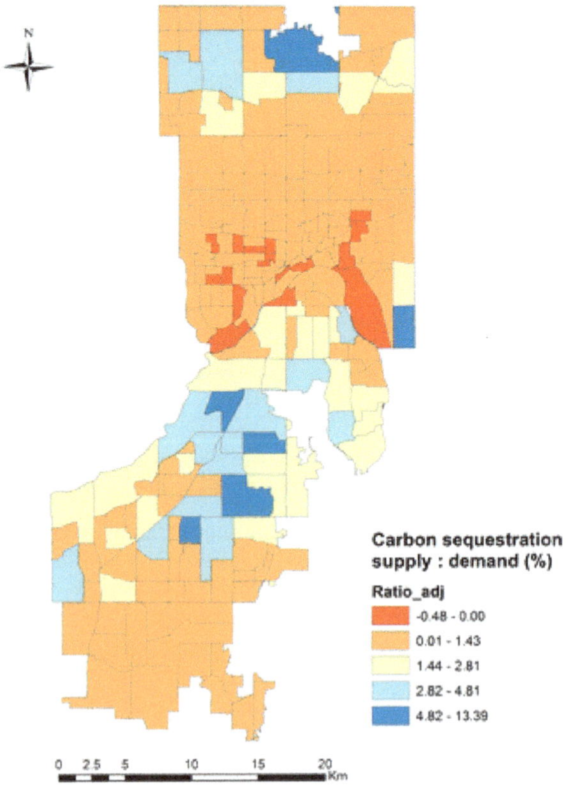

Figure 5: Carbon sequestration service supply-to-demand ratio map for Dakota and Ramsey County urban areas, representing the relative carbon sequestration balance. A natural breaks (Jenks) classification system was used to more clearly represent trends in the data due to uneven distributions of values. The first class was modified to show ratios below zero. http://dx.doi.org/10.1371/journal.pone.0136392.g005

DISCUSSION

Current assessments of ecosystem services tend to focus solely on the supply of these services. As this fails to consider a critical component of the ecosystem services dynamic, the demand or need for these services, these assessments do not fully represent the status of the services they assess. This strongly limits the use of these assessments in policy making as they fail to provide information on the full spectrum of ecosystem service delivery and consumption. Additionally, many existing studies are not conducted at scales relevant to policy making or lack a spatial component that could help policy makers identify the locations to which policies might be targeted or rely on data that are costly or difficult to acquire. In this study we presented a method for assessing both the supply and demand for an ecosystem service, carbon storage and sequestration by urban trees, as well as the relationship between this supply and demand, in a spatially explicit manner that could inform local policy making related to climate change mitigation. As such, this method begins to fill the existing gap in our ability to provide meaningful ecosystem service assessments to planners and policy makers. This method has numerous strengths that could link the ecosystem service more closely to planning and policy making as well as weaknesses that require further refinement.

A. Strengths and limitations of the LiDAR-based method for assessing ecosystem service supply

The supply of carbon storage and sequestration depends on ecological functions and processes that take place at the individual plant level, including tree growth, mortality and decomposition. As such, estimating carbon storage and sequestration requires a method that goes beyond traditional proxy-based approaches that only consider the capacity of ecosystem service provision by land cover type (e.g., InVEST [76]). Our approach uses LiDAR-based techniques that capture biophysical attributes of individual trees and thus provides the information needed to quantify and map final carbon storage and sequestration supply at a finer scale by considering ecological functions and processes. As opposed to traditional forest inventory-based methods (e.g., i-Tree model [38]), our LiDAR-based method has additional advantages related to reduced need for field surveys and greater cost-effectiveness in terms of data capture and processing.

Our findings indicate that LiDAR is a reliable method for assessing urban tree features. We did not conduct direct validation of LiDAR-derived dbh and tree height due to a lack of field measurements. However, since previous studies indicate that the dbh-height-crown width relationship for a given set of species and site conditions is relatively stable [42], the high accuracy indicated by our assessment for estimated crown width suggests that our dbh estimates are also highly accurate. Our estimated RMSE for crown width (RMSE = 0.43 m) is also less than those of previous studies that estimated crown width based on biometric relationships with typical field-collected forest inventory variables (i.e., dbh, height, height-to-crown base, crown class, basal area per hectare, and trees per hectare) (RMSE: 0.6081–1.48 m) [77] and that derived crown width by delineating tree shapes using a local maximum algorithm on a canopy height model (RMSE = 1.36–1.41 m) [42]. This suggests that feature extraction approaches using LiDAR data such as the approach employed here are able to measure crown width with higher accuracy than these models.

The accuracy of LiDAR feature extraction depends not only on the algorithms used, but also the characteristics of the LiDAR dataset (e.g., number of returns, density) used as well as study site conditions (e.g., tree canopy density, number of vegetation layers, slope, and tree species composition). Previous studies have suggested that the error associated with LiDAR-derived forest variables can be reduced using high density LiDAR data and an accurate DTM [78]. Such data allow more laser pulses to penetrate the upper canopy and reach the ground. This, in turn, results in greater accuracy in information such as tree locations and attributes derived from such datasets. Thus the high accuracy of tree attribute measurements indicated by our low RMSE likely results at least in part from the use of a high-density LiDAR dataset in this study.

We find the accuracy of tree features to vary with location. For example, feature extraction accuracy is relatively lower for trees located along building edges or sparsely distributed on agricultural lands, where ground features with rounded shapes (e.g., water towers, ethanol refineries) may be misclassified as trees. The lowest accuracy of LiDAR processing is observed on water which tends to absorb laser pulses and thus returns very weak signals to LiDAR sensors. This indicates that visual interpretation and validation of LiDAR-generated urban forest data is critical to ensuring the accurate extraction of tree features. Specific attention should be given to trees identified as located in water, on agricultural lands and in dense urban areas with complex building structures.

The inability of our LiDAR-processing technique to identify individual tree taxa could impact the accuracy of our carbon storage and sequestration estimates. The generalized allometric models that we used to generate these estimates depend on information related to tree taxa. Because the LiDAR processing in this study does not generate species data for individual trees, we assigned taxa to individual trees via random sampling using estimates of the relative abundance of dominant tree species from field data. Inaccuracy in these taxon assignments would lead to error in the calculation of above-ground biomass which in turn could affect the accuracy of carbon storage and sequestration estimates. Mapping tree species using a fusion of multispectral/hyperspectral and LiDAR data, a promising technique for identifying individual tree species, could minimize such error [79]. Although this approach has been shown to be capable of identifying tree species, it is challenging and not yet fully developed in urban areas due to their great native and exotic species richness and great spatial variation in tree species occurrence [79]. Field sampling of tree species at a finer resolution would also improve LiDAR-based allometric modeling, but at considerable cost to labor and time. Future studies might also avoid error related to inaccurate tree species identification by using methods other than allometric models in estimating carbon storage and sequestration. For example, such studies might investigate the use of LiDAR metrics with locally-adapted linear and nonlinear regression to estimate carbon storage and sequestration. In so doing, these studies would seek to predict carbon storage and sequestration based upon different combinations of LiDAR-derived forest variables and might improve upon the accuracy of estimates.

As mentioned above, uncertainty and error exist in our forest structure estimates, for example, in the estimates of dbh and crown width as well as in the use of generalized allometric models, which may impact estimated carbon storage and sequestration. These biases, however, are only minimally quantified in this study due to a lack of direct monitoring data for carbon storage and sequestration supply. To better assess these biases, future studies could: (1) systematically collect and use field data on tree attributes to validate LiDAR results; (2) identify sources of error and quantify error in the final estimates of carbon storage and sequestration supply based upon error propagation theory [80] or (3) compare the supply map generated using this method with independent studies for the same study region.

B. Implications for land-use planning and management

Quantifying the supply of carbon storage and sequestration by trees is increasingly important, given that such vegetation is a major terrestrial carbon pool that could help to mitigate climate change. Our assessment of carbon storage and sequestration by urban trees provides key insights that could inform climate change mitigation strategies at the local level, strategies that urban municipalities increasingly seek to implement. Climate change mitigation via GHG reduction, for example, is identified as a significant goal in the 2030 Comprehensive Plan for Dakota County and similar goals exist for cities worldwide, reflecting a growing interest in achieving carbon neutrality as a means to mitigate climate change.

Our assessment highlights the potential role of urban trees in reaching this carbon neutrality, indicating that urban forests can act as a carbon sink, albeit one that may be limited in terms of capacity. We find that trees in the study area store a significant amount of carbon (1735.69 million kgC). Our calculated carbon storage values per tree (mean = 238.13KgC/tree) are comparable to those estimated in a series of studies implemented in ten other US cities (mean = 227.00 kgC/tree, range: 91.81–638.95 kgC/tree) and are similar to estimates for densely-urbanized cities like New York, NY (235.07 kgC/tree), Philadelphia, PA (227.64 kgC/tree) and Boston, MA (244.97 kgC/tree) [35]. In line with other urban studies [44,81], our findings indicate that the direct net carbon sequestration by trees in urban areas makes only a small contribution to climate change mitigation (33.43 million kgC/year), offsetting approximately 1% of annual anthropogenic CO_2 emissions in the study area (3758.12 million kgC/year). This carbon offset rate is slightly greater than the rate obtained in other studies (Shengyang, China: 0.26%; Barcelona, Spain: 0.47%) [44,45], a difference which may result from actual differences in urban forest structure among locations or from different methodologies. This low offset rate suggests that city governments should recognize that the supply of carbon storage and sequestration service by urban trees is limited and that they should seek to not only maintain such services, but also to reduce demand via emissions reductions. This is not to say that trees are unimportant in climate change mitigation especially as trees provide additional benefits that could assist in reducing emissions, for example, through shading and evapotranspiration, which can reduce energy consumption for heating and cooling leading to avoided CO_2 emissions [82]. These additional benefits were not considered in this study.

Spatially explicit mapping is needed to quantify ecosystem services because the supply of and demand for ecosystem services vary geographically. The design and implementation of policies for achieving sustainability also depends on such spatially explicit information [83], which can improve the ability of planners and policy makers to target programs to locations where they may be most effective or needed [17]. The method demonstrated in this study makes the spatial variability of carbon storage and sequestration and its relationship with the underlying ecosystem components much clearer. The supply map our method produces illustrates how the supply of carbon storage and sequestration varies in space due to the spatial variation of urban tree resources (Fig 4). While large forested parks and reserves near MMRRA plays a major role in enhancing carbon storage, backyard and street trees within residential areas and trees on undeveloped land constitute a considerable proportion of the biomass carbon pool in more heavily urbanized areas. Spatial variability of carbon sequestration in urban trees mirrors that of storage, further indicating areas that are critical to the supply of this service. Such findings could help managers and policy makers to better understand where and how carbon storage and sequestration are provided in urban landscapes and to highlight the important role that urban tree protection ordinances and forest management practices can play in providing climate change mitigation benefits in urban areas.

Our method also facilitates the identification of spatially-explicit patterns in ecosystem service demand and the factors that influence these patterns. Our results indicate that demand for carbon sequestration is highest for the electricity utility sector, followed by the agricultural and transportation sectors. For example, in southern Dakota County and northern Ramsey County, with moderate populations, high vehicle numbers and more agricultural land, we find higher carbon emissions and thus greater demand (Fig 4). Conversely, lower demand for carbon sequestration occurs in highly urban St. Paul. This occurs because, although highly-urbanized census tracts have a high demand for carbon sequestration from residential, waste, industrial and commercial sectors, they have lower overall populations, vehicle numbers, and agriculture land areas than suburban areas with lower population densities and larger parcels. Additionally, although St. Paul is the most heavily urbanized part of the study area, important institutions and commercial and residential land uses also occur in other locations in the study area. This land use pattern further explains why most of St. Paul has relatively low carbon sequestration demand and why strategies for reducing demand might best be targeted to other locations. Our demand maps facilitate the identification of patterns such as these and can thus improve our understanding of the spatial variability in demand for ecosystem services and the factors that influence them. This in turn can improve our ability to identify and enact policies and management actions that enhance service delivery in the locations where demand is highest.

C. The supply-to-demand ratio: an index of ecosystem service status

An important issue with ecosystem service assessments is that measures of ecosystem service supply fail to consider how this supply compares with demand and thus cannot indicate the true status of an ecosystem service [84]. Assessments that compare supply with demand, as the present study does, better indicate the level of balance between benefits supplied by ecosystems and the societal need for these benefits. The supply-to-demand ratio we demonstrate in this study quantifies this balance, identifying the level of carbon storage and sequestration delivered locally from ecosystems to socio-economic systems. A particular strength of this ratio lies in its spatial explicitness, which facilitates consideration of the spatial heterogeneity of ecosystem service supply and demand in identifying areas with supply- demand tensions [22,84,85]. A previous study has used the supply-to-demand ratio of water provisioning service to indicate the level of water scarcity, mapping water balances under different global change and mitigation scenarios [85]. Our supply-demand ratio could be used in a similar manner, for example, to inform and thus improve local-level policy making related to carbon storage and sequestration, a service that is delivered globally, via an analysis of the supply-demand ratio under different scenarios.

Our supply-to-demand ratio map provides a location-specific overview of the status of carbon storage and sequestration across an urban area (Fig 5). In so doing, it not only identifies locations where carbon sequestration supply and demand are in greater or lesser balance, but, with closer examination, also helps to reveal the underlying factors that influence this supply and demand (e.g., number of trees, population distribution, number of vehicles). For example, the highest supply-to-demand ratio occurs in the MMRRA region where trees are dominant landscape features (i.e., high supply) and both population densities and traffic volumes are low (i.e., low demand). Conversely, landscapes characterized by agricultural lands or sparsely-distributed trees (i.e., low supply) where heavier traffic or higher populations at lower densities concurrently exist (i.e., high demand) typically have smaller ratios. Since St. Paul has relatively fewer trees than the rest of the study area, and southern Dakota County has relatively more agricultural land and higher populations, both areas have smaller supply-to-demand ratios. Such information could help policy makers and land managers to identify location-specific mechanisms for improving supply-to-demand ratios.

Implementing climate change mitigation policy at the local level requires us to answer questions associated with where and how to reduce service deficits to achieve a supply-demand balance. As discussed above, analyses such as the one described here that explicitly examine the spatial balance in supply and demand provide insights that could help cities to identify and spatially target practices to achieve carbon neutrality. For example, maps such as those generated by this study highlight key locations where supply and demand are out of balance as well as locations where the ability of urban trees to provide carbon sequestration or demand are relatively high or low. This indicates locations to which tree planting or protection ordinances to enhance carbon storage and sequestration or policies that encourage the use of public transit might best be targeted to mitigate climate change. As planners and policy makers seek to reduce the demand for carbon storage and sequestration while increasing the supply of these services, the method for quantifying, mapping and comparing supply and demand for carbon storage and sequestration illustrated in this study opens up a broad avenue for integrating the notion of carbon neutrality with urban forest management and carbon emissions reductions in policy making. As such supply-demand ratio maps in combination with basic supply and demand maps have great potential to improve the quality of policy and management decisions.

Because ecosystems provide multiple services at a given point in time and space, it is important to recognize that tradeoffs may exist among different services with regard to their supply and demand. However, it is difficult to assess these tradeoffs as units for the supply and demand for different services vary, depending on the type and dimension of services considered and the way in which they are quantified. Our supply-to-demand ratio provides a straightforward and standardized way to facilitate trade-off assessments, because this ratio is standardized and unitless. Thus such ratios could be used to directly compare multiple ecosystem services. Such ratios have the added benefits of increasing flexibility in indicator selection, given that they do not require measures of supply and demand for different services to have the same unit or order of magnitude. When measures of service supply and demand are not directly comparable, they are typically compared using semi-quantified values assigned through processes like normalization, standardization or expert-based ranking [86]. For example, based on a literature review and expert opinions, Burkhard et al. (2012) translated the provision of ecosystem services into five categories classified by numerical values: 0 = no relevant capacity, 1 = very low, 2 = low, 3 = moderate, 4 = high, and 5 = very high [26]. In generating such ordinal values, one is forced to make judgements as to what level of service supply or demand falls into a given category. Such judgements have well-recognized impacts on resulting estimates. Using supply-demand ratios to assess and compare tradeoffs among different services alleviates this effect as it results in a standardized measure that does not require such value judgments. Our study assessed only one ecosystem service and did not analyze tradeoffs among multiple services using the supply-to-demand ratio in the manner suggested here. We suggest that future efforts examine these tradeoffs among ecosystem services using the supply-to-demand ratio. The identification of these spatial tradeoffs could serve to further unravel the mechanisms behind the interrelationships among multiple services.

D. Caveats in assessing spatial mismatch between supply and demand

Special attention should be paid to the issue of scale when assessing in the balance between ecosystem service supply and demand spatially. The supply-to-demand ratio we demonstrate here could be implemented at any scale (spatial extent and grain), and the scale chosen should depend on the scale at which ecosystem services are produced, as well as the institutional scale at which they are received. The development of a spatial supply-demand ratio analysis for a given area of interest would first require identification of the SPA where ecosystem services are generated (e.g., population, ecosystem, landscape, biome, global) and the SBA where ecosystem services are demanded (e.g., individual, household, municipal, national, global) [87]. However, as ecosystem services are often supplied and consumed at multiple scales, identifying the proper scale for an analysis is challenging. For example, although carbon storage and sequestration is commonly taken as a global service, the supply of carbon storage and sequestration can be assessed at the plant, forest, landscape or global level. While beneficiaries of carbon storage and sequestration exist globally, demand for carbon sequestration occurs across local, regional and global scales and, indeed, can vary substantially with the scale and location considered. In identifying the best scale for an analysis one might consider the policy choices available and the scale at which they will operate, as well as the scale of information needed by policy makers and managers, who are often the end-users of ecosystem service assessments [88]. In our example, we sought to provide information illustrative of that required by policy makers within a metropolitan area. As such, we quantified carbon storage and sequestration service supply at the individual plant scale and aggregated this to the landscape scale for comparison with demand at a policy-relevant level. Future studies that use this methodology could aggregate to scales appropriate to specific planning and policy-making contexts. Additionally, although our study only considered a single scale in quantifying service demand, a multi-scale approach could be used to provide information of demand across multiple scales. For instance, demand for carbon sequestration could be assessed using participatory methods at the local scale [89], then analyzed using proxy or expert-based methods at the global scale [90]. Combining these methods would facilitate a wide range of ecosystem service assessments with purposes ranging from education to accounting for human well-being to specific landscape planning and management problems [91].

Linking supply and demand is another key issue in the assessment of ecosystem services. For ecosystem services with in situ spatial relationships (e.g., soil formation, erosion regulation) or proximity relationships (e.g., flood regulation, coastal protection) [4,92], spatial flow models that examine either the natural flow paths or anthropogenic flow corridors from SPA to SBA would be useful for exploring dynamic ecosystem service flows from supply sites to demand sites [10,13]. In this regard, processed-based models coupled with the ratio mapping method detailed here could be used to specifically identify service source, sink and benefit areas. Given carbon storage and sequestration is a singular service which is omnidirectional and spatially transferable, modeling the flow path of this service is rather challenging and not recommended. The spatial mismatch analysis presented in this study represents a simplified method for linking the supply of and demand for carbon storage and sequestration. This method focuses solely on the balance between carbon and sequestration service provision and demand within spatial units inside city boundaries and ignores the spatial import and export of carbon storage and sequestration at greater scales. Nevertheless, as discussed above, for a given region, and under the specific context of carbon neutrality and climate change mitigation, it is very useful and policy-relevant to assess the balance between the supply of and demand for this globally-delivered service in a spatially-explicit way at the local level.

CONCLUSIONS

Locations of high ecosystem service supply are often poorly matched with locations of high service demand. Quantifying, mapping and comparing such areas is critical to enhancing the ability of ecosystem service assessments to inform policy making as such information can help policy makers to visualize and thus to better understand service provision in their jurisdictions. This study demonstrates an approach to assessing carbon storage and sequestration provided by urban forests that distinguishes among the supply and demand for this service in a spatially-explicit manner. Assessing carbon storage and sequestration in this way improves both our understanding of the ecological capacity of urban trees to store and sequester CO_2 and our ability to consider relationships between this capacity and demand. In general, this spatially-explicit, supply-and-demand-driven approach to studying carbon storage and sequestration has great potential to improve assessments of this service in urban environments and the incorporation of such assessments into policy making associated with climate change mitigation and carbon neutrality in cities. This approach, if adapted for use with additional services, could additionally improve the general consideration of ecosystem services in local policy and land management decision making and the overall quality of environmental decision-making in local jurisdictions.

SUPPORTING INFORMATION

S1 Fig. Simple linear regression between LiDAR-derived crown width and crown width measured from the orthophotograph.

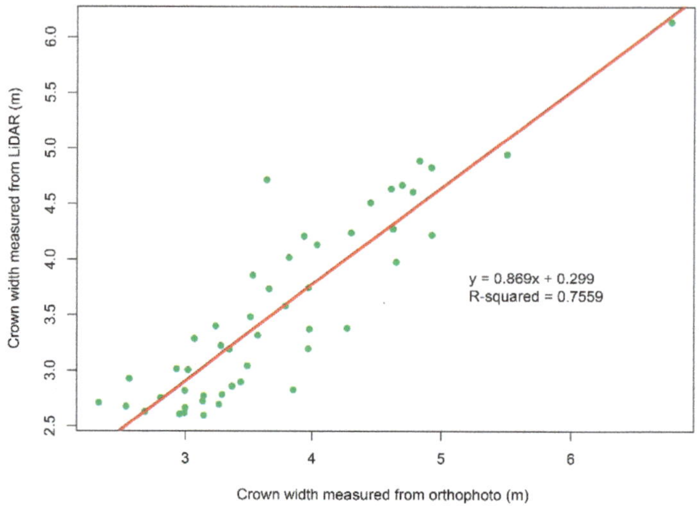

AUTHOR CONTRIBUTIONS

Conceived and designed the experiments: CZ HAS. Performed the experiments: CZ. Analyzed the data: CZ. Contributed reagents/materials/analysis tools: CZ HAS. Wrote the paper: CZ HAS.

REFERENCES

1. Millennium Ecosystem Assessment. Ecosystems and Human Well-being. Washington, DC: Island Press; 2005. doi: 10.1196/annals.1439.003.
2. De Groot RS, Alkemade R, Braat L, Hein L, Willemen L. Challenges in integrating the concept of ecosystem services and values in landscape planning, management and decision making. Ecol Complex. Elsevier B.V.; 2010;7: 260–272. doi: 10.1016/j.ecocom.2009.10.006.
3. Burkhard B, Kandziora M, Hou Y, Müller F. Ecosystem service potentials, flows and demands-concepts for spatial localisation, indication and quantification. Landsc Online. 2014;32: 1–32. doi: 10.3097/LO.201434.
4. Fisher B, Turner RK, Morling P. Defining and classifying ecosystem services for decision making. Ecol Econ. Elsevier B.V.; 2009;68: 643–653. doi: 10.1016/j.ecolecon.2008.09.014.
5. Haase D, Larondelle N, Andersson E, Artmann M, Borgström S, Breuste J, et al. A quantitative review of urban ecosystem service assessments: concepts, models, and implementation. Ambio. 2014;43: 413–33. doi: 10.1007/s13280-014-0504-0. pmid:24740614
6. De Groot RS, Wilson MA, Boumans RM. R. A typology for the classification, description and valuation of ecosystem functions, goods and services. Ecol Econ. 2002;41: 393–408. doi: 10.1016/S0921-8009(02)00089-7.
7. Boyd J, Banzhaf S. What are ecosystem services? The need for standardized environmental accounting units. Ecol Econ. 2007;63: 616–626. doi: 10.1016/j.ecolecon.2007.01.002.
8. Bastian O, Haase D, Grunewald K. Ecosystem properties, potentials and services–The EPPS conceptual framework and an urban application example. Ecol Indic. Elsevier Ltd; 2012;21: 7–16. doi: 10.1016/j.ecolind.2011.03.014.
9. Bagstad KJ, Johnson GW, Voigt B, Villa F. Spatial dynamics of ecosystem service flows: a comprehensive approach to quantifying actual services. Ecosyst Serv. 2013;4: 117–125. doi: 10.1016/j.ecoser.2012.07.012.
10. Serna-Chavez HM, Schulp CJE, van Bodegom PM, Bouten W, Verburg PH, Davidson MD. A quantitative framework for assessing spatial flows of ecosystem services. Ecol Indic. Elsevier Ltd; 2014;39: 24–33. doi: 10.1016/j.ecolind.2013.11.024.
11. Wolff S, Schulp CJE, Verburg PH. Mapping ecosystem services demand: A review of current research and future perspectives. Ecol Indic. 2015;55: 159–171. doi: 10.1016/j.ecolind.2015.03.016.
12. Schröter M, Barton DN, Remme RP, Hein L. Accounting for capacity and flow of ecosystem services: A conceptual model and a case study for Telemark, Norway. Ecol Indic. 2014;36: 539–551. doi: 10.1016/j.ecolind.2013.09.018.
13. Villamagna AM, Angermeier PL, Bennett EM. Capacity, pressure, demand, and flow: A conceptual framework for analyzing ecosystem service provision and delivery. Ecol Complex. 2013;15: 114–121. doi: 10.1016/j.ecocom.2013.07.004.
14. Geijzendorffer IR, Martín-López B, Roche PK. Improving the identification of mismatches in ecosystem services assessments. Ecol Indic. 2015;52: 320–331. doi: 10.1016/j.ecolind.2014.12.016.
15. Willemen L, Verburg PH, Hein L, van Mensvoort MEF. Spatial characterization of landscape functions. Landsc Urban Plan. 2008;88: 34–43. doi: 10.1016/j.landurbplan.2008.08.004.
16. Nelson E, Mendoza G, Regetz J, Polasky S, Tallis H, Cameron Dr, et al. Modeling multiple ecosystem services, biodiversity conservation, commodity production, and tradeoffs at landscape scales. Front Ecol Environ. 2009;7: 4–11. doi: 10.1890/080023.
17. Tallis H, Polasky S. Mapping and valuing ecosystem services as an approach for conservation and natural-resource management. Ann N Y Acad Sci. 2009;1162: 265–83. doi: 10.1111/j.1749-6632.2009.04152.x. pmid:19432652
18. Locatelli B, Imbach P, Vignola R, Metzger MJ, Hidalgo EJL. Ecosystem services and hydroelectricity in Central America: modelling service flows with fuzzy logic and expert knowledge. Reg Environ Chang. 2010;11: 393–404. doi: 10.1007/s10113-010-0149-x.
19. Boumans R, Costanza R, Farley J, Wilson MA, Portela R, Rotmans J, et al. Modeling the dynamics of the integrated earth system and the value of global ecosystem services using the GUMBO model. Ecol Econ. 2002;41: 529–560. doi: 10.1016/S0921-8009(02)00098-8.
20. Villa F, Ceroni M, Bagstad K, Johnson G, Krivov S. ARIES (Artificial Intelligence for Ecosystem Services): A new tool for ecosystem services assessment, planning, and valuation. 11th annual BIOECON conference on economic instruments to enhance the conservation and sustainable use of biodiversity. Venice, Italy; 2009.
21. .Paracchini ML, Zulian G, Kopperoinen L, Maes J, Schägner JP, Termansen M, et al. Mapping cultural ecosystem services: A framework to assess the potential for outdoor recreation across the EU. Ecol Indic. Elsevier Ltd; 2014;45: 371–385. doi: 10.1016/j.ecolind.2014.04.018.
22. Stürck J, Poortinga A, Verburg PH. Mapping ecosystem services: The supply and demand of flood regulation services in Europe. Ecol Indic. 2014;38: 198 211. doi: 10.1016/j.ecolind.2013.11.010.
23. Locatelli B, Imbach P, Wunder S. Synergies and trade-offs between ecosystem services in Costa Rica. Environ Conserv. 2013;41: 27–36. doi: 10.1017/S0376892913000234.
24. Schulp CJE, Lautenbach S, Verburg PH. Quantifying and mapping ecosystem services: Demand and supply of pollination in the European Union. Ecol Indic. 2014;36: 131–141. doi: 10.1016/j.ecolind.2013.07.014.
25. Nedkov S, Burkhard B. Flood regulating ecosystem services-mapping supply and demand, in the Etropole

municipality, Bulgaria. Ecol Indic. 2012;21: 67–79. doi: 10.1016/j.ecolind.2011.06.022.

26. Burkhard B, Kroll F, Nedkov S, Müller F. Mapping ecosystem service supply, demand and budgets. Ecol Indic. 2012;21: 17–29. doi: 10.1016/j.ecolind.2011.06.019.

27. Van Berkel DB, Verburg PH. Spatial quantification and valuation of cultural ecosystem services in an agricultural landscape. Ecol Indic. 2014;37: 163–174. doi: 10.1016/j.ecolind.2012.06.025.

28. Plieninger T, Dijks S, Oteros-Rozas E, Bieling C. Assessing, mapping, and quantifying cultural ecosystem services at community level. Land use policy. 2013;33: 118–129. doi: 10.1016/j.landusepol.2012.12.013.

29. Casado-Arzuaga I, Madariaga I, Onaindia M. Perception, demand and user contribution to ecosystem services in the Bilbao Metropolitan Greenbelt. J Environ Manage. 2013;129: 33–43. doi: 10.1016/j.jenvman.2013.05.059. pmid:23792888

30. Syrbe R-U, Walz U. Spatial indicators for the assessment of ecosystem services: Providing, benefiting and connecting areas and landscape metrics. Ecol Indic. Elsevier Ltd; 2012;21: 80–88. doi: 10.1016/j.ecolind.2012.02.013.

31. Liquete C, Zulian G, Delgado I, Stips A, Maes J. Assessment of coastal protection as an ecosystem service in Europe. Ecol Indic. Elsevier Ltd; 2013;30: 205–217. doi: 10.1016/j.ecolind.2013.02.013.

32. Egoh B, Reyers B, Rouget M, Richardson DM, Le Maitre DC, van Jaarsveld AS. Mapping ecosystem services for planning and management. Agric Ecosyst Environ. 2008;127: 135–140. doi: 10.1016/j.agee.2008.03.013.

33. Pataki DE, Alig RJ, Fung AS, Golubiewski NE, Kennedy CA, Mcpherson EG, et al. Urban ecosystems and the North American carbon cycle. Glob Chang Biol. 2006;12: 2092–2102. doi: 10.1111/j.1365-2486.2006.01242.x.

34. McPherson E, Nowak D, Rowntree R. Chicago's urban forest ecosystem: results of the Chicago Urban Forest Climate Project. USDA For Serv. 1994;186.

35. Nowak DJ, Crane DE. Carbon storage and sequestration by urban trees in the USA. Environ Pollut. 2002;116: 381–9. pmid:11822716 doi: 10.1016/s0269-7491(01)00214-7

36. Nowak DJ. Atmospheric carbon reduction by urban trees. J Environ Manage. 1993; 207–217. doi: 10.1006/jema.1993.1017

37. Nowak DJ, Greenfield EJ, Hoehn RE, Lapoint E. Carbon storage and sequestration by trees in urban and community areas of the United States. Environ Pollut. 2013;178: 229–36. doi: 10.1016/j.envpol.2013.03.019. pmid:23583943

38. Forest Inventory and Analysis National Program (FIA) [Internet]. 2014 [cited 10 Feb 2015]. Available: www.fia.fs.fed.us

39. Nowak DJ, Crane DE, Stevens JC, Ibarra M. Brooklyn's urban forest. USDA For Serv. 2002;290.

40. Phillips DL. Assessment of ecosystem services provided by urban trees: public lands within the urban growth boundary of Corvallis, Oregon. City. 2011; 1–18.

41. Myeong S, Nowak DJ, Duggin MJ. A temporal analysis of urban forest carbon storage using remote sensing. Remote Sens Environ. 2006;101: 277–282. doi: 10.1016/j.rse.2005.12.001.

42. Ã SCP, Popescu SC. Estimating biomass of individual pine trees using airborne lidar. Biomass and Bioenergy. 2007;31: 646–655. doi: 10.1016/j.biombioe.2007.06.022.

43. Brown S. Measuring carbon in forests: current status and future challenges. Environ Pollut. 2002;116: 363–72. pmid:11822714 doi: 10.1016/s0269-7491(01)00212-3

44. Liu C, Li X. Carbon storage and sequestration by urban forests in Shenyang, China. Urban For Urban Green. Elsevier GmbH.; 2012;11: 121–128. doi: 10.1016/j.ufug.2011.03.002.

45. Baró F, Chaparro L, Gómez-Baggethun E, Langemeyer J, Nowak DJ, Terradas J. Contribution of ecosystem services to air quality and climate change mitigation policies: the case of urban forests in Barcelona, Spain. Ambio. 2014;43: 466–79. doi: 10.1007/s13280-014-0507-x. pmid:24740618

46. Jenkins J, Chojnacky D. National-scale biomass estimators for United States tree species. For Sci. 2003;49.

47. Berland A. Twin Cities urbanization and implications for urban forest ecosystem services. University of Minnesota. 2012.

48. US Census Bureau. 2010 Census Urban and Rural Classification and Urban Area Criteria [Internet]. 2010.

49. Minnesota Geospatial Information Office [Internet]. 2015 [cited 1 Jan 2015]. Available:http://www.mngeo.state.mn.us/

50. Overwatch [Internet]. 2015. Available: http://www.geospatial.overwatch.com/

51. ESRI ArcGIS v.10.1 [Internet]. 2012. Available: http://www.esri.com/

52. Opitz D, Blundell S, Rajendran R, Morris M. An approach for collection of geospecific 3D features from terrestrial LiDAR. ASPRS 2008 Annual Conference. Portland, Oregon; 2008. pp. 0–7.

53. Minnesota Native Big Tree registry [Internet]. MNDNR; 2015. Available:http://www.dnr.state.mn.us/trees_shrubs/bigtree/index.html

54. Jenkins J, Chojnacky D, Heath L, Birdsey R. Comprehensive database of diameter-based biomass regressions for North American tree species. USDA For Serv. 2004;319.

55. Cairns MA, Brown S, Helmer EH, Baumgardner GA. Root biomass allocation in the world's upland forests. Oecologia. 1997;111: 1–11. doi: 10.1007/s004420050201.

56. Brown S, Lugo A. The storage and production of organic matter in tropical forests and their role in the global carbon cycle. Biotropica. 1982;14: 161–187. doi: 10.2307/2388024

57. Dixon R, Brown S. Carbon pools and flux of global forest ecosystems. Science (5144). 1994;263: 185–190. pmid:17839174 doi: 10.1126/science.263.5144.185

58. Iverson L, Brown S, Prasad A. Use of GIS for estimating potential and actual forest biomass for continental South and Southeast Asia. Effects of Land-Use Change on Atmospheric CO2 Concentrations. New York: Springer Verlag New York, Inc.; 1994.

59. Ecoregions of the United States [Internet]. USDA Forest Service; 2015. Available:http://www.fs.fed.us/rm/ecoregions/products/map-ecoregions-united-states/

60. Scheu S, Schauermann J. Decomposition of roots and twigs: effects of wood type (beech and ash), diameter, site of exposure and macrofauna exclusion. Plant Soil. 1994; 13–24.

61. Olson JS. Energy storage and the balance of producers and decomposers in ecological systems. Ecology. 1963;44: 322–331. doi: 10.2307/1932179

62. Chambers JQ, Higuchi N, Schimel JP, Ferreira LV., Melack JM. Decomposition and carbon cycling of dead trees in tropical forests of the central Amazon. Oecologia. 2000;122: 380–388. doi: 10.1007/s004420050044.

63. Harmon ME, Franklin JF, Swanson FJ, Sollins P, Gregory SV., Lattin JD, et al. Ecology of coarse woody debris in temperate ecosystems. Adv Ecol Res. 1986;15: 302. doi: 10.1016/s0065-2504(08)60121-x

64. Berendse F, Berg B, Bosatta E. The effect of lignin and nitrogen on the decomposition of litter in nutrient-poor ecosystems: a theoretical approach. Can J Bot. 1987;65: 1116–1120. doi: 10.1139/b87-155.

65. Boddy L, Owens E, Chapela I, Road N, Boddy L, Owens E, et al. Small scale variation in decay rate within logs one year after felling: effect of fungal community structure and moisture content. FEMS Microbiol Lett. 1989;62. doi: 10.1111/j.1574-6968.1989.tb03691.x

66. Russell MB, Fraver S, Aakala T, Gove JH, Woodall CW, D'Amato AW, et al. Quantifying carbon stores and decomposition in dead wood: A review. For Ecol Manage. Elsevier B.V.; 2015;350: 107–128. doi: 10.1016/j.foreco.2015.04.033.

67. David HA, Pastor J. Decomposition of aspen, spruce, and pine boles on two sites in Minnesota. Can J For Res. 1993;23: 1744–1749. doi: 10.1139/x93-220.

68. Fraver S, Milo AM, Bradford JB, D'Amato AW, Kenefic L, Palik BJ, et al. Woody debris volume depletion through decay: Implications for biomass and carbon accounting. Ecosystems. 2013;16: 1262–1272. doi: 10.1007/s10021-013-9682-z.

69. Lambert RL, Lang GE, Reiners WA. Loss of mass and chemical change in decaying boles of a subalpine balsam fire forest. Ecology. 2014;61: 1460–1473. doi: 10.2307/1939054

70. Mcpherson EG, Simpson JR. Through urban forestry: guidelines for professional and volunteer tree planters. USDA For Serv. 1999;171.

71. Teck R, Hilt D. Individual-tree diameter growth model for northeastern United States. USDA For Serv. 1991;649.

72. Minnesota Pollution Control Agency [Internet]. 2015. Available:http://www.pca.state.mn.us/

73. Xcel Energy Inc. Power Generation supply [Internet]. 2013 [cited 7 Jan 2015]. Available: http://www.xcelenergy.com/Company/Operations/Power_Generation

74. Minnesota Geographic Metadata. Metadata: Generalized Land Use 2010 for the Twin Cities Metropolitan Area [Internet]. 2011. Available:http://www.datafinder.org/metadata/GeneralizedLandUse2010.html

75. The terrestrial carbon cycle: managing forest ecosystems. U.S. EPA; 2007.

76. Conte M, Nelson E, Carney K, Fissore C, Olwero N, Plantinga AJ, et al. Terrestrial carbon sequestration and storage. In: Kareiva P, Tallis H, Ricketts TH, Daily GC, Polasky S, editors. Natural capital theory and practice of mapping ecosystem services. New York, NY: Oxford University Press; 2011. pp. 111–128.

77. Gill SJ, Biging GS, Murphy EC. Modeling conifer tree crown radius and estimating canopy cover. For Ecol Manage. 2000;126: 405–416. doi: 10.1016/S0378-1127(99)00113-9.

78. Estornell J, Ruiz LA, Velázquez-Martí B, Fernández-Sarría A. Estimation of shrub biomass by airborne LiDAR data in small forest stands. For Ecol Manage. Elsevier B.V.; 2011;262: 1697–1703. doi: 10.1016/j.foreco.2011.07.026.

79. Alonzo M, Bookhagen B, Roberts DA. Urban tree species mapping using hyperspectral and lidar data fusion. Remote Sens Environ. Elsevier Inc.; 2014;148: 70–83. doi: 10.1016/j.rse.2014.03.018.

80. Romaguera M, Hoekstra AY, Su Z, Krol MS, Salama MS. Potential of using remote sensing techniques for global assessment of water footprint of crops. Remote Sens. 2010;2: 1177–1196. doi: 10.3390/rs2041177.

81. Pataki DE, Carreiro MM, Cherrier J, Grulke NE, Jennings V, Pincetl S, et al. Coupling biogeochemical cycles in urban environments: ecosystem services, green solutions, and misconceptions. Front Ecol Environ. 2011;9: 27–36. doi: 10.1890/090220.

82. Simpson JR. Improved estimates of tree-shade effects on residential energy use. Energy Build. 2002;34: 1067–1076. doi: 10.1016/S0378-7788(02)00028-2.

83. Maes J, Egoh B, Willemen L, Liquete C, Vihervaara P, Schägner JP, et al. Mapping ecosystem services for policy support and decision making in the European Union. Ecosyst Serv. 2012;1: 31–39. doi: 10.1016/j.ecoser.2012.06.004.

84. Paetzold A, Warren PH, Maltby LL. A framework for assessing ecological quality based on ecosystem services. Ecol Complex. Elsevier B.V.; 2010;7: 273–281. doi: 10.1016/j.ecocom.2009.11.003.

85. Boithias L, Acuña V, Vergoñós L, Ziv G, Marcé R, Sabater S. Assessment of the water supply:demand ratios in a Mediterranean basin under different global change scenarios and mitigation alternatives. Sci Total Environ. Elsevier B.V.; 2014;470–471: 567–77. doi: 10.1016/j.scitotenv.2013.10.003. pmid:24176705

86. Schulp CJE, Burkhard B, Maes J, Van Vliet J, Verburg PH. Uncertainties in ecosystem service maps: a comparison

on the European scale. PLoS One. 2014;9: e109643. doi: 10.1371/journal.pone.0109643. pmid:25337913

87. Hein L, van Koppen K, de Groot RS, van Ierland EC. Spatial scales, stakeholders and the valuation of ecosystem services. Ecol Econ. 2006;57: 209–228. doi: 10.1016/j.ecolecon.2005.04.005.

88. Honey-Rosés J, Pendleton LH. A demand driven research agenda for ecosystem services. Ecosyst Serv. 2013;5: 160–162. doi: 10.1016/j.ecoser.2013.04.007.

89. Palomo I, Martín-López B, Potschin M, Haines-Young R, Montes C. National Parks, buffer zones and surrounding lands: Mapping ecosystem service flows. Ecosyst Serv. 2013;4: 104–116. doi: 10.1016/j.ecoser.2012.09.001.

90. Jacobs S, Burkhard B, Daele T Van, Staes J, Schneiders A. "The Matrix Reloaded": A review of expert knowledge use for mapping ecosystem services. Ecol Modell. Elsevier B.V.; 2014;295: 21–30. doi: 10.1016/j.ecolmodel.2014.08.024.

91. De Groot RS, Brander L, van der Ploeg S, Costanza R, Bernard F, Braat L, et al. Global estimates of the value of ecosystems and their services in monetary units. Ecosyst Serv. Elsevier; 2012;1: 50–61. doi: 10.1016/j.ecoser.2012.07.005.

92. Costanza R. Ecosystem services: multiple classification systems are needed. Biol Conserv. 2008;1: 8–10. doi: 10.1016/j.biocon.2007.12.020

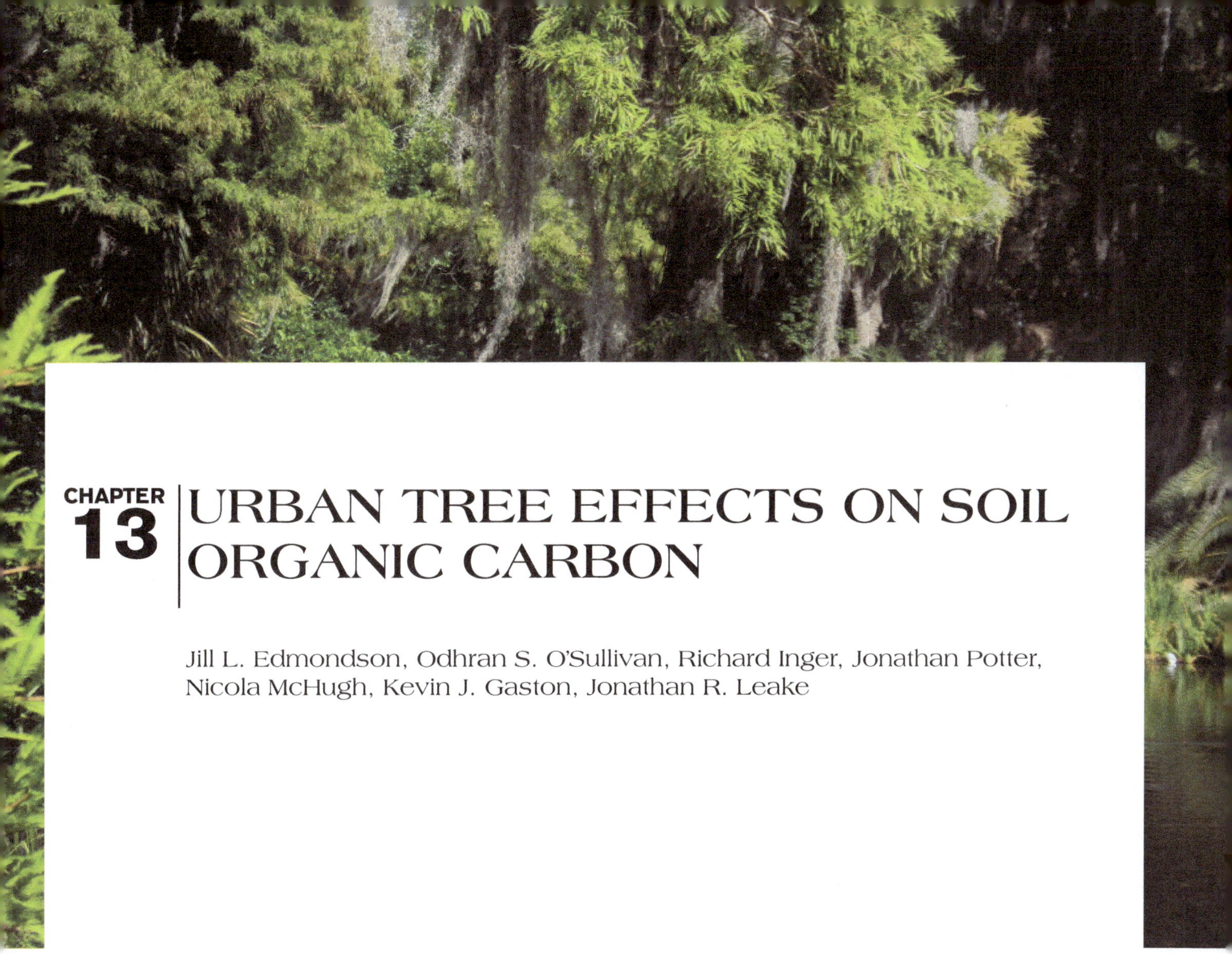

CHAPTER 13 URBAN TREE EFFECTS ON SOIL ORGANIC CARBON

Jill L. Edmondson, Odhran S. O'Sullivan, Richard Inger, Jonathan Potter, Nicola McHugh, Kevin J. Gaston, Jonathan R. Leake

INTRODUCTION

Urban ecosystems provide vital goods and services to the inhabitants of cities and towns [1]. Urban trees are especially important in providing a diverse range of ecosystem services. These include organic carbon storage [2]; flood mitigation and improved water quality [3]; filtration of atmospheric pollutants, especially removing health-damaging fine particulates such as PM10 [4], [5], and particulate-bound carcinogenic polycyclic aromatic hydrocarbons [6]; absorption of toxic gasses including O_3, SO_2 and NOX [7]; traffic noise pollution reduction [8]; and, amelioration of the urban heat island effect [9]–[11]. In addition, urban trees and greenspaces provide wildlife habitat and aesthetic values that further contribute benefits to human wellbeing, quality of life and health [12]–[14]. Trees are a ubiquitous part of citics and towns, and have been estimated to cover between 10–67% of urban and community areas in the USA [3] and 20% of Greater London, UK [5].

The recognized importance of trees for ecosystem service provision has stimulated global efforts to increase tree cover in urban areas, for example '*The Big Tree Plant*' in England run by DEFRA and the Forestry Commission [15], and the private-public run '*MillionTreesNYC initiative*' in New York, USA [16]. However, these initiatives have been conducted without a clear understanding of the effects of urban trees on provision of ecosystem services belowground. Given that approximately 75% of ecosystem carbon storage occurs in soils [17],[18], the net effect of trees on soil organic carbon (SOC) stores is particularly important since, in addition to providing a sink for atmospheric carbon dioxide fixed by photosynthesis, SOC is positively associated with regulating and supporting ecosystem services such as storm-water infiltration and nutrient holding capacity [19]. Without an assessment of the impact of tree planting on SOC and other soil properties that deliver ecosystem service benefits in an urban context it is not possible fully to understand the implications of this widespread management practice.

Trees can influence the biological, chemical and physical properties of soils directly through their deep roots and litter quality and quantity [20]. Changes in soil properties following afforestation are varied and dependent on former land-use and species, particularly whether broadleaved angiosperms or gymnosperms like pines [21], [22]. Previous studies on arable land, which has strongly depleted SOC stocks [23], have shown afforestation typically results in increased carbon sequestration both above and belowground [22]. Furthermore, the English national SOC inventory reports greater storage under woodland (primary and afforested) than pasture [24].

Citation: Edmondson JL, O'Sullivan OS, Inger R, Potter J, McHugh N, Gaston KJ, et al. (2014) Urban Tree Effects on Soil Organic Carbon. PLoS ONE 9(7): e101872. doi:10.1371/journal.pone.0101872

The UK Countryside Survey found lower soil bulk density (BD) values beneath woodland compared to improved grasslands [25], indicating a greater capacity of woodland soils to absorb sustained heavy rainfall and reduce run-off and flooding. As these countrywide survey habitats will include agricultural grasslands that have been ploughed and reseeded, they include soils that have previously been depleted in SOC and experienced increased BD prior to tree planting, as well as ancient woodlands.

Recent studies have shown that urban soils hold much higher concentrations of SOC than typical arable fields, into the same range as semi-natural grasslands and woodlands [26], [27]. The extent to which urban trees can increase SOC stocks over those of urban grasslands is currently unknown, as is the relative importance of different major urban tree genera. This knowledge-gap is strategically important with the rising pan-European threats to keystone urban tree species from virulent pests and diseases, including Ash dieback, Oak decline and Oak Processionary Moth [28]–[30]. Here, we use an observational study in urban parks to examine differences in soil properties (SOC, C:N ratio, and BD) beneath three tree genera (*Acer* spp., *Fraxinus excelsior*, and *Quercus robur*) selected for their abundance and capacity to grow into large specimens, and mixed urban woodlands in comparison to adjacent urban grassland soil. We test the hypothesis that SOC stocks and C:N ratios would be increased in soil under trees compared to grassland, paralleling and positively correlating with greater above-ground carbon storage, whereas soil BD would be reduced under trees compared to grassland as seen in semi-natural ecosystems [25].

METHODS

Study area

This research focussed on Leicester, a mid-sized UK city, located in the East Midlands of England (52°38′N, 1°08W). It has a human population of 310,000 [31], and covers an area of approximately 73 km^2. The region has a temperate climate, receiving 620 mm of precipitation each year and average annual daily minimum and maximum temperatures of 6.1°C and 13.9°C, respectively [32]. Soil types within the city are dominated by deep clays, deep loam and seasonally wet deep clays and loam, according to the National Soil Map for England and Wales produced by Cranfield University. The soil types sampled in the city were: Hanslope, Whimple, Salop, Beccles 3, Ragdale and Fladbury 1.

Sampling strategy

A GIS was used to select randomly urban parks for soil sampling within the city of Leicester and permission was granted for the work by the Leicester City Council. The land-use history of each park was assessed in a GIS using the series of historic Ordnance Survey maps dating back to 1887. No park had previously been built upon, one park and two country houses and grounds that went on to form parks were in existence in 1887, the remainder were agricultural land at this time. Over the following years to the present day the remainder of the parks were established as the city expanded into the surrounding agricultural landscape, with the two recent parks developed on farmland within the last 20–30 years (see Table S1 for site specific land-use history). At each park an initial assessment was made for the presence of individual trees that ranged in size from saplings to large mature specimens within our target genera, specifically isolated specimens of *Quercus robur*, *Fraxinus excelsior*, *Acer* spp. (comprised of*Acer pseudoplatanus* and *Acer platanoides*), or patches of mixed woodlands. The selected trees ranged in diameter at breast height (DBH) from 2.5 cm to 197 cm and in biomass from 1.3 kg to 61 tonnes for the largest mature *Q. robur* specimen (see Table S1). The tree genera were selected because of their importance in parks in Leicester and national abundance, *Q. robur* being the commonest tree and *F. excelsior* the second most common as an individual tree or within small patches of woodland in Great Britain [33]. *A. pseudoplatanus* is the fourth most common tree species in small patches of British woodland [33].

Where a park contained one or more of the tree genera and/or mixed woodland at least one grassland site was also identified for sampling. Each grassland site was situated in proximity to the tree sites identified, but was also over 50 m from any individual or patch of trees, to ensure that it was outside the influence of the trees. A GIS layer obtained from Leicester City Council was used to check that management at each grassland site was uniform, specifically that the regularity of mowing at all sites was approximately 25 times per year, these park grasslands were not irrigated nor did they receive any fertiliser input. The grassland sites were selected to act as a direct comparison (or paired sample) at each specific location (park) with the tree sample.

At each site, tree species, height and DBH were recorded within a 5×5 m quadrat centred on individual isolated trees within grassland or in mixed woodland. Soils were sampled in approximately 7 cm increments to 1 m depth [26], the reference depth for the national SOC inventory [21], [24], under target tree genera, mixed woodland and grasslands. Under isolated trees soil samples were taken within 1 m of the trunk to ensure that all samples were taken beneath the canopy of even the small immature trees. In total, soils were sampled beneath 12 specimens of *Quercus robur*, 11 of *Fraxinus excelsior*, 12 of *Acer* spp. In addition soils were sampled beneath mixed urban woodland at 8 sites, and urban grassland at 15 sites.

In addition to the three target tree genera specified a further six species; *Acer campestre*,*Corylus avellana*, *Crataegus monogyna*, *Salix caprea*, *Tilia* x *europaea* were identified in the mixed urban woodlands. Measured tree DBH and height were used to estimate aboveground biomass with allometric equations. For each species, where multiple equations were available they were combined to produce a generalised biomass prediction. Where species-specific equations did not exist genus level equations were used, following the methodology recently used to derive tree aboveground biomass across the city of Leicester [2]. However, allometric equations to predict biomass of urban trees specifically are scarce [34], thus those used were derived from European and North American forested ecosystems [2], [35]–[37].

Soil sample preparation and analysis

Soil samples were analysed for SOC, C:N ratio and BD using established procedures. In brief, soil samples were dried at 105°C for 24 hours, weighed, ball milled to homogenise, and passed through a 1 mm sieve [26]. Soil BD (g cm^{-3}) was calculated after removing the dry weight of matter greater than 1 mm [38]. Soils were analysed in duplicate for total N concentration (mg g^{-1}) in a CN analyser [26]. Inorganic carbon was removed from 2.5 g of soil sample by adding 10 ml 5.7 M HCl, samples were centrifuged at 1800 *g* for 10 minutes, supernatant discarded and dried at 105°C. Subsequent CN analysis in duplicate determined SOC concentration (mg g^{-1}) [26].

Statistical analysis

All dependant variables (SOC Concentration, SOC Density, Soil C:N & Soil Bulk density) were checked for normality and homoscedasticity, and log transformed where necessary. Each of these dependant variables was analysed using general linear mixed effects models. The maximal model included vegetation cover as a 5 level fixed factor (urban grassland, *Q. robur, F. excelsior, Acer* spp. and mixed woodland). Soil depth (at which the sample was taken) and the biomass of the individual tree in proximity to the soil sample or in mixed woodland the biomass of all trees within the 5×5 m quadrat were incorporated as covariates in the model, and urban park identity was included as a random (intercept) factor to account for differences in length of time since park establishment and geographic location within the city. Initial models included tree biomass as a variable, however this had no predictive power (as the confidence intervals spanned zero) for any measured soil property and therefore it was removed as a variable from all subsequent models. In all cases model simplification was attempted by comparing all possible simpler model subsets using Akaike's information criterion (AIC). The maximal model, containing all the predictor variables, was always found to be the top model (based on AIC) and used in the subsequent inferences. Pseudo R^2 values were calculated using the methods of Nagelkerke [39]. Mixed models were carried out using package lme4. F and p values were calculated using Satterthwaite approximations [40] to determine denominator degrees of freedom in package lmerTest. All tests were carried out in the R language and environment [41]. Raw data are available in Table S1.

RESULTS

There were significant differences in SOC concentration and amount per soil volume (SOC density) between the three tree genera, mixed urban woodland and grassland (F=9.95,p<0.001; and F=12.51, p<0.001, respectively). Soil depth had a strong effect on SOC concentration (F=609.89, p<0.001) and density (F=456.27, p<0.001). Median SOC concentration and density were greatest beneath *F. excelsior* throughout the depth profile, followed by *Acer* spp., with no difference between *Q. robur*, mixed woodland, and grassland (Fig. 1). Model pseudo R^2, including both soil depth and tree cover, explained 59% and 57% of variation in SOC concentration and density respectively. Estimates of total SOC storage, based on summed median values for each depth category, ranged from 14–26 kg SOC $m^{-2}$with lowest storage beneath *Q. robur*, mixed woodland, and grassland, and the highest under*F. excelsior* (Fig. 1). The explanatory power of the mixed effects model for C:N ratio was low, with a pseudo R^2 explaining only 11% of the variation in the data (soil depth F=20.74,p<0.001; vegetation cover F=7.579, p<0.001). The effect of trees on BD was not significant (F=1.24, p=0.295). Soil depth was the most important predictor of BD (F=244.35, p<0.001, model pseudo R^2=49%), with increased median BD with depth from 0.99–1.59 g cm^{-3} between 0–20 cm to 80–100 cm (Fig. 2). The mixed effects models including tree biomass as a variable (excluding grassland as this cover type has no woody biomass) revealed that the effect size of tree biomass on SOC, C:N ratio and BD was small, and as confidence intervals spanned zero had no predictive power (see Table S2 for model statistics).

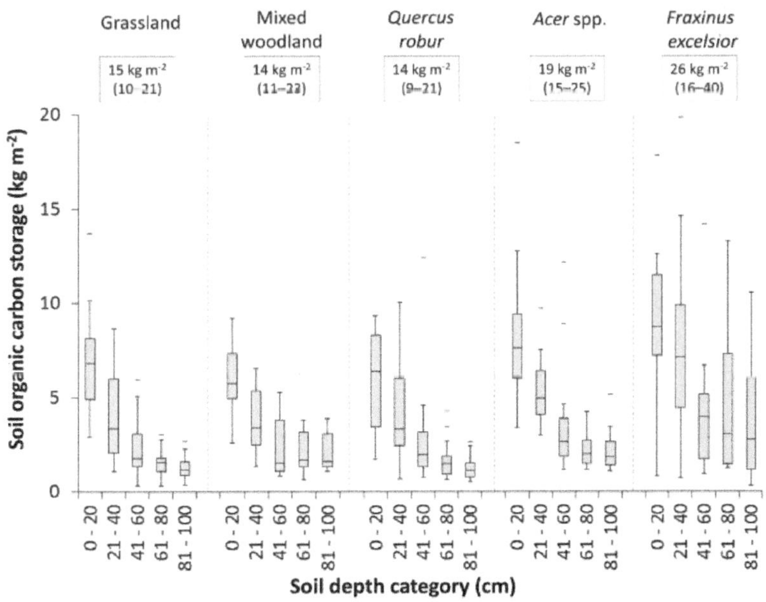

Figure 1: Soil organic carbon storage within each 20(summed median values are displayed in text boxes, values in parenthesis are total 25th and 75th percentiles), beneath ***Quercus robur*** (n=12), ***Fraxinus excelsior*** (n=11), ***Acer*** spp. (n=12), mixed woodland (n=8) and grassland (n=15) by depth class. The horizontal line within the box indicates median, box boundaries indicate 25th and 75th percentiles, whiskers indicate highest and lowest values, horizontal lines above or below whiskers indicate outliers. http://dx.doi.org/10.1371/journal.pone.0101872.g001

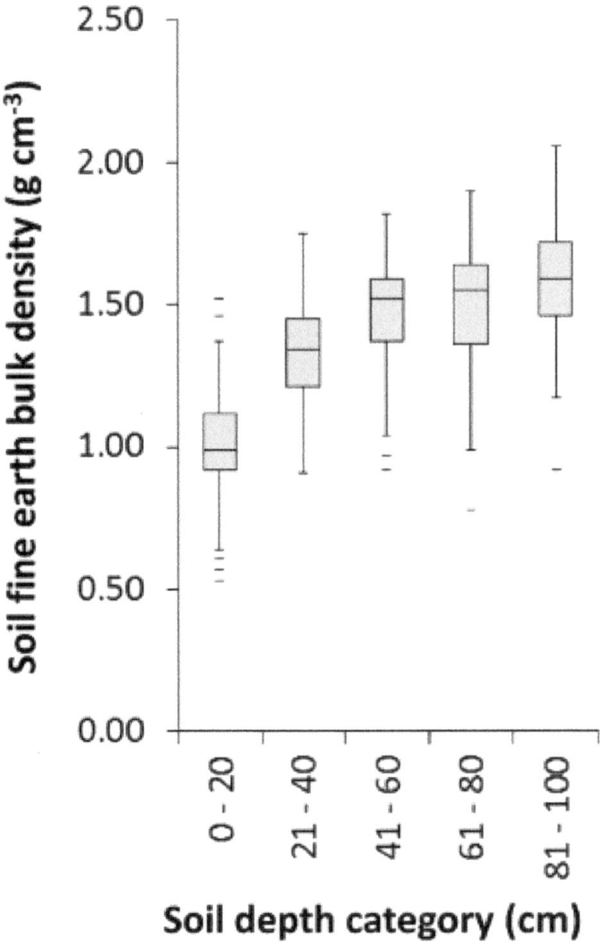

Figure 2: Soil bulk density in each 20 cm depth category. The horizontal line within the box indicates median, box boundaries indicate 25th and 75th percentiles, whiskers indicate highest and lowest values, horizontal lines above or below whiskers indicate outliers. http://dx.doi.org/10.1371/journal.pone.0101872.g002

DISCUSSION

Previous research in a typical UK city, Leicester, has shown that 97% of carbon present in aboveground ecosystem biomass is found in trees, with average storage increasing from 0.2 kg m^{-2} in herbaceous vegetation (most commonly urban grassland) to 28.5 kg m^{-2} in trees [2], affirming the importance of urban trees in aboveground carbon sequestration. In contrast we found that differences in SOC under trees compared to grassland were more modest and genus-specific and, surprisingly, did not occur under mixed woodland or *Q. robur*, even though several specimens of the latter had a trunk diameter of over 1.5 m. No significant differences in SOC concentration beneath urban forests and grasslands were found in Baltimore, USA[42], corresponding with our findings for mixed woodlands. Nonetheless, we did find increased SOC storage under *F. excelsior* and, to a lesser extent, *Acer* spp., with gains of 11 kg m^{-2} and 5 kg m^{-2} respectively compared to the adjacent grasslands (Fig. 1).

The effects of *F. excelsior* are of particular interest in light of current concerns about the impact of ash die-back disease on this widespread and abundant species in Europe and the UK, where it is the second most abundant tree in small woodland patches and the second most common individual tree in the countryside [33]. *A. pseudoplatanus* and *F. excelsior* are mesophilic species that thrive on well-watered alluvial soils, but the former is more drought susceptible [43]. The clear SOC enrichment throughout the soil profile under *F. excelsior*, especially from 40–100 cm depth, could be attributed to several factors. This species produces a greater root mass than *A. pseudoplatanus* and *Q. robur*, and establishes its extensive deep root network more quickly than other broadleaved species [44], and is especially well adapted to clay-rich soils such as those found in Leicester. Organic carbon introduced by tree roots into clay rich subsoils will experience long residence times [45].

Compared to park grassland, none of the tree genera or mixed woodland altered soil BD, which is consistent with previous reports of the absence of effects of broadleaved trees planted in pasture [46]. However, the urban trees could still aid storm-water drainage along root channels[20]. Similarly, soil C:N ratio, one of the major controls of N availability and leaching [47], was also unaffected by trees.

These data highlight the often overlooked importance of urban park grasslands as contributors to belowground ecosystem service provision, particularly SOC storage, which we show exceeds typical values for agricultural grassland by 23% [24]. Urban park management may reduce litter inputs from trees relative to grassland as mowings are not collected but autumn leaves are removed from beneath individual trees. However, it remains unclear how important these management practices are as leaves were not removed from beneath the stands of mixed woodland we sampled, yet these showed no significant increase in SOC storage compared to adjacent grassland. Carbon inputs into the soils under grassland and park trees are likely to be strongly influenced by roots. Mean residence time of root carbon in soil is typically 2.4 times that of shoot carbon, due to higher concentrations of the more recalcitrant components such as lignin, so that SOC is often mainly derived from root inputs [48].

This observational study aimed to provide a first indication of the long-term effect of trees in urban areas on soil properties. As our approach was non-experimental we cannot be certain whether the sampled trees were planted or naturally regenerated from seed, and we have to assume that any differences in soil conditions under trees and adjacent grassland are due to the trees. However, our strategy of sampling soil under trees of very different sizes enabled us to investigate if there were any relationships between tree size (as a proxy for tree age) and soil properties across a range of urban parks. Indeed, given the range of tree sizes we studied, the trees will likely have ranged in age from a decade to several centuries- a timespan difficult to achieve in experimental manipulation studies. Perhaps one of the most surprising findings arising from this was the absence of clear effects of tree size on soil carbon storage, especially for oaks where the largest individuals were 1.6 m–2.0 m DBH.

Tree planting within urban areas is one of a range of environmental management techniques used which impact, either intentionally or unintentionally, on ecosystem service provision [1]. Although tree planting is known to increase ecosystem service provision aboveground [2]–[14], we now show for the first time for urban trees that this benefit does not consistently extend into the soil system as, contrary to our original hypotheses, there is no direct relationship between aboveground tree biomass and SOC concentration, soil C:N ratio and soil BD.

We demonstrate the importance of targeted tree genus selection to maximise the long-term belowground ecosystem service benefits of urban tree planting with respect to SOC. Further research should seek to elucidate the belowground effects of other common tree species on a range of soil types and climate zones across urban areas globally, better to inform urban policy and planning. Our findings that *F. excelsior* makes an important contribution to enhancing urban SOC stocks coincides with the first record of ash dieback disease spreading into our study region (Leicestershire) [49]. This raises important questions about the likely future impacts of loss of this species on SOC stocks nationally. Furthermore, it highlights the importance of long-term planning in cities and towns to couple the ecosystem service benefits of tree planting with disease and climate change resilient urban tree populations in the future.

SUPPORTING INFORMATION

Table S1: Raw data file.

	1	2	3	4	5	6	7	8	9	10	11	12	13	14	15	16	17	18
1	Table S1													Land-use history using historical Ordnance Survey maps				
2	Sample code	Sample site	Park	x coordinate	y coordinate	Tree species/ grass	Tree DBH (cm)	Tree biomass (kg)	Soil depth (cm)	Soil bulk density (g cm-3)	Soil OC concentration(mg g-1)	Soil OC/N	Soil OC density (mg cm-3)	Map age 1887-1890	Map age 1930-1931	Map age 1955-1966	Map age 1971-1974	Map age 1982-1992
3	1	TD01	Braunstone Park	455610	303853	Quercus robur	8.276057040778557	22.222013549479527	10	0.7421	58.59873294830322	8.552674830297944	43.48611972093582	Grounds of country house	Park	Park	Park	Park
4	2	TD01	Braunstone Park	455610	303853	Quercus robur	8.276057040778557	22.222013549479527	16	0.9372	32.74851679801941	7.873886585891208	30.691909943103788	Grounds of country house	Park	Park	Park	Park
5	3	TD01	Braunstone Park	455610	303853	Quercus robur	8.276057040778557	22.222013549479527	27	1.3657	15.734044015407562	7.669363028761696	21.487983911842107	Grounds of country house	Park	Park	Park	Park
6	4	TD01	Braunstone Park	455610	303853	Quercus robur	8.276057040778557	22.222013549479527	35		3.916770815849304	4.760223409013056	5.103375567940265	Grounds of country house	Park	Park	Park	Park
7	5	TD01	Braunstone Park	455610	303853	Quercus robur	8.276057040778557	22.222013549479527	40		10.556161403656006	8.79927733867359	14.136338646977071	Grounds of country house	Park	Park	Park	Park
8	6	TD01	Braunstone Park	455610	303853	Quercus robur	8.276057040778557	22.222013549479527	51	1.5164000000000002	7.025592625141144	10.231974571951508	10.653608656764032	Grounds of country house	Park	Park	Park	Park
9	7	TD01	Braunstone Park	455610	303853	Quercus robur	8.276057040778557	22.222013549479527	62	1.7031999999999998	5.558382272720337	9.033797420845353	9.467036686897277	Grounds of country house	Park	Park	Park	Park
10	8	TD01	Braunstone Park	455610	303853	Quercus robur	8.276057040778557	22.222013549479527	74	1.7299	4.966342449188232	12.334413612629387	8.591275802850722	Grounds of country house	Park	Park	Park	Park
11	9	TD01	Braunstone Park	455610	303853	Quercus robur	8.276057040778557	22.222013549479527	86	1.6456	3.7698838114738464	10.374178736013992	6.203720800161361	Grounds of country house	Park	Park	Park	Park
12	10	TD01	Braunstone Park	455610	303853	Quercus robur	8.276057040778557	22.222013549479527	97	1.7431999999999999	4.8499225080013275	11.819639555288376	8.454384915947914	Grounds of country house	Park	Park	Park	Park
13	11	TD02	Braunstone Park	455805	303709	grassland			9	0.9901000000000001	29.540358781814575	8.887851186299217	29.24790922987461	Grounds of country house	Park	Park	Park	Park
14	12	TD02	Braunstone Park	455805	303709	grassland			16	1.081	17.371835708618164	8.061545231863334	18.778954401016236	Grounds of country house	Park	Park	Park	Park
15	13	TD02	Braunstone Park	455805	303709	grassland			26	1.2233	10.709753632545471	7.496429591917676	13.101241618692876	Grounds of country house	Park	Park	Park	Park
16	14	TD02	Braunstone Park	455805	303709	grassland			40	1.5326	4.379620403051376	7.1161004210297705	6.712206229716539	Grounds of country house	Park	Park	Park	Park
17	15	TD02	Braunstone Park	455805	303709	grassland			50		5.3816090524196625	9.913955005746653	7.532366079143965	Grounds of country house	Park	Park	Park	Park
18	16	TD02	Braunstone Park	455805	303709	grassland			56		1.2008900940418243	2.696128005441075	1.7177205783754192	Grounds of country house	Park	Park	Park	Park
19	17	TD02	Braunstone Park	455805	303709	grassland			65		1.5532540157437325	3.2429807813687708	2.284489332271385	Grounds of country house	Park	Park	Park	Park
20	18	TD02	Braunstone Park	455805	303709	grassland			69		1.1100919172167778	2.728443582905303	1.650669241271205	Grounds of country house	Park	Park	Park	Park
21	19	TD02	Braunstone Park	455805	303709	grassland			78		1.5664281696081161	3.355480782697066	2.381289997775173	Grounds of country house	Park	Park	Park	Park
22	20	TD02	Braunstone Park	455805	303709	grassland			87		2.868456691503525	5.691443736311351	4.445556241698657	Grounds of country house	Park	Park	Park	Park
23	21	TD02	Braunstone Park	455805	303709	grassland			94		2.0772135257720947	4.374930714620263	3.262860450091228	Grounds of country house	Park	Park	Park	Park

24	22	TD03	Braunstone Park	455759	303453	Quercus robur	16.870423967740905	127.35175569528008	9	0.9109999999999999	44.953999519348145	10.89590843774422	40.95309356212616	Grounds of country house	Park	Park	Park	Park
25	23	TD03	Braunstone Park	455759	303453	Quercus robur	16.870423967740905	127.35175569528008	16	1.1181	29.74682092666626	10.057459469394072	33.259920478105556	Grounds of country house	Park	Park	Park	Park
26	24	TD03	Braunstone Park	455759	303453	Quercus robur	16.870423967740905	127.35175569528008	22		7.85354346036911	4.494471690047268	9.244261573510773	Grounds of country house	Park	Park	Park	Park
27	25	TD03	Braunstone Park	455759	303453	Quercus robur	16.870423967740905	127.35175569528008	30	1.6766999999999999	7.027556896209717	6.39234455208412	11.783104647874831	Grounds of country house	Park	Park	Park	Park
28	25	TD03	Braunstone Park	455759	303453	Quercus robur	16.870423967740905	127.35175569528008	37		9.730928838253021	8.327802679388757	12.825557329977988	Grounds of country house	Park	Park	Park	Park
29	27	TD03	Braunstone Park	455759	303453	Quercus robur	16.870423967740905	127.35175569528008	48	1.5521	4.239840805530548	5.10988838306388	6.5806569142639635	Grounds of country house	Park	Park	Park	Park
30	28	TD03	Braunstone Park	455759	303453	Quercus robur	16.870423967740905	127.35175569528008	60	1.5167000000000002	5.4159800708293915	7.65311834177646	8.214416973426939	Grounds of country house	Park	Park	Park	Park
31	29	TD03	Braunstone Park	455759	303453	Quercus robur	16.870423967740905	127.35175569528008	80	1.6366	4.815860837697983	8.678542836108319	7.881637846976519	Grounds of country house	Park	Park	Park	Park
32	30	TD03	Braunstone Park	455759	303453	Quercus robur	16.870423967740905	127.35175569528008	90	1.7357	7.617310881614685	1.847283154697983	13.221366497218611	Grounds of country house	Park	Park	Park	Park
33	31	TD03	Braunstone Park	455759	303453	Quercus robur	16.870423967740905	127.35175569528008	100		5.046001672744751	11.87066627043028	8.010838678013638	Grounds of country house	Park	Park	Park	Park
34	32	TD04	Braunstone Park	455762	303433	Quercus robur	44.563384065730695	1426.6195293720973	10	0.8622	52.940380573272705	11.792392326592921	45.64519613027573	Grounds of country house	Park	Park	Park	Park
35	33	TD04	Braunstone Park	455762	303433	Quercus robur	44.563384065730695	1426.6195293720973	17		16.45482301712036	5.26438337728565	18.218519031137294	Grounds of country house	Park	Park	Park	Park
36	34	TD04	Braunstone Park	455762	303433	Quercus robur	44.563384065730695	1426.6195293720973	27	1.4531	11.481163501739502	6.065280769466021	16.683278684377672	Grounds of country house	Park	Park	Park	Park
37	35	TD04	Braunstone Park	455762	303433	Quercus robur	44.563384065730695	1426.6195293720973	39		2.5260156393051147	3.9514075216419977	3.3653893223975215	Grounds of country house	Park	Park	Park	Park
38	36	TD04	Braunstone Park	455762	303433	Quercus robur	44.563384065730695	1426.6195293720973	48		9.103280305862427	8.428806028786989	12.640656581543626	Grounds of country house	Park	Park	Park	Park
39	37	TD04	Braunstone Park	455762	303433	Quercus robur	44.563384065730695	1426.6195293720973	57		3.100506365299225	3.3008589597457307	4.449757436894307	Grounds of country house	Park	Park	Park	Park
40	38	TD04	Braunstone Park	455762	303433	Quercus robur	44.563384065730695	1426.6195293720973	67		2.3842164129018784	3.1567989522718847	3.526237404161842	Grounds of country house	Park	Park	Park	Park
41	39	TD04	Braunstone Park	455762	303433	Quercus robur	44.563384065730695	1426.6195293720973	80		2.120523750782013	3.5023428939139887	3.238182734555146	Grounds of country house	Park	Park	Park	Park
42	40	TD04	Braunstone Park	455762	303433	Quercus robur	44.563384065730695	1426.6195293720973	89		4.029257297515869	5.586255976675632	6.269400736653492	Grounds of country house	Park	Park	Park	Park
43	41	TD04	Braunstone Park	455762	303433	Quercus robur	44.563384065730695	1426.6195293720973	94		2.939334958791733	4.468345615943071	4.617069775264081	Grounds of country house	Park	Park	Park	Park
44	42	TD05	Braunstone Park	455721	303504	Quercus robur	161.06480240899808	36855.995140934894	7		14.409145712852478	4.816354402840259	12.48748788249957	Grounds of country house	Park	Park	Park	Park
45	43	TD05	Braunstone Park	455721	303504	Quercus robur	161.06480240899808	36855.995140934894	17		16.92640244960785	7.774506975955501	18.74064429839342	Grounds of country house	Park	Park	Park	Park
46	44	TD05	Braunstone Park	455721	303504	Quercus robur	161.06480240899808	36855.995140934894	23		19.743744730949402	7.0304049642507875	23.47792817467192	Grounds of country house	Park	Park	Park	Park
47	45	TD05	Braunstone Park	455721	303504	Quercus robur	161.06480240899808	36855.995140934894	30		9.120108485221863	5.24528988632565	11.501958062460268	Grounds of country house	Park	Park	Park	Park
48	46	TD05	Braunstone Park	455721	303504	Quercus robur	161.06480240899808	36855.995140934894	36		7.878369987010956	5.453641820707405	10.32532855366015	Grounds of country house	Park	Park	Park	Park
49	47	TD05	Braunstone Park	455721	303504	Quercus robur	161.06480240899808	36855.995140934894	40		4.376338869333267	4.240745146116998	5.860597041401568	Grounds of country house	Park	Park	Park	Park
50	48	TD05	Braunstone Park	455721	303504	Quercus robur	161.06480240899808	36855.995140934894	47		2.7991360425949097	3.3599194750495847	3.8708552963967002	Grounds of country house	Park	Park	Park	Park

Table S2: Results from mixed effect model and model averaging for models including tree biomass.

Variable	Pseudo R^2	z-value	*p* value	Relative Importance
SOC Concentration	0.58	0.802	0.42	0.33
SOC Density	0.51	0.61	0.53	0.54
Soil C:N	0.16	1.49	0.13	0.51
Soil Bulk Density	0.49	0.85	0.39	0.33

Pseudo R^2 values are calculated using the methods of Nagelkerke (1991). Z & p values are based on the final averaged model. Relative importance is the proportion of all the top models which contained the term for tree biomass. For comparison all top models for all variables contained soil depth in all top models (relative importance =1). Vegetation cover was retained in all the top models for all variables apart from the model for soil bulk density where the relative importance was 0.21.

ACKNOWLEDGMENTS

We gratefully acknowledge Leicester City Council for access to their land.

AUTHOR CONTRIBUTIONS

Conceived and designed the experiments: JLE NM KJG JRL. Performed the experiments: JLE OSO JP. Analyzed the data: RI. Contributed to the writing of the manuscript: JLE RI OSO JP NM KJG JRL. Designed, arranged permissions: JLE JP OSO. Carried out fieldwork: JLE JP OSO.

REFERENCES

1. Gaston KJ, Ávila-Jiménez ML, Edmondson JL (2013) Managing urban ecosystems for good and services. J Appl Ecol 50: 830–840. doi: 10.1111/1365-2664.12087
2. Davies ZG, Edmondson JL, Heinemeyer A, Leake JR, Gaston KJ (2011) Mapping an urban ecosystem service: quantifying above-ground carbon storage at a city-wide scale. J Appl Ecol 48: 1125–1134. doi: 10.1111/j.1365-2664.2011.02021.x
3. Nowak DJ, Greenfield EJ (2012) Tree and impervious cover in the United States. Landscape Urban Plan 107: 21–30. doi: 10.1016/j.landurbplan.2012.04.005
4. Tiwary A, Sinnett D, Peachey C, Chalabi Z, Vardoulakis S, et al.. (2009) An integrated tool to assess the role of new planting in PM_{10} capture and the human health benefits: a case study in London.
5. Tallis M, Taylor G, Sinnett D, Freer-Smith P (2011) Estimating the removal of atmospheric particulate pollution by the urban tree canopy of London, under current and future environments. Landscape Urban Plan 103: 129–138. doi: 10.1016/j.landurbplan.2011.07.003
6. Joraeva VA, Johnson DL, Hassett JP, Nowak DJ (2002) Differences in accumulation of PAHs and metals on the leaves of *Tilia* x *euchlora* and *Pyrus calleryana*. Environ Pollut 120: 331–338. doi: 10.1016/s0269-7491(02)00121-5
7. Nowak DJ, Crane DE, Stevens JC (2006) Air pollution removal by urban trees and shrubs in the United States. Urban For. Urban Gree 4: 115–123. doi: 10.1016/j.ufug.2006.01.007
8. Maleki K, Hosseini SM (2011) Investigation of the effects of leaves, branches and canopies of trees on noise pollution reduction. Annals of Environmental Science 5: 13–21.
9. Akbari H (2002) Shade trees reduce building energy use and CO_2 emissions from power plants. Env Pollut 116: S119–S126. doi: 10.1016/s0269-7491(01)00264-0
10. Hamada S, Ohta T (2010) Seasonal variations in the cooling effect of urban green areas on surrounding urban areas. Urban For Urban Gree 9: 15–24. doi: 10.1016/j.ufug.2009.10.002
11. Armson D, Stringer P, Ennos AR (2012) The effect of tree shade and grass on surface and globe temperatures in an urban area. Urban For Urban Gree 11: 245–255. doi: 10.1016/j.ufug.2012.05.002
12. McPherson EG, Nowak D, Heisler G, Grimmond S, Souch C, et al. (1997) Quantifying urban forest structure, function, and value: the Chicago urban forest climate project. Urban Ecosystems 1: 49–61. doi: 10.1023/a:1014350822458
13. Nowak DJ, Crane DE, Stevens JC, Hoehn RE, Walton JT, et al. (2008) A ground-based method of assessing urban forest structure and ecosystem services. Arboriculture and Urban Forestry 34: 347–358.
14. White MP, Alcock I, Wheeler BW, Depledge MH (2013) Would you be happier living in a greener urban area? A fixed-effects analysis of panel data. Psychol Sci 24: 920–928. doi: 10.1177/0956797612464659
15. DEFRA. The Big Tree Plant. Available: http://www.defra.gov.uk/bigtreeplant/Accessed: 2014 Jan.
16. MillionTressNYC. Million Trees NYC. Available:http://www.milliontreesnyc.org/html/about/about.shtml Accessed: 2014 Jan.
17. Batjes NH (1996) Total carbon and nitrogen in soils of the world. Eur J Soil Sci 47: 151–163. doi: 10.1111/ejss.12115
18. Lal R (2008) Carbon sequestration. Philos T Roy Soc B 363: 815–830. doi: 10.1098/rstb.2007.2185
19. Franzluebbers AJ (2002) Soil organic matter stratification ratio as an indicator of soil quality. Soil Till Res 66: 95–106. doi: 10.1016/s0167-1987(02)00018-1
20. Day SD, Wiseman PE, Dickinson SB, Harris JR (2010) Tree root ecology in the urban environment and implications for a sustainable rhizosphere. Arboriculture and Urban Forestry 36: 193–205.
21. Jobbagy EG, Jackson RB (2000) The vertical distribution of soil organic carbon and its relation to

climate and vegetation. Ecol Appl 10: 423–436. doi: 10.1890/1051-0761(2000)010[0423:tvdoso]2.0.co;2

22. Guo LB, Gifford RM (2002) Soil carbon stocks and land use change: a meta analysis. Global Change Biol 8: 345–360. doi: 10.1046/j.1354-1013.2002.00486.x
23. Lal R (2004) Soil carbon sequestration impacts on global climate change and food security. Science 304: 1623–1627. doi: 10.1126/science.1097396
24. Bradley RI, Milne R, Bell J, Lilly A, Jordan C, et al. (2005) A soil carbon and land use database for the United Kingdom. Soil Use Manage 21: 363–369. doi: 10.1079/sum2005351
25. .Carey PD, Wallis S, Chamberlain PM, Cooper A, Emmett BA, et al. (2008) Countryside Survey: UK Results from 2007. NERC/Centre for Ecology & Hydrology (CEH Project Number:C03259).
26. Edmondson JL, Davies ZG, McHugh N, Gaston KJ, Leake JR (2012) Organic carbon hidden in urban ecosystems. Scientific Reports 2: 963. doi: 10.1038/srep00963
27. Edmondson JL, Davies ZG, McCormack SA, Gaston KJ, Leake JR (2014) Land-cover effects on soil organic carbon stocks in a European city. Sci Total Environ 472: 444–453. doi: 10.1016/j.scitotenv.2013.11.025
28. DEFRA (2013) Chalara management Plan. Department for Environment Food and Rural Affairs, London.
29. Forestry Commission. Bleeding Canker of Horse Chestnut. Available:http://www.forestry.gov.uk/fr/infd-6kybgv Accessed 2013 Dec.
30. Tree Health and Plant Biosecurity Taskforce (2013) Tree Health and Plant Biosecurity Task Force. Department for Environment Food and Rural Affairs, London.
31. Leicester City Council. Population of Leicester 1801–2011. Available:http://www.leicester.gov.uk/your-council-services/council-and-democracy/city-statistics/other-statistics/population-statistics/populationsince1801/ Accessed 2014 Jan.
32. Met Office. 1981–2010 averages. Available:http://www.metoffice.gov.uk/climate/uk/averages/19812010/ Accessed 2014 Jan.
33. Maskell L, Henrys P, Norton L, Smart S, Wood C (2013) Distribution of ash trees (Fraxinus excelsior) in Countryside Survey Data. Centre for Ecology and Hydrology
34. McHale MR, Burke IC, Lefsky MA, Peper PJ, McPherson EG (2009) Urban forest biomass estimates: is it important to use allometric relationships developed specifically for urban trees? Urban Ecosystems 12: 95–113. doi: 10.1007/s11252-009-0081-3
35. Ter-Mikaelian MT, Korukhin MD (1997) Biomass equations for sixty-five North American tree species. Forest Ecol Manag 97: 1–24. doi: 10.1016/s0378-1127(97)00019-4
36. Zianis D, Muukkpnen P, Mäkipää R, Mencuccini M (2005) Biomass and stem volume equations for tree species in Europe. Silva Fennice Monographs 4: 63.
37. Snorrason A, Einarsson SF (2006) Single-tree biomass and stem volume functions for eleven tee species used in Icelandic forestry. Icelandic Agricultural Science 19: 15–24.
38. Edmondson JL, Davies ZG, McCormack SA, Gaston KJ, Leake JR (2011) Are soils in urban ecosystems compacted? A citywide analysis. Biol Letters 23: 771–774. doi: 10.1098/rsbl.2011.0260
39. Nagelkerke NJD (1991) A note on a general definition of the coefficient of determination. Biometrika 78: 691–692. doi: 10.1093/biomet/78.3.691
40. Satterthwaite FE (1946) An approximate distribution of estimates of variance components. Biometrics Bull 2: 110–114. doi: 10.2307/3002019
41. R Core Team (2012) R: A language and environment for statistical computing. R Foundation for Statistical Computing, Vienna, Austria. ISBN 3-900051-07-0, Available:http://www.R-project.org/.
42. Pouyet RV, Yesilonis ID, Golubiewski NE (2009) A comparison of soil organic carbon stocks between residential turf grass and native soil. Urban Ecosystems 12: 45–62. doi: 10.1007/s11252-008-0059-6
43. Lemoine D, Peltier J-P, Marigo G (2001) Comparative studies of the water relations and the hydraulic characteristics in Fraxinus excelsior, Acer pseudoplatanus, and A.opalus trees under soil water contrasted conditions. Ann For Sci 58: 723–731. doi: 10.1051/forest:2001159
44. Kerr G, Cahalan C (2004) A review of site factors affecting early growth of ash (Fraxinus excelsior L.). Forest Ecol Manage 188: 225 234. doi: 10.1016/j.foreco.2003.07.016
45. Krull ES, Baldock JA, Skemstad JO (2003) Importance of mechanisms and processes of the stabilisation of soil organic matter for modelling carbon turnover. Funct Plant Biol 30: 207 222. doi: 10.1071/fp02085
46. Berthrong ST, Jobbagy EG, Jackson RB (2009) A global meta-analysis of soil exchangeable cations, pH, carbon and nitrogen with afforestation. Ecol Appl 19: 2228–2241. doi: 10.1890/08-1730.1
47. Dungait JAJ, Cardenas LM, Blackwell MSA, Wu L, Withers PJA, et al. (2012) Advances in the understanding of nutrient dynamics and management in UK agriculture. Sci Total Environ 434: 39–50. doi: 10.1016/j.scitotenv.2012.04.029
48. Rasse DP, Rumpel C, Dignac M-F (2005) Is soil carbon mostly root carbon? Mechanisms for a specific stabilization. Plant Soil 269: 341–356. doi: 10.1007/s11104-004-0907-y
49. DEFRA. Ash Die Back in Leicestershire Press Release. Available:http://www.gov.uk/government/news/ash-disease-found-in-Leicestershire Accessed 2013 Dec.

CHAPTER 14 STREET TREES IN THE URBAN FOREST CANOPY

Joseph Poracsky and David Banis

ACKNOWLEDGMENTS

This study was made possible by a grant from the Urban and Community Forestry Program of the Oregon Department of Forestry, administered by Paul Ries, Urban and Community Forestry Program Manager. We greatly appreciate Paul's assistance in the implementation and completion of this work.

Thanks go to Wayne Larimer, (B.S. Geography, Portland State University, 2005) for preparation of the shaded relief map of Portland.

Invaluable inspiration during the many hours of writing was provided by:

T. G. Albinoni

C. P. E. Bach

J. S. Bach

L. Boccherini

D. Buxtehude

J. Desprez

F. Danzi

J. Haydn

B. Joel

W. A. Mozart

A. Vivaldi

SUMMARY

Attempts to identify the contribution of street trees to the overall urban forest of a city have been rare and lack consensus on how to measure that contribution – percentage of trees, percentage of canopy cover, or percentage of leaf area. The actual numeric values presented in the literature also vary over a broad range and often are based on estimates, extrapolations from aggregated data, or simply stated with no empirical data referenced. This study was undertaken to evaluate the contribution of street trees to canopy in Portland, Oregon. The study involved both visual and digital analysis of multi-band aerial images, field observation, and GIS analysis involving several ancillary data themes. The strategy was to calculate the street tree component (STC) of the urban forest canopy within a sample of areas in the city and generalize that pattern to the whole city.

The analysis indicates that street trees constitute 3.4% of the whole study area. In commercial / industrial areas the value drops to only 1.1% and in residential areas rises to 4.8%. For the overall study area the STC of the canopy is 17.2%, a value in the middle of the range of values found in the published

Citation: Joseph Poracsky, David Banis, Street Trees in the Urban Forest Canopy, http://pdxscholar.library.pdx.edu/cgi/viewcontent.cgi?article=1023&context=geog_fac

literature. In commercial / industrial areas we found a value of 23.7% and a value of 16.6% for residential areas.

Residential areas are overall better treed than commercial / industrial areas, but do not rely on street trees to contribute as large a percentage to overall canopy as do commercial / industrial areas. Available planting space in the right-of-way (ROW) is much more nearly constant (or independent of land cover) than the amount of planting space available inside a block. Therefore, the STC tends to vary inversely with the total amount of vegetated area. As the vegetated (read, plantable) area decreases, the ROW becomes a more significant space for planting and the STC increases. Conversely, as the vegetated area increases, the ROW becomes a less significant space for planting and the STC decreases.

For both residential and commercial / industrial areas, the percentage of street tree area within the total canopy varies from locale to locale. The relatively small sampling used in this study suggests that neighborhood age and physical characteristics influence the STC in a given neighborhood, particularly for residential areas. Analysis allowed division of the study area into five residential neighborhood groups that attempt to reflect local differences. A larger / more detailed sampling would allow for improved delineation, perhaps below the level of neighborhood. Three generalized spatial patterns emerge for Portland.

1. Street tree canopy as a percent of area seems to follow an east-west gradient, with lower values to the east and higher values to the west.
2. Street trees as a component of the total canopy takes on an inverted-U shape, with higher values in the Inner Eastside and lower values to the east and to the west.
3. The overall tree canopy category takes on just the opposite pattern – a standard-U – with lower values in the Inner Eastside and higher values to the east and west.

This study represents a useful starting point for understanding the dynamics of an important and highly visible component of the urban forest.

Trees in the Urban Environment

The past several decades have seen an increased recognition and appreciation of the roles played by trees in the urban environment. Research has shown that trees provide a wide variety of aesthetic, ecological, economic, social and human health benefits, ranging from traffic calming, to providing wildlife habitat, to reduction of runoff, to creation of neighborhood pride and improved property resale value. Extended discussions of these and other urban tree values may be found in a variety of sources (e.g., Ebenreck 1989; Miller 1997). Here we will simply note that there are many positive values to trees and these values serve to make a strong case for proactive management of urban trees

In order to make intelligent decisions regarding tree management, urban forest researchers and managers need to have data on such items as numbers of trees, species composition, age profiles, health conditions, and maintenance needs. Data on counts and characteristics are most useful when they include locations that allow them to be presented in map form. The map format is invaluable in helping to identify spatial patterns and to point to where management attention should be directed. In collecting spatial data about the urban forest, two distinct approaches are used – inventory and mapping of the trees as individual point features and mapping the extent of the aggregated tree crowns or tree canopy. Enumerating individual trees and their locations produces a wealth of detailed data, but requires a high cost of time and effort. While the increasing availability of digital tools that make the process easier and faster to complete has led to having inventories performed more often, point-based inventories of trees are still not widely-performed. And, since the cost of inventorying individual trees is directly proportional to the number of trees and the area covered, when individual tree inventories are performed, they usually cover only limited areas, such as a neighborhood or a small community.

In contrast to tree inventory, tree canopy mapping produces less-detailed data. This is done by not considering individual trees but viewing groups of trees in the aggregate as an interwoven surface of branches and leaves. Mapping can be completed more quickly and at less expense than tree inventory by using aerial or satellite imagery and established techniques in remote sensing data analysis. Canopy mapping has been used for areas the size of cities or regions (Rowntree 1984; Newman 1997; Iverson and Cook 2001). It offers a more economical approach than mapping tree locations, especially for large areas, where the synoptic view provided by the canopy extent map can provide assistance in making strategic management decisions

Since they provide different kinds of information, the choice between an inventory and a canopy mapping obviously depends on the types of information desired, as well as cost and time constraints. Since funds for urban forestry programs are usually very limited, canopy mapping is usually more practical to perform than a tree inventory. Not only is a canopy mapping less expensive and more quickly produced, but it can provide overview data that is invaluable in facilitating the planning and implementation of an inventory.

Tree Canopy and Street Trees

A recent study of tree canopy in Portland, Oregon over the period 1972-2002 employed satellite imagery at a 25 meter resolution to map differences in canopy density and to identify a recommended canopy cover target level for tree planting projects (Poracsky and Lackner 2003). The study segmented the city into four of the urban land environments (ULEs) identified in the City's Urban Forest Management Plan. While nominally based on land uses, the ULE categories are intended to recognize that different land uses offer different sets of constraints on, and opportunities for, tree planting. As a consequence, different ULEs are likely to be associated with different canopy cover percentages.

The canopy study made recommendations for target canopy cover in two ULEs – residential and commercial / industrial. No recommendation was made for the ULE of right-of-way (ROW) street trees – plantings along the streetside – since the narrowness of the street rightof-way makes it impossible for medium resolution satellite data to separate out the street trees from adjacent trees in other ULEs in which they are embedded.

Although the canopy study did not address the street tree issue, it was recognized that street trees are an important component of the urban forest. One reason is that the right-of-way environment is where regulatory forestry programs normally have jurisdiction and can therefore directly impact tree management. A second reason for the importance of street trees is their location along the streetscape. Given their size and prominent location between street and sidewalk on the everyday routes along which urban residents live and travel, street trees are the most highly visible component of the urban forest.

From a canopy cover analysis viewpoint, an important question for urban tree management is, "What is the street tree component (STC) of the urban forest?" A second related question is "Does the STC vary over the whole canopy and, if so, what kind of variation pattern is exhibited?" In an initial attempt to answer these questions, a brief review of the urban forestry literature was undertaken. Four conclusions emerged from the review.

First, while there are a number of canopy mapping studies, there have not been a great number of attempts documented in the literature to identify the contribution of street trees to the total urban forest. A total of six references were identified, and these are summarized in Figure 1. Second, the literature reflects variability in the approach to evaluating street trees as a component of the urban forest. Sometimes what is reported are percentages based on counts of trees (Moll 1989) and at other times percentages based on canopy cover (Grey and Deneke 1986). Sometimes two figures are presented, one for percentages of trees and one for canopy area. For example, McPherson, Nowak and Rowntree (1994) report that 10% of the trees (by count) in Chicago are street trees, but also note that the street trees constitute 24% of the leaf area (a canopy measure that incorporates the third dimension component of tree height). Third, the actual numeric values presented in the literature vary over a broad range. For example, while McPherson, Nowak and Rowntree (1994) report that 10% of the trees in Chicago are street trees, they report a much smaller figure of 2.7% for suburban Cook County. In 1989 Moll (16) cites a figure of 33% for cities in general, but four years later (1993) reports a figure of 10%. In neither case does he cite an empirical study. It is possible, and perhaps likely, that, rather than representing discrepancies, this wide range of numbers (2.7% - 33%) represents real differences that result from variations in defining individual study area criteria, differences in the level of development of study areas, and variation in local environmental conditions. Fourth, there is often a reliance in the literature on "estimates" and calculations from aggregated data. For example, Moll (1989 16) says "about 1/3 of the trees within the city limits are street trees" without citing any specific studies. Grey and Deneke (1986 17) present a figure of 30% for "trees in urban areas [that] are publicly owned" but that figure must be reduced an unknown amount for street trees, since the publicly owned land category includes parks, railroad rights-of-way, and public buildings and grounds. In 1999 McPherson et al found 21.4% of the canopy of Modesto, CA was composed of street trees. (This figure was not presented directly, but was derived by calculation from other figures presented in the study.) What can we conclude based on the four points drawn from this brief literature review? Simply stated, there is no consensus either in terms of definition or numerical value as to what street trees contribute to the urban forest canopy. In fact, there is strong indication that a moderately wide range of values – not a single universal value – may be the rule, the actual number depending on the characteristics of the area of concern. It may not be possible to transfer street tree canopy values from one area and apply them to another. Understanding a particular area is likely to require a local study.

Percent		Source	Comments
Trees	Area		
<30	-	Grey and Deneke 1986 17	Estimate of trees in urban areas that are publicly owned is 30%; minus parks, public buildings and
33	-	Moll 1989 16	"about 1/3 of the trees within the city limits are street trees"
10	-	Moll and Kollin 1993 (quoted in Miller 1997 31)	10% of urban trees are on publicly owned land
10.1	-	Nowak 1994 13	City of Chicago
-	24	Nowak 1994 13	City of Chicago – leaf area (not canopy)
2.7	-	Nowak 1994 13	Suburban Cook County
-	9.5	Nowak 1994 13	Suburban Cook County – leaf area (not canopy)
-	10	American Forests 1997	Source not physically viewed; cited on website (www.edc.uri.edu/personal/greg/atlas/) August 2,
-	21.4	McPherson et al 1999 241	Figure not cited directly in the article; calculated by authors from other figures presented in the article.

Figure 1: Summary of literature on percent of urban forest canopy that is street trees. Six citations were found that listed numbers, though one (Nowak 1994) listed multiple values, as described in the "Comments;" each of these values is listed separately in the table. (Source: complied by authors from cited sources.

Based on these conclusions, this study was undertaken with the intention to: (1) evaluate the contribution of street trees to canopy in Portland, Oregon; (2) identify possible spatial patterns associated with the street contribution; and (3) explore some of the factors that might help explain the pattern. Isolating the issue of street trees as a factor separate from the rest of the urban canopy can improve our understanding of how urban forest processes operate. The results will help identify differences between segments of the city, and can be helpful in performing comparative studies between Portland and other cities.

The Portland Landscape

The City of Portland lies at the confluence of the Columbia and Willamette Rivers (Figure 2). The Columbia River forms the northern boundary of the city, while the Willamette River flows through the city and marks a distinctive border between two very different parts of the city

Most of the city landscape west of the Willamette River is a system of steep-sided hills that rise 500 to 1,000 feet above the narrow strip of river flood plain adjacent to the river. The West Hills are heavily forested with a mix of native trees dominated by second-growth bigleaf maple, Douglas-fir, and alder. The flat lands below are heavily developed and have a much less dense canopy. The pattern of development on the low-lying areas may be divided into four roughly equivalent-sized zones. The northern two-quarters is predominantly industrial land, the next one-quarter south includes the historic downtown / central business district, and the southernmost one-quarter is largely industrial and commercial.

Land on the east side of the river is generally flat to slightly rolling, except for several remnants of isolated volcanic buttes. Much of the Eastside along the south shore of the Columbia River is industrial, as is most of the area on the east bank of the Willamette, together creating a swath of commercial / industrial land adjacent to the rivers that takes on the general shape of a northwesterly-pointing "V" (Figure 3). The width of the industrial-V varies and engulfs the largely residential area of the Eastside that is home to the majority of Portland's population. Throughout the residential area are small pockets and strips of commercial development, the most prominent being the Lloyd District, a high-rise development that represents the leap-frogging of the downtown across the river from its

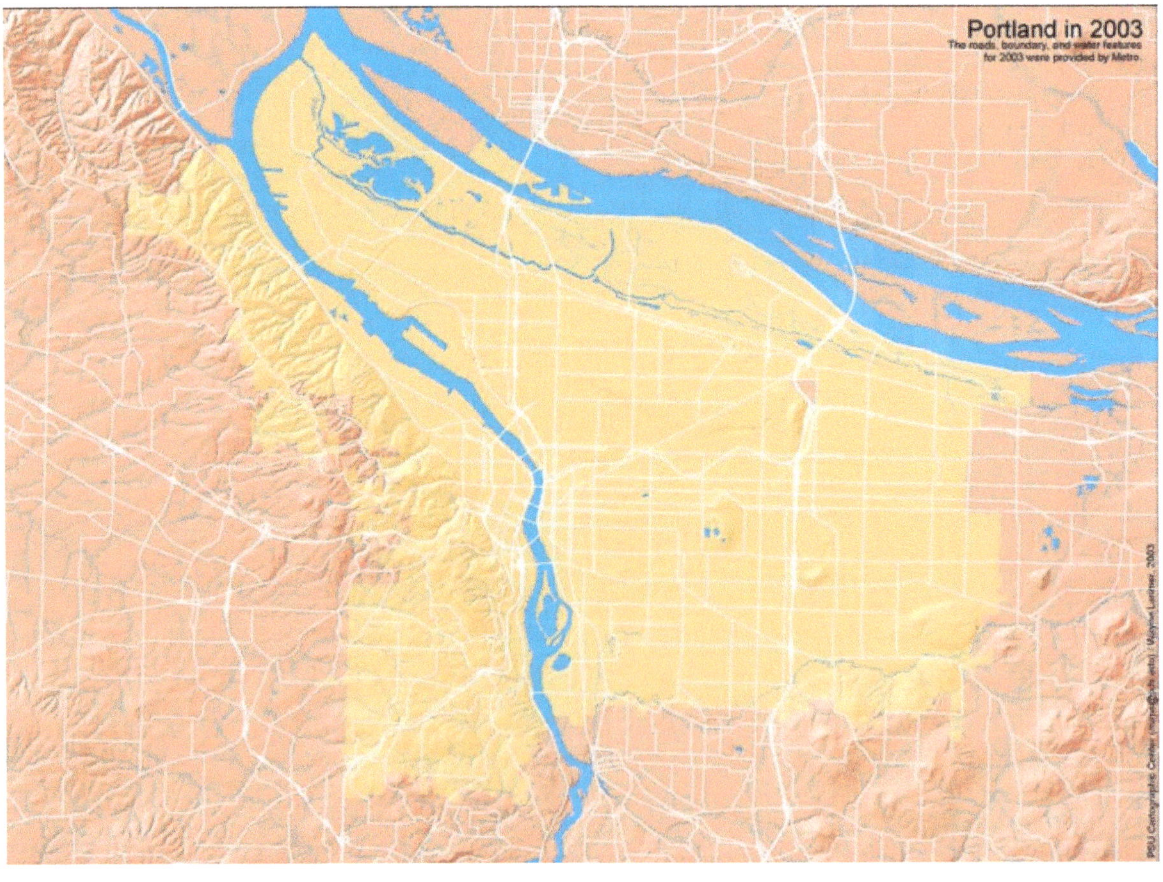

Figure 2: The City of Portland has two distinct physical characteristics that are illustrated by this map. First is its situation at the confluence of the Columbia and Willamette Rivers. The Columbia River forms the northern boundary of the city, while the Willamette River flows through the city dividing it into two distinct, though unequal-sized, portions

This division is echoed by the City's second critical physical characteristic, its site character relative to topography. The area west of the Willamette River is dominated by a north-northwest trending linear system of hills rising steeply about 1,000 feet above the narrow, heavily-developed river terrace. The east side of the river is flat to slightly rolling, but punctuated by several remnant volcanic buttes. (Map by Wayne Larimer.)

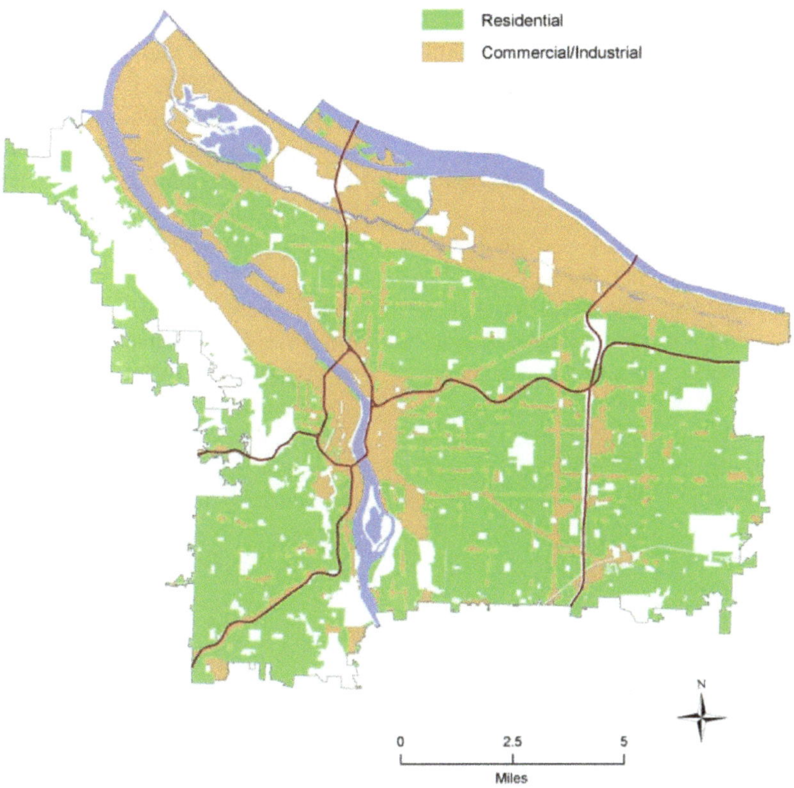

Figure 3: The study area examined in this report is portrayed in color to indicate the associated urban land environment. Only residential and commercial / industrial areas were examined. The study area thus consists of the City of Portland minus areas in parks, rights-of-way associated with Interstate highways, and rivers within the city limits – a total of 77% of the city area. Note the northwesterly-pointing "V" of commercial / industrial land adjacent to the rivers confining historic location west of the Willamette.

The pattern of relief in the city – with the linear feature of the West Hills dominating the western portion, and the smaller but visually prominent isolated hills on the east side – plays a potentially important role in the pattern of tree canopy over the city. Much of the densest canopy in the city is found on the steeper, landslide-prone, and therefore more difficult to develop parcels of the West Hills. In contrast, the buttes to the east have more stable slopes and are being rapidly developed as view-home properties, thus losing their previous dense tree cover.

Methodology

The basic methodology of this study involved calculating the street tree component (STC) within a sample of small areas throughout the city, identifying a generalized pattern in the results, and considering explanations for the pattern. The processes involved included visual image interpretation, digital image analysis, field observation, development of several GIS layers, mapping and statistical tabulation of data. Image processing and GIS analysis was performed using Idrisi Kilimanjaro software and ArcGis was employed for map composition. The methodology is discussed here in the text as eight major steps:

1. Preparing an Urban Land Environments map
2. Selecting sample patches
3. Visually interpreting street trees
4. Assessing the visual interpretation accuracy
5. Digitally interpreting total canopy of the city
6. Assessing accuracy of the digital interpretation
7. Calculating street tree contribution to canopy
8. Identifying patterns and generalizing results

Step 1: Preparing an Urban Land Environments (ULE) map

Metro, the regional government for the Portland area, provides basic land use data through its Regional Land Information System (RLIS). The classifications of land in their data base were used to delineate ULEs for this study. Two ULEs were evaluated in this study – residential, constituting 51.5% of the city, and commercial/industrial, constituting 25.5% of the city. We expected that the percentage of street trees comprising the canopy within these two classes would be substantially different. Land that was part of the parks ULE and the natural areas ULE was not considered in the analysis, nor were several broad strips of rightof-way associated with the Interstate highways. Together with portions of the rivers within city limits, these excluded areas constituted 23% of the city's physical area. Figure 3 portrays the portions of the city actually included in the study.

Step 2: Selecting sample patches

The narrow linear character of street tree plantings means that they are not distinguishable at the moderate-to-coarse spatial resolution of scanning systems such as the Landsat MSS and

Thematic Mapper. To obtain the higher resolution that was called for, one-meter resolution digital aerial camera images of the city, acquired in June of 2002 by the City of Portland Bureau of Environmental Services, became the basis for the study. Three bands – green, red, and near infrared – were used for the analysis. At one-meter resolution, the remote sensing imagery constituted a large data set, which necessitated a sampling approach to the analysis. A sample patch size of 200 meters x 200 meters (40,000 pixels) was decided on. This area represents about 5-7 blocks, depending on the block size in the neighborhood. A grid of 200 meter squares was overlaid on the digital data and 200 random sample patches selected. (See Appendix 1 for a discussion of digital image size issues.) The sample patches were then overlaid with the ULE delineations and those patches that included just residential land were retained to constitute the residential area sample. For commercial areas a somewhat different approach was used. Most commercial strips are too narrow to completely fill the sample patch size of a 200 meter x 200 meter square. Accordingly, an allowance was made to keep samples in which the patch included a large proportion of a commercial strip. In these instances only the commercial portion of the sample patch was considered and included in the analysis, making those patches smaller than the 200 meter x 200 meter size. Any sample patches that fell outside the City limits, within parks, in water, or were split by large highways or freeways, were eliminated from the sample. The final result was a total of 62 sample patches – 37 residential areas and 25 commercial / industrial areas – that fit the sampling criteria. The distribution of sample patches by neighborhood is portrayed in Figure 4. The total area included in the 62 sample patches comprised about 0.5% of the study area. These sample patches constituted the foundation for the visual interpretation and analysis. (Appendix 2 contains the sample data.)

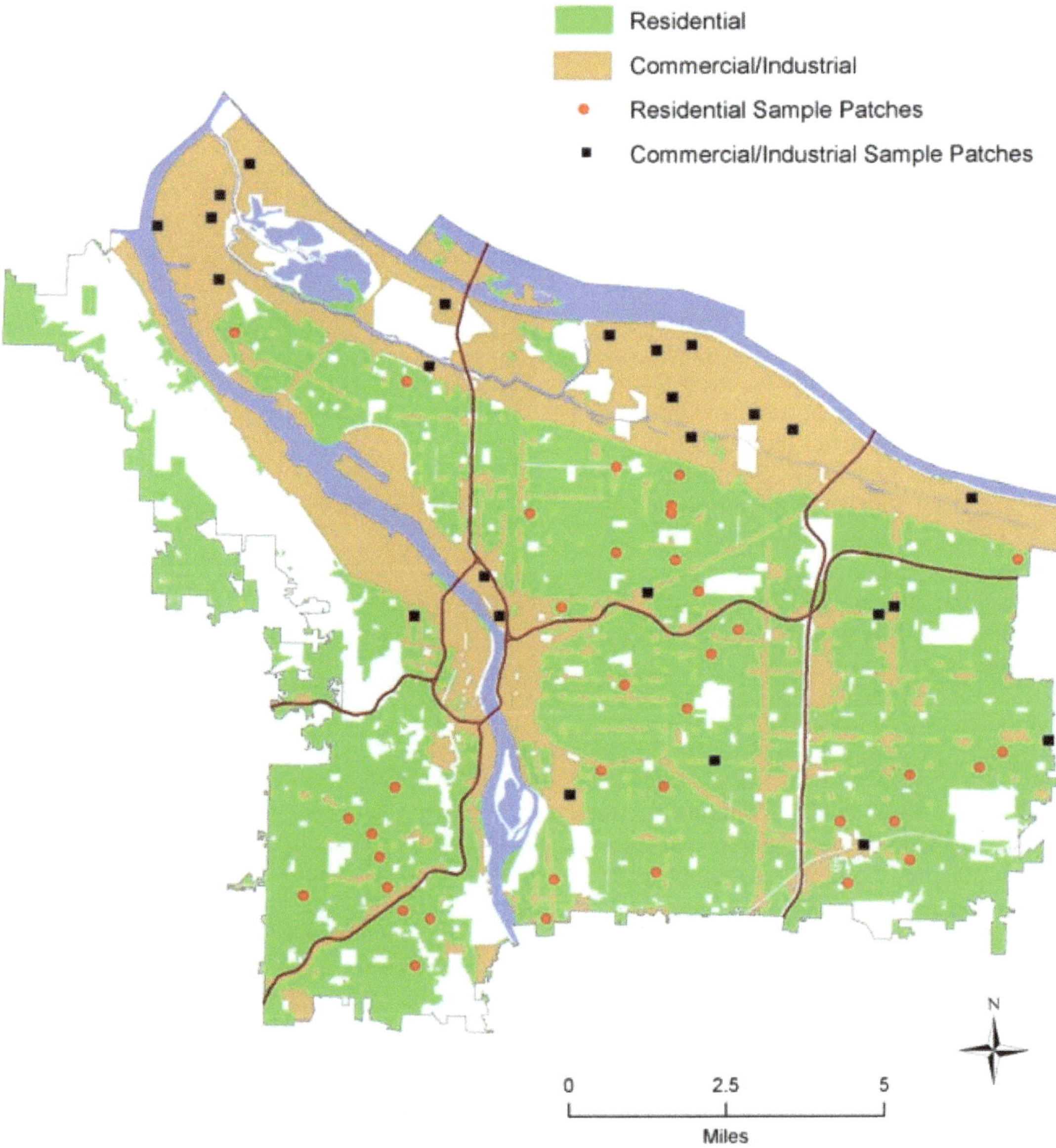

Figure 4: The 62 sample patches used to visually interpret and digitize canopy cover. Patches represent 200 meters x 200 meters on the ground but are shown here at an exaggerated size for legibility.

Step 3: Visually interpreting street trees

The canopy for street trees was interpreted visually from false color composite displays on the computer screen and digitized as polygons. Figure 5 portrays two examples of the results of the visual interpretation process. This visual interpretation of street tree canopy was later used in conjunction with data on the total area covered by tree canopy to calculate the STC – the percentage of total canopy area that is represented by street tree canopy – for each of the 62 sample patches. Distinguishing vegetation from non-vegetation areas is a generally straightforward visual interpretation process when using color infrared data. Likewise, trees can usually be separated from grass by a combination of tone and texture. However, when attempting to distinguish street trees as a sub-class of woody vegetation there are several persistent conditions that introduce subjectivity into the interpretation. First, there is no clear spectral division that separates what is a shrub or bush from what is a tree. In general, trees can be distinguished from large bushes by shape and from small bushes by shadows, but these shape and size clues are not always conclusive. A second problem is that shadows cast by the trees obscure edges and can make outlining the exact canopy boundary difficult. This condition is especially vexing for large trees, which cast broad shadows: the edge of the tree is somewhere in the shadow but, depending on the shape / form of the tree, the shadow may be mostly on the ground or mostly on the tree, or somewhere in between. In such cases the interpreter can do nothing but make their best subjective determination. A third problem is that there are areas where the canopies of street trees and yard trees have grown together and are continuous. Here it is sometimes difficult to discern that the canopy includes both street and yard trees, and, in instances where it is clear that both occur, it is sometimes difficult to identify where to separate the two. None of these three problems is overwhelming, but together they serve to complicate the interpretation process and introduce a potential source of error. In industrial areas, the generally small amounts of vegetation – both in terms of total area and in terms of the sizes of individual canopy patches – works to make the visual identification of street trees easier. The same holds true for commercial strips, where street trees tend to be the largest segment of what is usually a small amount of vegetation. Most plantings in both commercial and industrial areas are in orderly rows, making identification easier by bringing in the added visual clue of pattern. Of particular help in commercial and industrial areas is the frequently high proportion of impervious surface, which provides a spectral background against which trees strongly contrast

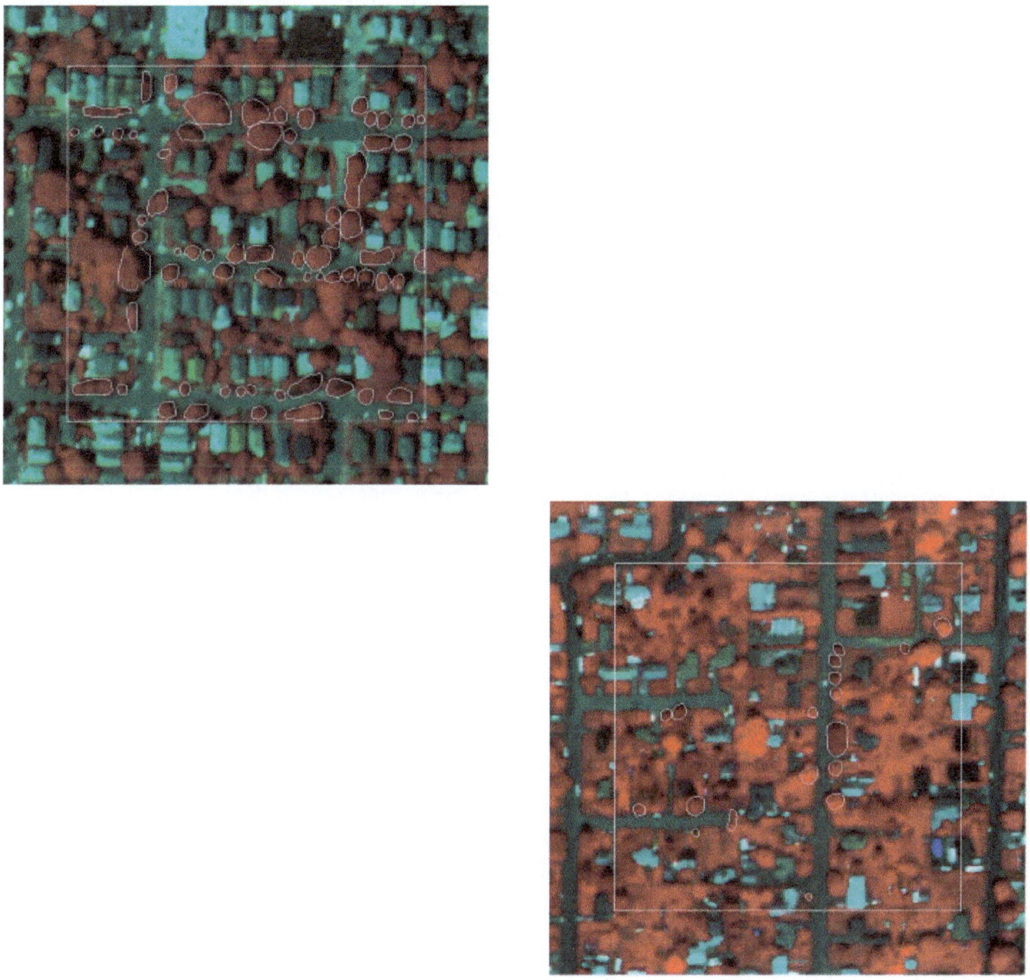

Figure 5: Examples of image interpretation of street trees in two different residential neighborhoods of Portland. Each white box represents a 200 meter x 200 meter square sample patch within which street trees were delineated. For each sample patch the total area of delineated street trees was then calculated.

For residential areas, the ease with which right-of-way areas and associated street trees could be distinguished from all other non-street trees varied by neighborhood. In this regard, the presence or absence of sidewalks was the most important factor. Three geographically distinct types of residential area could be identified, and each provided its own unique challenges during the interpretation process. First, the North Portland and the Inner Eastside neighborhoods have a conventional grid layout and sidewalks. Here, there was little confusion between street trees and yard trees. The challenge in these neighborhoods was the substantial numbers of small street trees that were difficult to detect on the imagery. The second type of residential area was found in Southwest Portland. This area is still heavily forested, even in developed portions with many winding streets and few sidewalks. In these neighborhoods, it could be difficult to distinguish between street and yard trees. The third type of residential area was found in the Outer Eastside of Portland. Here there is a conventional grid layout, but frequently no sidewalks and a higher proportion of undeveloped properties than the Inner Eastside. The right-of-way can be difficult to define, and distinguishing street trees from yard trees could be difficult. Despite the subjectivity introduced by these persistent conditions, the visual interpretation process proceeded smoothly and was completed with a high degree of confidence.

Step 4: Assessing visual interpretation accuracy

The accuracy of the visual interpretation method was checked by conducting an on-theground tree inventory for 12 of the 62 sample patches. A total of four commercial / industrial and eight residential patches were selected, providing data on 349 residential street trees and 133 commercial / industrial street trees. Two sources of error were identified. First, even with the high resolution imagery, some young or columnar street trees were too small to be identified. In some sample areas, these small trees were as much as one third of the count of street trees. The accuracy assessment indicated that the visual interpretation process missed 21% of the total count of trees that were observed in the field. Note that this error rate refers to individual trees, not canopy. Partially mitigating this problem is the fact that these trees are so small that, even in total, they comprise a very small percentage of the total canopy cover. Thus, this 21% missed tree figure does not imply a 21% canopy area error.

The second source of error was occasional confusion between street trees and yard or private property trees. This number varied by type of neighborhood, with a 76% accuracy rate in neighborhoods without sidewalks and an 87% accuracy rate in neighborhoods with sidewalks. This problem was most common in Southwest Portland and the Outer Eastside, both areas having many streets without sidewalks. Additional analysis of the accuracy of the visual interpretation is provided by examining the accuracies with regard to ULE. Figure 6 presents a graphic portrayal that assists in understanding this analysis. For residential areas, 79% of the street trees observed in the field were identified by the interpretation. And, of all the residential area trees that were identified by the interpretation as street trees, 83% were correctly identified as street trees. In commercial / industrial areas, 90% of the street trees observed in the field were identified by the interpretation. And, of all the commercial / industrial area trees that were identified by the interpretation as street trees, 92% were correctly identified. These accuracy figures are conservative, since they are based on counts of trees, not extent of canopy, and almost all the street trees that were missed or misidentified were small. As would be expected from raster sampling theory, the threshold for being able to detect small trees was about 2 meters x 2 meters, or four times the area of the pixel. Trees below this size were frequently not distinguishable on the one-meter pixel images. Figure 7 presents the accuracy data in tabular form and makes the case that the visual interpretation likely results in a slight overestimation of the amount of street tree canopy. The actual amount of likely overestimation is unknown, but is likely to be in the range of 3 to 6%.

Step 5: Digitally interpreting total canopy

The next step was to perform a digital interpretation of the one-meter digital data and produce a map of land cover types. The necessity to use high resolution data is well recognized in the remote sensing literature. Jensen and Cowen (1999) note that high resolution imagery is required in order to effectively study particular urban attributes. However, due to the relatively recent availability of high resolution digital imagery, urban land use classification with high resolution digital imagery has not been done a great deal

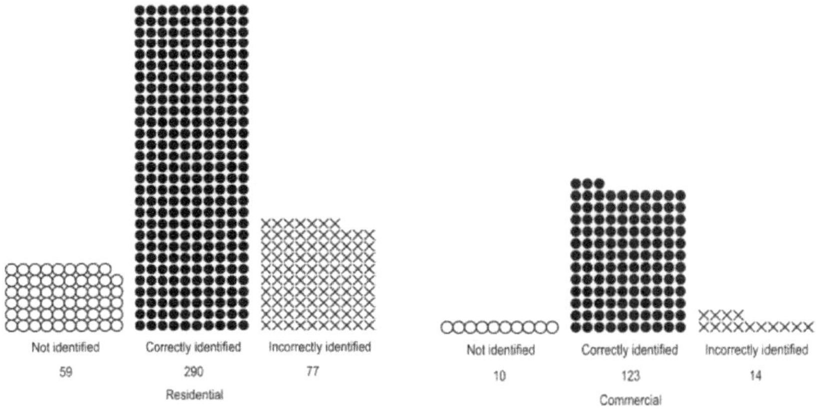

Figure 6: Accuracy of the visual interpretation of street trees as compared with field observations. Black circles represent street trees that were correctly identified through visual interpretation. Open circles represent street trees that were missed / not identified through visual interpretation. The x's represent non-street trees that were incorrectly identified through visual interpretation as street trees.

In residential areas, the visual interpretation correctly identified 83% (290/349) of the street trees observed on the ground. The visual interpretation failed to identify 17% (59/349) of the street trees. Of those trees that the visual interpretation identified as street trees, 79% (290/367) actually were, producing an incorrect identification in 21% (77/367) of the cases. In commercial / industrial areas, the visual interpretation correctly identified 92% (123/133) of the street trees observed on the ground. The visual interpretation failed to identify 8% (10/133) of the street trees. Of those trees that the visual interpretation identified as street trees, 90% (123/137) actually were, producing an incorrect identification in 10% (14/137) of the cases. Note that these figures refer to counts of trees, not canopy area. Many of the errors involve small trees that, in the aggregate, do not constitute a large portion of the canopy

	Residential	Commercial / Industrial	Combined / Total
# Trees - Total	349	133	482
# Correctly ID'd	290	123	413
% Correctly ID'd	83%	92%	86%
# Omitted / Missed	59	10	69
% Omitted / Missed	17%	8%	14%
#ID'd as Street Trees - Total	367	137	504
# Incorrectly ID'd	77	14	91
% Incorrectly ID'd	21% (77 / 367)	10% (14 / 137)	18% (91 / 504)

Figure 7: Summary of the accuracy and error rates for the visual interpretation of street trees. In both the case of the residential category and the commercial / industrial category there is a higher rate of commission errors (i.e., incorrectly identifying a non-street-tree as a street tree) than omission errors (i.e., failing to identify a street tree).

From this we would conclude that the visual interpretation leads to a slight over-estimation of street tree canopy, since (a) there are more trees incorrectly identified as street trees than there are trees omitted / missed, and (b) the trees that are incorrectly identified are larger and contribute more to canopy than the missed smaller trees.

(Zhang 2001; Myeong et al 2001.) Going to higher resolution imagery introduces some problems that are not obvious beforehand but quickly become apparent once the analysis is undertaken. The critical issues are: (a) the high resolution of the data means there likely are different technical issues to consider compared to using medium resolution imagery; and (b) there is a very limited body of work to rely upon for urban land use classification with high resolution imagery. To calculate the percentage of street trees, we first performed a digital classification of total canopy for the entire digital data set. If our goal was simply to identify the vegetated areas of the data set, a vegetation index using a ratio of the near infrared band to the red band would likely suffice (Lillesand and Kiefer 2000). However, in order to distinguish between different categories of vegetation a different method was needed. A supervised approach using a minimum distance to means classifier and the near-infrared, red and green bands was employed. A total of 67 training sites were used for the classification. Three land use/cover classes were identified: trees comprised one class, grass and shrubs a second class, and all other land use types (impervious surfaces, water, bare soil) were combined into a third class, here called non-vegetated surfaces for convenience. An example of the resulting classification map for a sample patch is shown in Figure 8. The results of the final classification are summarized in Figure 9. For the whole study area, non-vegetated surfaces predominate with 59% of the area. Breaking this figure into commercial / industrial (76.8%) and residential (48.2%) clearly illustrates the difference in potential tree planting space between these two land covers. And that difference in potential clearly shows up in the tree category where the 29.2% residential tree canopy represents a value 6.2 times the 4.7% of the commercial / industrial category.

Step 6: Assessing accuracy of the digital interpretation

Accuracy assessment relied on a random set of pixels and the use of visual interpretation to determine the "true" identification of each pixel. From that data, an error matrix was prepared and used to calculate the accuracy of the digital classification. The results are presented in Figure 10. The overall accuracy for all of the digitally identified classes is 90%. With only a single composite class for pavement, buildings, water, and bare soil, it is not surprising that the accuracy for identifying that class is a high value of 96%. It is clear that vegetated features can be very reliably distinguished from non-vegetated (provided they are large enough to be captured by the pixel resolution).

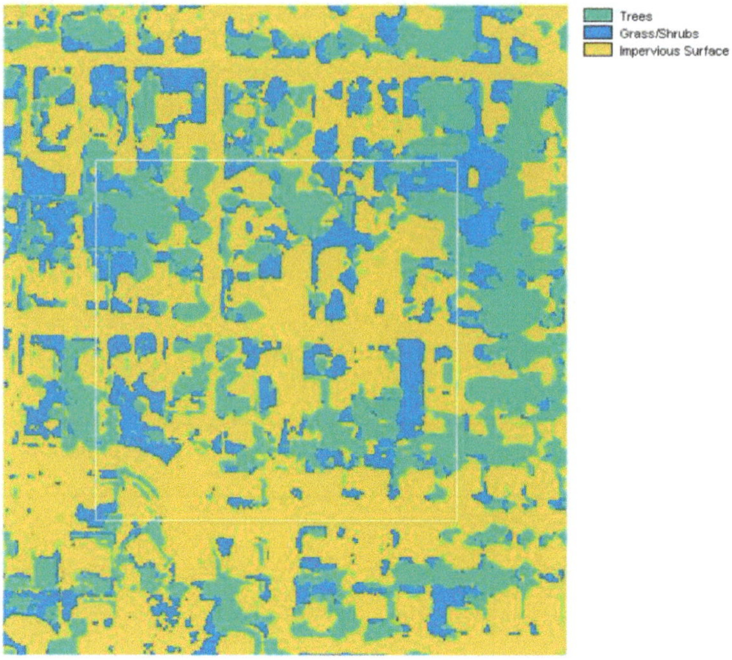

Figure 8: Land cover classification for a sample patch in Northeast Portland. The white box represents a 200 meter x 200 meter square sample patch within which three different land covers were delineated.

	Tree Canopy	Grass/Shrub	Nonveg Surface
Study Area - Residential	29.2	22.6	48.2
Study Area - Commercial/	4.7	18.5	76.8
Study Area - Total	19.9	21.0	59.0

Figure 9: Summary of the digital classification for the whole study area. Note the high percentage of non-vegetated surface in the commercial / industrial areas compared to residential areas – 76.8% versus 48.2% – which represents a 1.6 times difference in the potential for tree planting. This can be compared to the 29.2% versus 4.7% tree canopy cover for residential versus commercial / industrial areas – a 6.2 times difference. When also compared to the figures for grass, it is clear that residential areas exhibit not only a greater area of overall vegetation cover, but also a much higher disposition to plant trees than do commercial industrial areas.

		Visual Interpretation				
		Trees	Grass	Non-Veg	Row Total	**User's Accuracy**
Digital Classification	Trees	53	7	1	61	**87%**
	Grass	6	25	0	31	**81%**
	Non-Veg	3	1	92	96	**96%**
Column Total		62	33	93	188	Overall = 90%

Figure 10: Error matrix for the digital land cover classification. Numbers represent the count of individual pixels examined in each category. The shaded diagonal cells represent agreement between the digital interpretation and the visual interpretation. The User's Accuracy represents the reliability of the digital interpretation performed on the sample of patches, as compared with the visual interpretation. This ranged from 81% for grass, to 87% for trees and a high of 96% for non-vegetated surfaces.

The accuracies of 87% and 81% for trees and grass, respectively, while not as high as for the non-vegetated class, are still good. The error pattern for both of these classes indicates the lowered reliability of distinguishing sub-classes of vegetation. The signatures for trees covered a wide range of values and were occasionally confused with the signature for grass / shrub, which had a much smaller range of values. To a smaller degree there was also some conflation between non-vegetated surfaces and shadow areas of trees, especially for conifers, which are frequently much taller than deciduous trees and cast longer shadows. Note that there are few firm guidelines on what is an "acceptable" accuracy for a remote sensing problem. One figure frequently quoted in the literature for land use / land cover analysis is 85% (Anderson et al 1976). Using this figure as a standard, the 81-96% range for accuracy in this situation would appear to be acceptable. Commercial / industrial areas had varying amounts of canopy cover vegetation, but a consistently small percentage of these areas were vegetated and frequently the trees were small and potentially not identified. For two of the sample patches, total tree canopy was determined using the same visual identification and digitization method that was used for street trees. One case was when the total tree canopy was 1-2% of the total number of pixels in the sample patch. The digital interpretation tended to underestimate total tree canopy (due to the shadow problem) for commercial / industrial, where it overestimated total tree canopy for residential (the grass problem is bigger than the shadow problem). The other case where visual interpretation was used was when looking at a commercial strip instead of the entire sample patch. Thus, for the commercial / industrial areas included in the table, it is a combination of the digitally and manually digitized values that are included in the table.

Step 7: Calculating street tree contribution to canopy

Once we had digitally classified the land cover for the study area, we were then able to calculate an estimate of the street tree contribution to canopy. For each of the 62 sample patches the area of the patch delineated as street trees was divided by the area of total tree canopy from the digital land cover analysis. A breakdown of the findings for tree canopy into street trees and non-street trees is presented Figure 11. Four key observations can be made.

1. A total of 3.4% of the study area consists of a land cover of street trees and 16.5% nonstreet trees. While it is possible to add these two figures to obtain a total for trees of 19.9%, it must be remembered that the total figure – like the street tree and non-street tree figures – refers only to the study area of residential and commercial / industrial land, not to the entire city. This study did not include park areas, but if they were added in, the city-wide canopy figure would be higher than 19.9%. It is important that all of these figures be used with caution, and accompanied by a clear description of what they represent.

	Street Trees	Non-Street Trees	All Trees *(Street plus Non-Street)*	**Street Tree Component**
Study Area - Residential	4.8	24.4	29.2	**16.6**
Study Area - Commercial/	1.1	3.6	4.7	**23.7**
Study Area - Total	3.4	16.5	19.9	**17.2**

Figure 11: Percentages of the study area in street trees and non-street-trees. Note that although residential areas have a much higher percentage of overall tree canopy than do commercial / industrial areas (29.2% versus 4.7%), the street tree component of the urban canopy is higher for commercial / industrial areas (23.7%) than for residential (16.6%).

2. The percentage of the study area that consists of street tree canopy varies with the land cover category. The city-wide figure for commercial / industrial is only 1.1 % and for residential 4.8%. While neither figure is very high, note that there is more than a four times difference between them. Residential area street trees clearly represent a greater portion of the city area than do commercial / industrial area street trees.
3. The percentage of total tree canopy contributed by street trees varies with the land cover category. In residential areas 16.4% (4.8 / 29.2) of tree canopy comes from street trees, while in commercial / industrial areas the figure is 23.4% (1.1 / 4.7). Thus, street trees are proportionally a greater contributor to overall canopy in commercial / industrial areas than in residential areas.
4. The percentage of street trees as part of the total study area canopy cover – 17.2% – is in the middle of the 9.5 to 24% range quoted in the literature.

Step 8: Identifying patterns and generalizing results

Figure 12 presents three different approaches to viewing tree canopy – street trees as a percent of total area, street trees as percent of total canopy, and total canopy as a percent of area. Some notable variation in street trees is seen between different residential areas. The Southwest group of neighborhoods with 6.1% of their area in street trees is the highest, while Outer Southeast with 2.8% is the lowest. The general pattern of tree densities follows an east-west gradient, with lower values to the east and higher values to the west.

Visual examination of the data and field observation experiences led to dividing the residential areas into the five neighborhood groups indicated on Figure 13: Inner Northeast, Inner Southeast, Outer Eastside, Southwest, and Northwest. These divisions closely follow the locally recognized but informal terminology for grouping neighborhoods of the City. Interstate 205 serves generally as the dividing line between Inner and Outer Eastside. The underlying logic of these groupings was to keep individual neighborhoods intact and to group them together based on similar canopy densities. Since they exhibit similar

		Street Tree Component (STC)		Total Tree Canopy	
	Street Trees - Percent of the Area	Percent	Max-Min Values	Percent	Max-Min Values
Inner NE - Residential	5	22.6	6.8 - 42.8	22.1	12.7 - 30.4
Inner SE - Residential	5	23.2	7.0 - 38.6	21.6	12.1 - 35.4
Outer East - Residential	2.8	10.4	6.4 - 14.6	27.2	10.5 - 49.7
SW - Residential	6.1	12.9	3.0 - 26.8	47.1	21.0 - 61.5
NW - Residential	--	--	--	--	--
Study Area - Residential	4.8	16.6	3.0 - 42.8	29.2	10.5 - 61.5
Study Area - Commercial/	1.1	23.7	0 - 84.3	4.7	0 - 31.1
Study Area - Total	3.4	17.2	--	19.9	--

Figure 12: Three perspectives on tree canopy percentages in residential and commercial / industrial areas of Portland. No values are given for residential in NW because the amount of residential land is small and much of it is a mixed use combination with commercial dominating on the streetscape. In this unique situation it was felt that residential figures would not be meaningful.

The first data column represents street trees as a percentage of total area, characterized by a relatively narrow range of single-digit figures. The second column provides the figures for street trees as a percent of total tree canopy, with a range form 10.4 to 23.7%, depending on the geographical area being considered. The third data column represents total canopy cover in the area, with a very wide range of 4.7 to 47.1%. This column serves as the denominator when calculating street trees as a percentage of the total canopy.

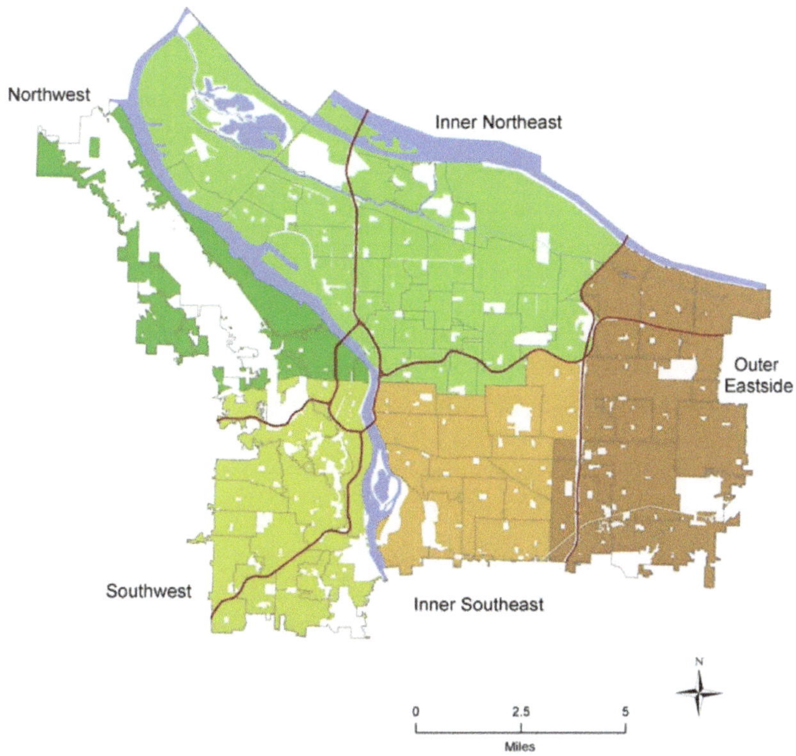

Figure 13: Based on a variety of factors examined during this study, a regionalization was made of Portland's residential areas as five neighborhood groups. The regions closely follow locally recognized, informal terminology for grouping neighborhoods of the City. The underlying logic of these groupings was to keep individual neighborhoods intact and to group them together based on similar canopy densities. The divisions are based primarily on the data examined in this study, but also include both a degree of subjectivity and a degree of generalization. While not definitive, this regionalization provides a useful initial attempt at understanding the pattern of street trees in Portland.

Composite Inner Eastside for the sake of discussion. The northwest section of Portland is dominated by Forest Park and industrial areas with relatively small residential clusters, and there were no residential samples from this area.

The Inner Eastside residential areas have the highest street tree contribution to the urban forest canopy – 22.6% in Inner Northeast and 23.2% in Inner Southeast. These areas also exhibited high variation in street tree planting, with some blocks dominated by large street trees and other blocks almost devoid of street trees. Sidewalks typically delineate the right of way, but the width of the right of way is variable. The mean tree canopy itself was fairly consistent and uniform throughout. The Outer Eastside residential areas have the lowest street tree contribution to the urban forest canopy (10.4%). There was quite low variation in the percentage of street trees although the total tree canopy was variable. These two findings can be attributed to the combination of relatively new developments and a larger number of undeveloped parcels compared to the Inner Eastside. These residential areas rarely have a well-defined right of way. The Southwest residential areas also have a fairly low street tree contribution to the urban forest canopy (12.9%), but for different reasons than the Outer Eastside. These neighborhoods are much more heavily forested than the others, with many developments carved out of closed canopy stands. As might be expected for such an environment, street trees are a small percentage of the total canopy. The general pattern of street trees as a component of the total canopy takes on an invertedU shape, with higher values in the center of the map and lower values to the east and to the west. The total tree canopy category takes on just the opposite pattern – a standard-U – with the lowest values in the center of the map and the higher values to the east and west. Industrial land exhibits a variety of spatial configurations but is, by definition, dominated by non-vegetated surfaces such as pavement and buildings. There is no consistent repeating pattern in the proportions of treed areas to other vegetation covers. In some areas, tree plantings, especially street trees, are the only vegetation. This ULE also incorporates undeveloped land, which varies from bare soil to heavily vegetated, and that vegetation is much more likely to be grass or shrub than tree. Given the extensive areas associated with this ULE, the sample unit size is probably too small for detailed analysis of this type of land cover.

Commercial properties are of two basic types: strip development along streets, and large properties such as malls with a minimal street component. For the strips, street trees are an important component. For commercial / industrial areas, unlike for residential areas, there was no apparent relationship between percentage of street trees and location within the city

RESULTS

This study set out to do three things:

(1) evaluate the contribution of street trees to canopy in Portland, Oregon;

(2) identify possible spatial patterns associated with the street contribution;

(3) explore some of the factors that might produce and help explain the patterns.

The analysis indicates that street trees constitute 3.4% of the whole study area. In commercial / industrial areas the figure drops down to only 1.1% and in residential areas rises to 4.8%.

For the overall study area the street tree component of the canopy is 17.2%, a figure that falls in the middle of the range of figures found in the published literature. Going a step further and considering the land cover characteristics, we found a figure of 23.7% in commercial / industrial areas and a figure is 16.6% in residential areas. Clearly we can see that in residential areas – which are overall better treed than commercial / industrial areas – street trees contribute proportionally less to overall canopy than they do in commercial / industrial areas.

In general, as the amount of non-vegetated surface increases, the percentage of street trees within the total tree canopy also increases. Logically, if there is less area in general available for planting, the right-of-way zone becomes a more significant area for planting. Conversely, more vegetation in general would mean that street trees would be a smaller component of that vegetation, since they are limited to the narrow right-of-way zones. While available planting space in the right-of-way may vary, it is much more constant than the highly variable amount of planting area available inside the block. A related trend indicates that as non-vegetated surface area increases, the total tree canopy as a percentage of vegetation increases. Conversely, as the amount of vegetation increases in a given area, the contribution of trees becomes smaller. No trends were noted related to the percentage of street trees or tree canopy to the amount of grass / shrub in a given area.

The percentage of street trees as part of the total canopy cover varies substantially from area to area for both residential and commercial / industrial land uses, although the latter has much greater variation. The results of this relatively small sampling of the city suggest that neighborhood characteristics influence the number of street trees as a percentage of total canopy cover in a given neighborhood. A part of this influence comes from topography as factor, where steep slopes tend to limit the kind of development to residential, and to reduce the potential density to lower figures than on flatter land. Based on numerical data and field observation experiences, the study area was divided into five residential neighborhood groups that reflect notable regional differences throughout the city. Three very generalized patterns emerge from the groupings:

1. The general pattern of street tree canopy as a percent of area seems to follow an eastwest gradient, with lower values to the east and higher values to the west.
2. The general pattern of street trees as a component of the total canopy takes on an inverted-U shape, with higher values in the center of the map and lower values to the east and to the west.
3. The overall tree canopy category takes on just the opposite pattern – a standard-U – with the lowest values in the center of the map and the higher values to the east and west.

Concluding Comments

Results of this study only apply to Portland (and, in fact, only to the 77% of the city included as part of the study area). The numbers found would likely be different depending on where the boundaries were drawn and the size of the delineated areas. The regionalization is clearly somewhat subjective, highly generalized and prone to variation based on even small changes in the boundaries. However, it represents a useful starting point for understanding the dynamics of an important and highly visible component of the urban forest. Isolating the issue of street trees as a factor separate from the rest of the urban canopy adds to our understanding of the character of the urban forest and how urban forest processes operate.

Where might this research go next? Topically, the relationship of street trees to income levels, age of the neighborhood, and percent non-vegetated surface are just a few of the variables that could be explored in detail at a more local scale. The potential role of topography in street tree canopy in Portland is an avenue likely worth pursuing. Steep areas of the city tend to have higher overall percentages of canopy, but also lower densities of roads and therefore a smaller percentage of street trees.

Methodologically, a larger / more detailed sampling by residential neighborhood would allow for improved estimation of trends, perhaps below the level of neighborhood. Commercial and industrial areas could also be further evaluated following a different sampling plan from residential areas. The method of digital image interpretation used in this study is fairly labor intensive, but necessary to accurately estimate something as detailed as street tree canopy. It is also quite accurate. Delineating street trees by computer algorithm is not realistic, since simply distinguishing between grass and trees during supervised classification is difficult. Previous detailed analyses of street trees have relied upon field inventories, which are much more labor intensive than the method employed in this study. Incorporating existing field inventories of street trees into the study could supplement image interpretation, improving its accuracy and reducing its labor-intensive nature. In terms of management, further study could focus on identifying neighborhoods that would be best served by tree planting programs, especially in the Outer Eastside. The high resolution imagery can be used to identify specific blocks within neighborhoods that could be targeted for planting (see Appendix 3 for additional comments on remote sensing issues). The results of this research could be helpful in performing comparative studies between Portland and other cities. Street trees are defined as those trees planted in the right-of-way, either in the parking strip delineated by the sidewalk, or in the absence of a sidewalk, a similar distance from the street (usually about 6 feet but often up to 12 feet). By virtue of their being planted in the right-ofway they (in most communities) come under the jurisdiction of the city forestry program. All other trees on private property are considered yard trees and are under control of the property owner. However, it is quite common for trees planted on private property close to the street to serve as functional street trees, i.e., the canopy of these trees covers part of the street, serving a similar role in providing canopy as would a definitional street tree – one planted in the right-of-way. This thought raises an interesting management issue: If what we are concerned with is canopy over the sidewalk and street, does it really matter whether the tree stem resides in the ROW or on private property? Pondering the answer to this question may provide some food for thought on how street trees are administered and how their planting is promoted.

Appendix 1: Image Data Tiles

The small pixels of the one-meter data resulted in large files, so the data were stored as five image tiles representing different flight lines and covering different sections of the city. Each tile was processed individually, resulting in five separate analysis files for the city.

Portions of the imagery outside the city were masked out of the images, as were parks and Interstate ROWs, resulting in five subsets of data files with the following sizes:

North & Northwest Portland & Forest Park	24,585 acres
Southwest Portland	10,852 acres
Downtown Portland & the inner eastside	42,591 acres
Outer eastside	32,127 acres
East Portland	3,340 acres

The sum of the tile areas is 113,495 acres (177.3 square miles), a total area that is several tens of square miles larger than the city. Although parks, interstate rights-of-way and the non-city areas around the edges of the tiles were masked out, there was substantial overlap between the tiles, and these overlap areas are included in the above tile size figures. When selecting the 62 sample patches, they were distributed among the five image tiles in proportion to the city area covered by that image tile. Since reflectance values varied somewhat between the five image tiles, each was classified separately. Between 12 and 16 training sites were applied to each image tile. The results of the final classification are summarized in Figure 9. Although the image tiles were analyzed separately, the samples were pooled to produce a composite accuracy assessment for the five image classifications. Random pixels were selected for each of the five image tiles and the accuracy of the digital classification determined by visual interpretation. The results are presented in Figure 10.

Appendix 2: Sample Patch Data

ID	Neighborhood	Cross Streets	Area	Area - in pixels						STC - %	Description / Notes
				Street Trees	Non-Street Trees	Total Trees	Grass	Total Vegetated	Non-Vegetated		
1_15	St. Johns	N Kellogg/ St. Johns	NE	1,475	4,444	5,919	12,978	18,897	21,103	24.9	grid
3_10	King	NE 8th/Skidmore	NE	3,288	4,760	8,048	9,749	17,797	22,203	40.9	grid
3_11	Beaumont- Wilshire	NE 35th/Siskiyou	NE	3,019	6,485	9,504	10,490	19,994	20,006	31.8	mix of grid and winding streets
3_12	Cully	NE 50th Pl/Going	NE	1,214	10,800	12,014	10,351	22,365	17,635	10.1	grid- dead end streets- no sidewalks
3_13	Cully	NE Prescott	NE	797	8,714	9,511	12,864	22,375	17,625	8.4	grid- no cross streets- sidewalk on arterial
3_14	Concordia	NE 34th/Ainsworth	NE	3,020	5,807	8,827	12,106	20,933	19,067	34.2	grid
3_15	Cully	NE 52nd/Simpson	NE	1,106	6,726	7,832	12,831	20,663	19,337	14.1	grid- dead end streets- no sidewalks
3_16	Sullivan's Gulch	NE 19th/ Clackamas	NE	2,765	7,292	10,057	6,384	16,441	23,559	27.5	grid
3_9	Kenton	N Hunt/Endicott	NE	549	7,486	8,035	10,572	18,607	21,393	6.8	grid
4_13	Rose City Park	NE 53rd/Siskiyou	NE	2,178	2,911	5,089	10,163	15,252	24,748	42.8	grid
4_14	Rose City Park	NE 60th/Tillamook	NE	2,515	9,631	12,146	8,390	20,536	19,464	20.7	grid
3_17	Mt. Tabor	SE 57th	SE	693	9,274	9,967	8,551	18,518	21,482	7.0	grid- no cross streets
3_19	Creston-Kenilworth	SE 33rd/Francis	SE	791	6,816	7,607	8,318	15,925	24,075	10.4	grid
3_20	Creston-Kenilworth	SE 50th/Cora	SE	2,322	5,019	7,341	9,653	16,994	23,006	31.6	grid
3_22	Sellwood	SE 17th/Lambert	SE	1,799	6,574	8,373	7,218	15,591	24,409	21.5	grid
3_23	Brentwood/ Darlington	SE 48th/Ogden	SE	1,368	5,003	6,371	9,649	16,020	23,980	21.5	grid
3_25	Sellwood	SE 16th/Marion	SE	3,263	6,748	10,011	9,044	19,055	20,945	32.6	grid
3_30	Buckman	SE 37th/Taylor	SE	5,464	8,676	14,140	6,088	20,228	19,772	38.6	grid
3_31	Center/Mt. Tabor	NE 65th/Burnside	SE	1,661	7,675	9,336	10,691	20,027	19,973	17.8	grid
4_4	Montavilla	NE 72nd/Pacific	SE	747	4,071	4,818	7,858	12,676	27,324	15.5	grid- half of area is school
4_8	Lents-Powellhurst-Gilbert	SE 108th/ Mitchell	OE	886	8,572	9,458	14,115	23,573	16,427	9.4	large properties with undeveloped land- no sidewalks
4_9	Powellhurst-Gilbert	SE Steele	OE	1,547	9,201	10,748	15,623	26,371	13,629	14.4	one street- several dead ends- no sidewalks
5_1	Wilkes	NE 162nd/Fargo	OE	1,818	11,940	13,758	5,006	18,764	21,236	13.2	includes apartment complex, new construction and undeveloped land
2_1	Hillsdale	SW Seymour	SW	1,885	18,970	20,855	8,870	29,725	10,275	9.0	winding streets- includes ~40% undeveloped land- no sidewalks
2_2	Hayhurst	Cullen / SW 39th	SW	3,204	21,399	24,603	4,838	29,441	10,559	13.0	winding streets- includes undeveloped land- no sidewalks
2_3	Hayhurst	Illinois/SW 30th	SW	3,409	13,471	16,880	7,072	23,952	16,048	20.2	grid- no sidewalks
2_4	Multnomah	Texas/SW 28th	SW	2,856	12,147	15,003	6,696	21,699	18,301	19.0	grid- includes church and undeveloped land- no sidewalks
2_5	Ash Creek	Garden Home/ SW 52nd	SW	3,475	18,384	21,859	7,823	29,682	10,318	15.9	mix of grid and winding streets- large properties- no sidewalks
2_6	Multnomah	Multnomah / SW 25th	SW	2,253	6,158	8,411	5,534	13,945	26,055	26.8	includes school- no sidewalks
2_7	Markham	Marigold / SW 19th	SW	2,112	14,054	16,166	8,062	24,228	15,772	13.1	grid- no sidewalks
3_24	Marshall Park	SW 8th Dr	SW	1,938	20,551	22,489	8,717	31,206	8,794	8.6	winding streets- includes ~30% undeveloped land- no sidewalks
3_32	Marshall Park	SW 14th Dr / Kari Ln	SW	698	22,572	23,270	6,248	29,518	10,482	3.0	winding streets- large lots- no sidewalks

Commercial / Industrial Sample Patches

Sample	Neighborhood	Cross Streets	AREA - Pixels				STC - %	Description / Notes
			Street Trees	Total Trees	Vegetated	Non-Vegetated		
1_1	St. Johns	None	-	-	-	40,000		open space- no trees
1_2	St. Johns	None	522	651	3,016	36,984	80.2	industrial- riverside
1_3	St. Johns	N Ramsey	68	335	4,282	35,718	20.3	industrial- includes only part of one street
1_5	St. Johns	None	-	-	896	39,104		industrial parking area- no trees
1_14	St. Johns	N Marine Dr/ Leadbetter	102	241	1,461	38,539	42.3	industrial- half undeveloped with road intersection
3_18	Reed	SE 24th/Pardee	-	-	-	40,000		no trees
3_26	Northwest District	NW 22nd/ Lovejoy	2,564	3,043	7,489	32,511	84.3	commercial strip- most of sample area is a hospital but zoned residential
3_27	Eliot	N Ross/Page	579	1,110	3,224	36,776	52.2	industrial
3_28	Lloyd District		217	827	2,892	37,108	26.2	Rose Garden
3_29	Hollywood	NE 44th/Sandy	1,614	2,233	6,119	33,881	72.3	commercial strip- includes ~10% residential
3_4	Kenton	Expo TC	1,736	4,737	7,245	32,755	36.6	mostly undeveloped
3_5	Airport	None	-	-	39,939	61		open space- no trees
3_6	Airport	None	-	-	-	40,000		no trees
3_7	Airport	None	-	-	5,438	34,562		no trees
3_8	Kenton	N Wilbur/ Newark	37	217	643	39,357	17.1	industrial- one street- few trees
4_15	Parkrose Heights/ Hazelwood	NE 118th/ Halsey	355	866	X	X	41.0	commercial strip- includes ~30% residential
4_16	Russell	NE 122nd	336	1,603	1,156	38,844	21.0	commercial
4_19	Powellhurst- Gilbert	Unnamed roads	-	12,457	25,987	14,013	0.0	mostly open space- no street trees
4_2	Airport	NE Cascades Pkwy	1,685	2,645	27,524	12,476	63.7	industrial- mostly undeveloped- one road with trees
4_20	South Tabor	SE 67th/Powell	1,085	1,725	X	X	62.9	commercial strip- most of area is residential
4_21	Cully	NE Skyport Wy	511	9,202	17,642	22,358	5.6	industrial- includes portion of Columbia Slough and open space- only part of one street
4_22	Airport	Perimeter Rd	-	-	21,271	18,729		mostly open space- no trees
4_3	Airport	Perimeter Rd	-	-	17,522	22,478		mostly undeveloped- no trees
4_5	Argay	NE 148th	-	2,683	10,306	29,694	0.0	industrial- no roads- mostly undeveloped
5_2	Centennial	SE 174th/Division	871	1,753	X	X	49.7	commercial strip- includes ~20% residential

Appendix 3: Remote Sensing Issues

Although the use of high spatial resolution imagery is necessary for the type of detailed analysis this project entailed, there are a number of challenges inherent in the use of highspatial resolution imagery (compared with moderate or low-resolution imagery) for land use classification. The complex heterogeneous nature of the urban environment is a problem for classification processes in general, but even more so at one-meter resolution. A large number of edges and mixels / mixed-pixels are inevitable, especially along the many changes between vegetated and non-vegetated areas. Although post-classification filtering can be used to decrease the misclassification that results from the high number of edges, those processes were not applicable to the detailed nature of this study. Identifying isolated and small groups of pixels of a distinct and different class was the goal and these "salt and pepper" effects could not to be removed.

The number of pixels corresponding to an individual tree often numbers in the single digits. The resulting preponderance of edges leads to misclassification of individual pixels. However, each sample patch was comprised of 40,000 pixels, and the classification exhibits improved accuracy over this larger area as errors of commission tend to balance errors of omission. For example, some grass areas are classified as deciduous trees and increase canopy cover. However, some types of deciduous tree have signatures coincident with grass and are not counted as part of the canopy.

The same kind of balance is found in shadow areas. In high resolution imagery, shadow effects are pronounced. Compared with lower resolution satellite images where the shadows comprise a small percentage of any given pixel, high resolution imagery will have many individual pixels completely encompassed by shade. The shadow areas for buildings conflate with the signatures for conifer trees, and this tends to increase canopy cover. On the other hand, shadowed portions of deciduous tree canopy may be misclassified as non-vegetated surface and not counted as part of the canopy.

Part of what allowed this study to be performed was the availability of one-meter resolution digital remote sensing data. Ten years ago such data was virtually non-existent and processing it in a reasonable timeframe would require a specialized computer system. Ten years from now it is possible that one-meter data will be considered medium resolution or at least be superceded by "hyper-resolution" data measured in single-digit (or perhaps even decimal) inches. Computing power will likely allow even faster processing of the necessarily larger "hyper-resolution" data sets than technology allows today with relatively smaller data sets. This study is thus not definitive from a remote sensing viewpoint, but instead represents the current state-of-the-art of a rapidly-evolving technology

REFERENCES

1. Anderson, J. R., E. E. Hardy, J. T. Roach, & R. E. Witmer. 1976. A Land Use and Land Cover Classification System for Use with Remote Sensor Data. USGS Professional Paper 964; Washington, DC: USGPO.
2. City of Portland. 2004. Portland Urban Forestry Management Plan. Portland, OR.
3. Ebenreck, Sara. 1989. The Values of Trees; pp. 49-57 in Moll, Gary and Sara Ebenreck, eds. Shading Our Cities. Washington, DC: Island Press.
4. Grey, Gene W. and Fred J. Deneke. 1986. Urban Forestry, 2nd ed. New York: Wiley.
5. Iverson, L. R., and E. A. Cook. 2001. Urban Forest Cover of the Chicago Region and its Relation to Household

Density and Income. Urban Ecosystems 4: 105-124.

6. Jensen, John R. and Dave C. Cowen. 1999. Remote Sensing of Urban and Suburban Infrastructure and Socio-Economic Attitudes. Photogrammetric Engineering and Remote Sensing 65(5):611-622.

7. Kielbaso, James J. 1990. Trends and Issues in City Forests. Journal of Arboriculture 16(3):69–76.

8. Maco, Scott E., and E. Gregory McPherson. 2002. Assessing Canopy Cover Over Streets and Sidewalks in Street Tree Populations. Journal of Arboriculture 28(6): 270-276.

9. McPherson, E. Gregory. 1993. Monitoring Urban Forest Health. Environmental Monitoring and Assessment, 26, 165-74.

10. McPherson, E. Gregory and Craig W. Johnson. 1988. A Community Forestry Planning Process: Case Study of Citizen Participation. Landscape and Urban Planning 15:185-94.

11. McPherson, E. Gregory, James R. Simpson, Paula J. Peper and Qingfu Xiao. 1999. Benefit-Cost Analysis of Modesto's Municipal Urban Forest. Journal of Arboriculture 25(5):235-248.

12. Miller, Robert W. 1997. Urban Forestry: Planning and Managing Urban Greenspaces, 2nd ed. Upper Saddle River, NJ: Prentice-Hall.

13. Moll, Gary. 1989. In Search of an Ecological Urban Landscape; pp. 13-24 in Moll, Gary and Sara Ebenreck, eds. Shading Our Cities. Washington, DC: Island Press.

14. Moll, Gary and Cheryl Kollin. 1993. A New Way to See Our City Forests. American Forests 99(9–10):29–31.

15. Moll, Gary. 1997. A Recipe for Sustainability. American Forests 103(3):5.

16. Myeong, Soojeong, David Nowak, Paul F. Hopkins, and Robert H. Brock. 2001. Urban Cover Mapping Using Digital Imagery. Urban Ecosystems 5:243-256.

INDEX

A

Akaike's information criterion (AIC) 181
Animals injure trees 83
Australian Capital Territory (ACT) 2

B

Bloomington Urban Forestry Research Group (BUFRG) 67
Bloomington Urban Forestry Research Group [BUFRG] 62
Bulk density (BD) 180

C

Canopy height model (CHM) 47
Categories of Interest (COI) 27
Census block groups (CBG) 132
Coarse woody debris (CWD) 148
Combined management approach 8, 10
Community Accomplishment Reporting System, (CARS) 113
Compaction 81
Continuous Change Detection and Classification (CCDC) 147

D

Department of Conservation and Recreation (DCR) 114
Diameter at breast height (DBH 49
Diameter at breast height (DBH) 180
Digital surface model (DSM) 52
Digital Surface Model (DSM) 163
Digital Terrain Model (DTM) 50

E

Emphasise 10
Environmental justice (EJ) 129
Environmental stress 82

F

Few soils 81

G

Global positioning system (GPS) 62
Greenspace 2, 4, 5, 6, 7, 8, 9, 10, 11

H

Habitat structures 1, 2, 9, 10, 11
Healthy root system 81

K

Keep Indianapolis Beautiful, Inc. (KIB) 62

L

Landscape 75, 76, 77, 79, 86, 87, 88, 100, 101
Large old trees 1, 9
Least squares regression (OLS) 131
Light Detection and Ranging (LiDAR) 161
Low density residential (LDR) 147, 149
Low temperature 82

M

Medium density residential (MDR) 147, 149
Metropolitan Statistical Area (MSA) 146
Minnesota Department of Natural Resources (MNDNR) 164
Mississippi River Corridor Critical Area (MMRRA) 170

N

National Agriculture Imagery Program (NAIP) 131
Newly planted street trees 61
New York City Department of Parks and Recreation (NYC DPR) 37, 38
New York City (NYC) 37, 38
Normalized Difference Vegetation Index (NDVI) 50

O

Other development (OTH) 147

P

Performance Measurement Accountability System (PMAS) 113
Peroxyacetylnitrate 77

R

Radiation 77
Regional Land Information System (RLIS) 193
Right-of-way (ROW) 190

S

Social-ecological system (SES) 64
Social-spatial 17
Soil organic carbon (SOC) 179
Soil stress 83
Spatial autoregression (SAR) 130, 131
Street tree component (STC) 189, 191, 193
Street Tree Resource Analysis Tool for Urban Forest Managers (STRATUM) 37, 38, 116
Structural integrity 9, 10
Sustainability 121, 122, 123, 124, 125, 126, 127
Sustainable forest management (SFM) 121

T

Terra Remote Sensing Inc. (TRSI) 49
Twin Cities Metropolitan Area (TCMA) 162

U

Urban environments. 19
Urban Forest Effects Model (UFORE) 37, 38
Urban forest management 125, 126
Urban forests 121, 122, 123, 124, 125, 126, 127, 128
Urban land environments (ULEs) 190
Urban landscape 17
Urban landscapes 19
Urban tree canopy (UTC) 21
Urban Tree Canopy (UTC) 21, 130
Urban Tree Growth and Longevity (UTGL) 61, 74

Y

Youth Tree Team (YTT) 67